03-20-06

$85.00

Science in the Contemporary World

Other Titles in ABC-CLIO's

History of Science

Series

Science in the Contemporary World
An Encyclopedia

Eric G. Swedin

A B C ● C L I O

Santa Barbara, California   Denver, Colorado   Oxford, England

Library of Congress Cataloging-in-Publication Data

Swedin, Eric Gottfrid.
Science in the contemporary world : an encyclopedia / Eric G. Swedin.
      p. cm.   (ABC-CLIO's history of science series)
Includes bibliographical references and index.
ISBN 1-85109-524-1 (hardback : acid-free paper)–ISBN 1-85109-529-2
(eBook)
1. Science—Encyclopedias.   I. Title. II. Series:
ABC-CLIO's history of science series.
Q121.S94 2005
503—dc22   2004026950

07 06 05 04 10 9 8 7 6 5 4 3 2 1

This book is available on the World Wide Web as an e-book. Visit abc-clio.com for details.

ABC-CLIO, Inc.
130 Cremona Drive, P.O. Box 1911
Santa Barbara, California 93116-1911

This book is printed on acid-free paper ∞.
Manufactured in the United States of America

Dedicated to my wife, an unabashed science enthusiast,
Betty Jean Harris Swedin

Contents

Science in the Contemporary World: An Encyclopedia

Preface and Acknowledgments

This encyclopedia is organized so that approximately half of the entries are biographies of people and the other half cover topics. Alas, I could not include everything. My goals have been both breadth and selected depth. Individual scientists have been selected for inclusion both based on how important their discoveries were and in an attempt to have a broad representation of the different scientific disciplines. Topics are examined in a depth in proportion to how much change has occurred in that scientific area since 1950.

A book this substantial is not done in isolation. I want to thank all the scientists and historians who have created the research that this book draws upon. I also want to thank the series editor, William E. Burns; my editors at ABC-CLIO, Simon Mason and Carla Roberts; and my copy editor, Silvine Farnell. My family also supported me with their enthusiasm for science and allowing me too much time squirreled away in my study.

Eric G. Swedin
Weber State University

Introduction

World War II neatly divides the twentieth century in almost every historical way imaginable. Few people on Earth were unaffected by that titanic struggle. Old empires ended, ideologies gained fresh power, and science and technology gained a sharp boost. World War II has also been called the physicists' war because scientists and engineers developed many important weapons that fundamentally changed the course of the war, including rockets, radar, proximity fuses, the electronic computer, and the mightiest weapon of all, the atomic bomb. (*See* Nuclear Physics; Physics.)

The American economy boomed during World War II for many reasons, including a sense of shared purpose, the expansion of military forces that solved unemployment problems, and massive armaments spending that kept factories running at full capacity. Significant scientific and technological innovation occurred during the war, and that innovation, combined with prudent government economic policies, helped fuel the postwar economic boom. Multiple economic studies have found that scientific and technological research lead to greater productivity and economic growth. At the end of the war, the United States dominated the world, its gross national product making up half of the world economy. This enabled Americans to expand their extensive higher education system and to pour money into scientific and technological research. (*See* National Institutes of Health; Universities.)

The United States became scientifically assertive during the postwar years. Many scientists, especially Jewish scientists from Germany, had fled to the United States and continued their work in their new home. The electrical engineer Vannevar Bush (1890–1974) directed the American Office of Scientific Research and Development during World War II. As the war drew to an end, Bush wrote an influential report arguing for continued federal spending to develop new science and technology. His report led Congress to create the National Science Foundation (NSF) in 1950. (*See* National Science Foundation.)

The destruction of the war in Europe and Japan retarded the practice of science in those areas for at least a decade after the war. After economic recovery, the practice of science was revitalized in Europe and Japan. The United States, the Soviet Union, the other developed nations in Europe, China, and Japan—all engaged in big science projects, mostly related to developing nuclear weapons or nuclear energy. (*See* Big Science.) Despite the continuing prestige of the U.S. graduate education system, by the end of the century, thirteen other countries, including Canada, Japan, South Korea, Taiwan, and a number of European nations, produced more college graduates with science, engineering, or technology degrees on a per capita basis than the United States. (*See* Third World Science.)

In the United States and other nations, the number of women and minorities going into science blossomed after World War II, as prejudices within the scientific community declined and the social movements of the 1960s and 1970s opened up opportunities for women and minorities. Nevertheless, at the end of the century, women and minorities were not represented in any of the scientific disciplines in proportion to their numbers in the general population, even though more women than men were going to college, and more minorities than ever before were going to college. In physics, only 10 percent of doctorates went to women in 2000, and less than 2 percent went to minorities.

Scientific and technological research requires funding. Some funding comes from their own pockets, but most scientists must turn to other resources: a commercial enterprise, an educational institution, a private foundation, local government, national government, or international organizations. (*See* Foundations.) The history of science is more than just the work of scientists; it includes the work of science popularizers, attitudes toward science held by the lay public, and the scholarly examination of science. Science popularizers use books, science museums, the media, and education to translate complex scientific concepts for the lay audience. They also try to convey the wonder of science. (*See* Asimov, Isaac; Clarke, Arthur C.; Cousteau, Jacques; Gould, Stephen Jay; Hawking, Stephen; Hoyle, Fred; Media and Popular Culture; National Geographic Society; Sagan, Carl; Wilson, Edward O.) Science fiction also partially serves this function. (*See* Science Fiction.) Historians, philosophers, and sociologists have also studied the actual practice of science. (*See* Big Science; History of Science; Kuhn, Thomas S.; Philosophy of Science; Popper, Karl; Social Constructionism; Sociology of Science.) Feminist scholars have critiqued both the practice and conclusions of science. (*See* Feminism.)

Science since World War II has been dominated by nine themes: the cold war, space exploration, undersea exploration, physics, new tools, the life sciences, earth science, medicine, studying humans, and environmentalism.

The Cold War

The cold war pitted the ideology of democracy and free markets, supported by the United States and its allies, against communism and Marxist economics, supported by the Soviet Union and its allies. The conflict led directly to the hydrogen bomb in 1952 and the continuing nuclear arms race. (*See* Cold War; Economics; Hydrogen Bomb; Oppenheimer, J. Robert; Sakharov, Andrei; Soviet Academy of Sciences; Soviet Union; Teller, Edward; War.) Both the superpowers poured funding and talent into scientific and technological research, seeking strategic military advantage and each striving to show that its political ideology was best suited to the goal of scientific progress. The launch of *Sputnik I,* the first artificial satellite, in 1957 opened a new frontier in the cold war ideological struggle. (*See* Satellites.)

Space Exploration

Earth is a very small part of the universe. In 1950, only science fiction writers and a few far-sighted scientists recognized the value of artificial satellites or dreamed of sending human explorers beyond the atmosphere. The International Geophysical Year of 1957–1958 provided an opportunity for military ballistic rocket programs in the United States and Soviet Union to turn their attention briefly to placing an artificial satellite in orbit. (*See* International Geophysical Year.) The Soviets succeeded first in 1957 with *Sputnik I,* which shocked the Americans and launched a space race, fueled by the ideological competition of the cold war. (*See* Satellites.) The United States founded the National Aeronautics and Space Administration (NASA) in 1958 and transferred the

Jet Propulsion Laboratory (JPL) in Pasadena, California, from army control to NASA. (*See* National Aeronautics and Space Administration.) JPL runs all of NASA's lunar and interplanetary unmanned scientific missions. (*See* Jet Propulsion Laboratory.)

In response to the Soviets placing the first person in orbit in 1961, the United States created the Apollo project to put the first person on the Moon by the end of the decade. *Apollo 11* succeeded in 1969. (*See* Apollo Project.) In 1969, men walked on the Moon, the highlight of the manned space program. Only twelve astronauts visited the Moon before the Apollo project foundered on budget cuts and the lack of planned sequels that could sufficiently motivate the country to continue the effort.

Although the manned space program remained the priority for both superpowers, enough money has been set aside for unmanned spacecraft programs that these ventures have been able to revolutionize planetary astronomy. (*See* Space Exploration and Space Science.) By the end of the century, NASA had sent spacecraft to every planet except Pluto. The Soviets sent landers or spacecraft to the Moon, Venus, and Mars. The Europeans became involved after the founding of the European Space Agency (ESA) in 1975, sending spacecraft on solar exploration missions and on joint ESA-NASA efforts. (*See* European Space Agency.)

These spacecraft have discovered that Venus is a hellish place with melting surface temperatures caused by a runaway greenhouse effect. Mercury is covered with craters like Earth's Moon. Mars also has numerous craters, along with a massive extinct volcano higher than anything on Earth and a canyon stretching partway around the planet that puts the Grand Canyon in Arizona to shame. The thin atmosphere of Mars has failed to quickly erode these natural wonders, unlike the thick atmosphere of Earth. Although Mars now contains small amounts of water vapor and ice, scientists have found intriguing hints that flowing water once shaped the Martian landscape, raising hopes that while life may not currently exist on Mars, bacterial life might once have existed millions or billions of years in the past. A meteorite recovered from Antarctica shows possible evidence of microfossils from bacteria, but the evidence is hotly contested.

The gas planets have been visited by multiple spacecraft, including the twin Voyager spacecraft in the late 1970s and 1980s. Photographs transmitted back from the spacecraft show numerous new moons, erupting volcanoes on Jupiter's moon Io, and possible frozen oceans on the larger moons, and revealed that all the gas giants have systems of rings girdling the planet. (*See* Voyager Spacecraft.) The search for possible life has provided a major motivation for all the interplanetary exploration by robotic spacecraft, though no scientist expects to find anything more advanced than bacterial life. Scientists do hope to find evidence of intelligent extraterrestrial life around other stars, and in 1960 they began to search in earnest for radio signals from other stars or galaxies as part of SETI, or the search for extraterrestrial intelligence. (*See* Search for Extraterrestrial Intelligence.) So far their efforts have found nothing.

The geologist Eugene Shoemaker spent a lifetime identifying meteor impact craters on the surface of Earth and searching the solar system via a telescope for asteroids or comets that might hit Earth. His effort resulted in the 1993 discovery of the Shoemaker-Levy 9 comet, which then collided with Jupiter in a spectacular display watched by astronomers from around the world. (*See* Shoemaker, Eugene; Shoemaker-Levy Comet Impact.) The father-son team of Luis W. Alvarez and Walter Alvarez, a physicist and a geologist, advanced the theory that an asteroid or large comet had hit Earth 65 million years ago, causing climatic change and the extinction of the dinosaurs. The evidence for this collision is found in a worldwide layer of iridium at the same stratigraphic level, a layer of iridium that comes from extraterrestrial sources; the actual crater was later identified in Mexico. (*See* Alvarez, Luis W., and Walter Alvarez.) In 1908 a comet or asteroid exploded in the air over Tunguska, Siberia, devastating a large expanse of forest. Scientists now realize that catastrophic events happen more often than previously suspected, with dramatic impact on the natural environment.

Undersea Exploration

Outer space has revealed new wonders, and the oceans, covering two-thirds of the planet, also hide their own wonders, now being revealed. (*See* Undersea Exploration.) The French naval officer Jacques Cousteau invented the Aqua-Lung in 1943 and opened up the near-surface environment to ready scientific exploration. Cousteau later traveled the world in his research ship, the *Calypso,* making documentaries while engaged in oceanographic research. (*See* Cousteau, Jacques; Oceanography.) In 1960, the bathyscaphe *Trieste* descended to the deepest point in any of the oceans at the Challenger Deep of the Mariana Trench in the Pacific Ocean. In 1977, the deep-sea submersible *Alvin* discovered, in the Galapagos Rift near the Galapagos Islands at about 2,700 meters, deep-sea hydrothermal vents, geological marvels that sustain a unique new ecology not based on photosynthesis. Some scientists argue that life on Earth first emerged at such sites. (*See* Deep-Sea Hydrothermal Vents; Geology.) The International Geophysical Year also provided an opportunity for oceanographic studies, especially of the role of the oceans in meteorology (*See* International Geophysical Year; Meteorology.) Studies of the ocean floor, finding rifts and magnetic patterns in cooled magma, directly led to the theory of plate tectonics. (*See* Plate Tectonics.) Despite tremendous advances in technology and our growing understanding of the oceans, much beneath the waves remains a mystery.

Physics

The explosion of the atomic bomb in 1945 symbolized the new discoveries of relativity and quantum mechanics. Catapulted to prominence by nuclear weapons, physicists also continued trying to find the fundamental building blocks of matter, as well as reaching out into the farthest regions of space, producing theories about the beginning of time, using computers to create sophisticated models of chaotic systems, and seeking a grand unified theory to explain all physics. (*See* Chaos Theory; Grand Unified Theory; Mathematics; Physics.) There were occasional disappointments, including the failed effort to develop commercial fusion reactors, and the cold fusion debacle of 1989 showed that scientists should verify their experimental results and follow the normal process of announcing their breakthroughs at conferences and publishing in peer-reviewed journals. (*See* Cold Fusion; Nuclear Physics.)

Since World War II, nuclear physicists have used cloud chambers and then particle accelerators to smash particles into ever smaller particles, tracking the subatomic particles that persist briefly before decaying. (*See* Particle Accelerators; Particle Physics). In 1953, the physicist Murray Gell-Mann proposed the property of strangeness to organize the numerous subatomic particles that had been discovered. In 1964, Gell-Mann proposed a new type of fundamental matter, called quarks and anti-quarks, as forming subatomic particles. Quark theory is now part of the standard model that attempts to combine the four basic forces (weak nuclear force, strong nuclear force, electromagnetism, and gravity). (*See* Gell-Mann, Murray; Neutrinos). Another foundation of the standard model came in the 1950s through the work of Richard P. Feynman, Julian Schwinger, Shinichiro Tomonaga, and Freeman Dyson to reform the theory of quantum electrodynamics (QED). (*See* Dyson, Freeman; Feynman, Richard; Schwinger, Julian.) Yoichiro Nambu began the development of string theory in 1970 when he proposed that quarks act as if they were connected by strings. Developed extensively during the 1980s, string theory deals with multidimensional mathematics and hypothetical strings that are near the Planck length in size, and thus too small for instruments to detect them. (*See* Nambu, Yoichiro.)

Since 1950, physicists have continued to discover phenomena that no one had suspected: not only quarks and other fundamental building blocks of matter, but quasars, pulsars, and black holes. Astronomers have even found direct evidence of planets orbiting other stars. (*See* Astronomy; Black Holes; Extrasolar Planets; Pulsars; Quasars.) The theorists who developed the Big Bang theory asserted that the universe had been created from a primordial concentration of mass and energy in a single point, a singularity. The astronomer Fred Hoyle developed the steady state theory, which argued that matter was being continuously created as the universe expanded. During the 1950s, debate raged between supporters of the two theories, but the discovery in 1964 of a constant background radiation to the universe fit the Big Bang theory, and most astronomers came to support that theory. (*See* Big Bang Theory and Steady State Theory; Gamow, George; Hoyle, Fred.)

New Tools

The most important technological advances since 1950 have been computers and space travel. Digital electronic computers came into being as a result of projects in World War II to create machines to help in code breaking and to create artillery ballistic tables. The mathematical genius John von Neumann took this early work and laid the theoretical basis for modern computing. (*See* Computers; Neumann, John von.) John Bardeen, William Shockley, and Walter H. Brattain invented the transistor in 1947 at Bell Telephone Laboratories, which led to

the development of integrated circuits in the late 1950s. (*See* Bardeen, John; Integrated Circuits.) Integrated circuits led to microprocessors, which are complete computers on a single chip.

Engineers developed networks in the late 1960s to enable computers to communicate with other computers, with ARPAnet coming online with four nodes in 1969. Founded by the United States defense department, ARPAnet grew into the Internet, a worldwide network of networks. The programmer Tim Berners-Lee developed the World Wide Web in 1991, an easy way to use the Internet. (*See* Berners-Lee, Timothy; Internet.) Computers and networks made it easier for scientists to communicate with each other, as well as to delve into new fields where computers made the onerous burden of repetitive mathematics manageable. Chaos theory and the theories of complexity could not have arisen without computers. (*See* Chaos Theory.)

Another important new tool for scientists is deep-core drilling to retrieve samples from deep in the earth, under the seabed, or from ice sheets. Ice cores can reveal past climatic conditions and help scientists understand the current phenomenon of global warming. (*See* Deep-Core Drilling; Global Warming.) Lasers have made long-distance communications easier and helped scientists understand chemical reactions through laser spectroscopy. (*See* Lasers.) Particle accelerators are the basis of advances in particle physics, where various forms of accelerators accelerate particles and smash them against target atoms, creating smaller particles, whose short lives, less than a fraction of a second, are recorded and analyzed, allowing physicists to identify ever smaller particles. (*See* Particle Accelerators.) Advanced ground-based telescopes and space-based telescopes have extended what astronomers can see. Optical telescopes have been joined by telescopes looking at other bands in the electromagnetic spectrum, including ultraviolet, infrared, x-ray, and radio. Orbiting satellites have contributed to understanding the oceans, assessing environmental damage and land use, finding archaeological sites, and providing precise locations through the Global Positioning System (GPS). (*See* Satellites; Telescopes.)

The Life Sciences

One of the most important scientific discoveries since World War II occurred in 1953, when James D. Watson and Francis Crick discovered the double helix molecular structure of deoxyribonucleic acid (DNA). The x-ray diffraction photographs of DNA molecules by Rosiland Franklin provided a key element of Watson and Crick's discovery. The science of genetics leapt forward and molecular biology came to rapidly dominate the other fields in the life sciences. (*See* Biology and the Life Sciences; Crick, Francis; Franklin, Rosiland; Genetics; Watson, James D.)

The biochemists Herbert Boyer and Stanley Cohen developed the techniques of genetic engineering in the early 1970s, spawning the biotechnology industry. (*See* Biotechnology; Boyer, Herbert; Cohen, Stanley N.) The Human Genome Project, running from 1990 to 2003, found that the final human genome contained 3.1 billion pairs, making up 35,000 to 40,000 genes. (*See* Human Genome Project). In 1996, Dolly the sheep was born, the successful result of cloning a mammalian adult somatic cell. (*See* Cloning.) Genetic engineering and cloning raised ethical and bioethical questions about the ecological effects of new species, as well as whether species developed in the laboratory can be owned, whether people should be able to choose the genetic characteristics of their children as if ordering from a catalogue, and the like. (*See* Bioethics; Ethics.)

Scientists like the ornithologist Ernst Mayr and the biologist Theodosius Dobzhansky created the theory of evolution known as the modern synthesis by combining Mendelian genetics with the Darwinian theory of natural selection. (*See* Dobzhansky, Theodosius; Evolution; Mayr, Ernst.) The famous 1953

Miller-Urey experiment showed that organic molecules could have emerged spontaneously from the early primeval atmosphere. Later scientists expanded on the theory of evolution, arguing that evolution may occur in bursts rather than gradually, and that species should be classified by genetic and evolutionary relationships rather than the classical physiological differences. (*See* Evolution; Gould, Stephen Jay; Miller, Stanley L.) Scientists such as Elso Barghoorn found microfossils of ancient bacteria, pushing back the first known life on Earth into the billions of years. (*See* Barghoorn, Elso.) Anthropologists such as those in the Leakey family sought the elusive missing link connecting extinct primate species with modern human beings, and showed that humans evolved in Africa first. (*See* Anthropology; Leakey Family.)

Lynn Margulis revived the older theory of symbiosis to create her serial endosymbiosis theory, which posited that eukaryotic cells, which contain nuclei, evolved when nonnucleated bacteria fused together billions of years ago. She also showed that microbes could transmit induced characteristics, a Lamarckian notion that came to be accepted as part of the mechanism of natural selection within bacteria. (*See* Margulis, Lynn; Microbiology; Symbiosis.) James Lovelock developed the Gaia hypothesis in the late 1960s, which proposes that Earth functions like a living organism, maintaining surface and atmospheric conditions so that the chemistry and temperature will sustain life. (*See* Lovelock, James.) The theory is still controversial.

Some evolutionists applied natural selection to not just biology, but to human behavior and psychology. Edward O. Wilson, the world's foremost authority on ants, caused a storm of controversy when his influential *Sociobiology: A New Synthesis* was published in 1975. Richard Dawkins went even further by promoting evolutionary psychology. (*See* Dawkins, Richard; Sociobiology and Evolutionary Psychology; Wilson, Edward O.)

Earth Science

Geologists in 1950 were certain of two things: that gradualism explained the stratigraphical geological layers found on Earth and that continents were fixed in place. Then, in the 1960s, scientists like Harry Hess, John Tuzo Wilson, and others used evidence from the ocean floor to create the theory of plate tectonics. (*See* Geology; Hess, Harry; Plate Tectonics; Wilson, John Tuzo.) According to this now generally accepted theory, which has transformed the practice of geology, the continents actually move around on a mantle of magma, pushing up mountain ranges where the plates meet and forming rift valleys where plates pull apart. Earthquakes and volcanic activity cluster around plate boundaries. (*See* Earthquakes; Volcanoes.) The foundations of the prejudice against gradualism, originally developed partially as a reaction to biblical creation theories that argued that Earth was a scant six thousand years old, were shaken with the discovery that dinosaurs had been driven into extinction by the collision of an asteroid with Earth in Mexico 65 million years ago. (*See* Alvarez, Luis W., and Walter Alvarez; Creationism.)

Medicine

Medical research has regularly introduced new wonders in the last century. (*See* Medicine.) Some of the accomplishments since 1950 have included the polio vaccines introduced in the 1950s by Jonas Salk and Albert Sabin and new wonder drugs to treat ever more obscure diseases. (*See* Pharmacology; Sabin, Albert; Salk, Jonas.) Robert Edwards and Patrick Steptoe applied in vitro fertilization to create the first human test-tube" baby in 1978. (*See* Edwards, Robert; In Vitro Fertilization; Steptoe; Patrick.) The oral contraceptive, introduced for general use in 1960, helped usher in a sexual revolution in the 1960s and 1970s. (*See* Birth Control Pill.) Organ transplants begin in 1954 with a kidney transplant and by the end of the century included successful transplants of numerous different organs, and even a single (failed)

example of transplanting a hand. Efforts to use xenotransplants from pigs and baboons have begun amid much opposition on ethical and moral grounds. (*See* Organ Transplants.)

Institutions such as the American Centers for Disease Control and Prevention (CDC) and National Institutes of Health (NIH) fund multibillion-dollar efforts in continued research and prevention. (*See* Centers for Disease Control and Prevention; National Institutes of Health.) The World Health Organization successfully led a worldwide effort to completely eradicate the disease of smallpox, declaring victory in 1979. (*See* Smallpox Vaccination Campaign; World Health Organization.) Even in the midst of so many advances in industrialized countries, leading to ever higher life expectancy rates, new threats have emerged. The excess use of antibiotics has led, through the process of natural selection, to drug-resistant strains of bacteria. In 1981, gay-related immunodeficiency disease (GRID), later renamed acquired immunodeficiency syndrome (AIDS), was found in California, and it has grown into a major health threat, especially in sub-Saharan Africa. (*See* Acquired Immunodeficiency Syndrome.) Proteinaceous infectious particles were discovered in the 1970s and are now thought to be the cause of bovine spongiform encephalopathy (BSE), or mad cow disease, scrapie in sheep, and the rare kuru and Creutzfeldt-Jakob diseases in humans. (*See* Medicine; Prions.)

Studying Humans

Neuroscientists and psychologists have made enormous strides in understanding how people think and behave, though much still remains unknown. Neuroscientists now understand more fully the nature of neurons, how neurotransmitters work, and the locations within the brain of certain cognitive functions. Neuroscientists have contributed to the study of cognitive science, which is the study of how humans think. (*See* Cognitive Science; Neuroscience.) These studies have also contributed to the effort to achieve artificial intelligence, an effort begun in the 1950s that has shown only incremental progress. (*See* Artificial Intelligence.)

Psychology grew as a discipline after World War II, vigorously expanding into clinical psychology, challenging psychiatry for the privilege of treating patients. (*See* Psychology.) The psychologist B. F. Skinner promoted a theory of strict behaviorism that was later rejected by mainstream psychology. (*See* Skinner, B. F.) The psychiatrist Elisabeth Kübler-Ross developed a theory explaining the five stages of grief and acceptance that the terminally ill and bereaved go through. (*See* Kübler-Ross, Elisabeth.) The psychologist Carl R. Rogers promoted client-centered therapy, helping to break the dogmatic paradigms of behaviorism and Freudian psychoanalysis, and helping to create humanistic psychology and the subsequent growth of eclecticism in the techniques used by psychotherapists. (*See* Rogers, Carl R.) The experimental psychologist Harry F. Harlow used studies of affection deprivation in primates to show that love is a primary motivator in both primates and humans. (*See* Harlow, Harry F.) The study of primates in the wild has also illuminated behaviors that are shared with our closest cousins in the animal kingdom, showing that many behaviors are not unique to humans. (*See* Fossey, Dian; Goodall, Jane; Sociobiology and Evolutionary Psychology.) The study of past humans and other cultures by anthropologists and archaeologists has flourished, assisted by the development of carbon-14 dating by Willard F. Libby, and helped to date artifacts at archaeological sites. (*See* Anthropology; Archaeology; Libby, Willard F.)

Sexologists in Europe and the United States struggled with the sensitive nature of their topic, persisting in the face of those who thought their studies unseemly. The sociological work of Alfred Kinsey, published in the 1940s and 1950s, changed how Americans thought about sexuality, though his studies are now seen as flawed by methodology and bias.

(*See* Kinsey, Alfred C.) The team of William H. Masters and Virginia Johnson used both physiological and psychological studies to examine sex in a more rigorously scientific way. (*See* Masters, William H., and Virginia Johnson.)

Environmentalism

Human population increased from 2.53 billion in 1950 to 6 billion in 1999. (*See* Population Studies.) Only the Green Revolution allowed farmers to feed such a dramatic increase in people. (*See* Agriculture; Green Revolution.) Increased population also combined with increasing industrialization to strain the planet's resources. In 1962, the biologist Rachel Carson published *Silent Spring,* warning of the dangers from overusing pesticides. Carson's book became a major impetus, reviving the conservation movement as a vigorous environmental movement. (*See* Carson, Rachel; Environmental Movement.) Environmental activists drew on the growing science of ecology to understand the interconnections among the different elements of the environment. (*See* Ecology.) More support for environmentalism emerged with the publication in 1968 of *The Population Bomb* by Paul R. Ehrlich. Movements to limit growth and attain zero population growth emerged, which helped slow population growth in the industrialized countries and China, but had less impact on poorer countries. (*See* Ehrlich, Paul R.)

In 1974, the chemists Mario Molina and F. Sherwood Rowland, drawing on the earlier work of Paul Crutzen, found that man-made chlorofluorocarbons (CFCs) were eroding the ozone layer in the upper atmosphere. (*See* Chemistry; Crutzen, Paul; Molina, Mario; Ozone Layer and Chlorofluorocarbons; Rowland, F. Sherwood.) In one of the most significant successes of the environmental movement, international agreements banned the manufacture of chlorofluorocarbons. Global warming, species depletion, pollution, and loss of wilderness and open spaces—all are continuing issues for the environmental movement. (*See* Global Warming.)

The Future

The practice of science has expanded with unprecedented vigor since 1950. At least 80 percent of all the scientists in history were alive in the last half of the twentieth century. By the end of the century, as measured by total pages in scientific journals, scientific knowledge was doubling every ten to fifteen years. That is a crude measure, since a single groundbreaking scientific article can have an impact outweighing thousands of other articles, but at least it gives us a way to measure scientific output. Most of the published output is confined to an elite group of scientists concentrated at an elite set of well-funded institutions—universities, government research labs, and corporate research labs. The vast majority of scientists publish only a handful of articles in their lifetime, while a few publish hundreds of articles. Big Science— that is, large government-funded efforts like the Human Genome Project, the building of particle accelerators, large-scale oceanic research, space exploration, the creation of large telescopes, and well-funded medical research—has come to dominate much scientific output.

In his 1996 book, *The End of Science: Facing the Limits of Knowledge in the Twilight of the Scientific Age,* the journalist John Horgan described the heroic age of science as ending, with all the important discoveries already made. All that remained for scientists, he argued, was filling in the details of a framework already discovered. Sir John Maddox, who retired in 1995 after twenty-three years as editor of the journal *Nature,* published a book in 1998 that answered Horgan, calling it *What Remains to Be Discovered: Mapping the Secrets of the Universe, the Origins of Life, and the Future of the Human Race.* Maddox boldly predicted exciting new discoveries in particle physics, genetics, molecular biology, and artificial intelligence. (*See* Artificial Intelligence; Biology and the Life Sciences; Biotechnology; Genetics.) Around 1900, ironically enough, many pundits feared that physics had explained everything. Even physicists believed this, at times encouraging their brightest students to find other fields of study. Within a few years came the discoveries about the nature of the atom and electron that eventually led to quantum mechanics. Einstein's relativity also completely changed the landscape of physics. In truth, in 1900, physics was on the verge of exciting changes, not faced with the stagnation of scientific innovation. It is possible that the same is true today.

So what other scientific advances does the future hold? Undoubtedly, nanotechnology will continue to make strides, giving us a greater degree of control on the molecular level that will yield new scientific instruments and industrial advances. (*See* Nanotechnology.) The twentieth century can be characterized as the century of the physical sciences, with the revolutions in relativity, quantum mechanics, chemistry, and particle physics, but it seems likely that the twenty-first century will be the century of the life sciences, bringing to fruit the sometimes frightening promises of genetics and biotechnology. Some future wonders may be gleaned from the pages of science fiction, a literature of wonder, just as science often evokes a sense of wonder in scientists and students. (*See* Science Fiction.)

Topic Finder

Scientific Disciplines
Agriculture
Anthropology
Archaeology
Artificial Intelligence
Astronomy
Biology and the Life Sciences
Biotechnology
Chemistry
Cognitive Science
Ecology
Economics
Evolution
Genetics
Geology
Mathematics
Medicine
Meteorology
Microbiology
Nanotechnology
Neuroscience
Nuclear Physics
Oceanography
Particle Physics
Pharmacology
Physics
Psychology

Theories and Ideologies
Big Bang Theory and Steady State Theory
Chaos Theory
Creationism

Grand Unified Theory
Plate Tectonics
Social Constructionism
Sociobiology and Evolutionary
    Psychology
Symbiosis

New Vistas
Apollo Project
Search for Extraterrestrial Intelligence
    (SETI)
Space Exploration and Space Science
Undersea Exploration
Voyager Spacecraft

Tools of Science
Computers
Deep-Core Drilling
Integrated Circuits
Internet
Lasers
Particle Accelerators
Satellites
Telescopes

Institutions and Projects
Associations
Centers for Disease Control and
    Prevention
European Space Agency
Foundations

A

Acquired Immunodeficiency Syndrome

In 1981, physicians in the United States identified a new disease, which they initially named gay-related immunodeficiency disease (GRID). Homosexual men were developing rare diseases, such as Pneumocystis carinii pneumonia and Kaposi's sarcoma, as a result of suffering severely depressed immune systems. Two years later, once it became obvious to the Centers for Disease Control (CDC) in Atlanta that the disease had nothing to do with homosexuality, the CDC officially changed the name of the disease to acquired immunodeficiency syndrome (AIDS). The disease took years to run its course in an individual. The disease particularly devastated the American homosexual community, and gay activists mobilized to demand government funding for scientific research, funding for medical care, and respect for the civil rights of AIDS patients.

The teams of Robert Gallo (1937–) at the National Institutes of Health in the United States and Luc Montagnier (1932–) at the Pasteur Institute in France both discovered in 1984 the retrovirus, human immunodeficiency virus (HIV), that caused AIDS. A bitter dispute over priority ensued, not only for scientific glory, but also for patent rights to the blood test developed two years later on

the basis of the discovery. Although a few scientists still dispute that AIDS is caused by HIV, the consensus is that HIV causes AIDS. Later studies of stored blood showed that AIDS had probably entered the U.S. population in the mid-1970s and had its origin in the vicinity of Lake Victoria in Africa.

As a blood-borne disease, HIV spread fastest among high-risk populations such as intravenous drug users, hemophiliacs, and promiscuous homosexuals. These populations were also recognized as being vulnerable to other blood-borne viruses and diseases, such as hepatitis B. Sexually transmitted diseases in general reached epidemic proportions in the United States after the sexual revolution of the 1960s, but effective antibiotics muted the consequences of those diseases. (The number of syphilis infections reported to public health authorities in America quadrupled from 1965 to 1975, and the number of reports of gonorrhea tripled during the same period.)

HIV/AIDS appeared nearly simultaneously in at least twenty nations in North America, Europe, and Africa. Most virulent diseases burn themselves out, as they kill their carriers faster than the disease can be transmitted. HIV, however, lingers in the lymph nodes for up to a decade, spreading to other carriers, before finally causing AIDS and killing the host. As with other viruses,

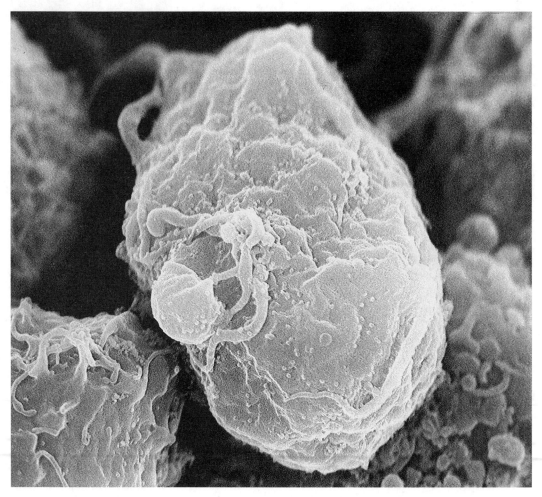

Scanning electron micrograph of human immunodeficiency virus (HIV), the virus believed to be responsible for AIDS (CDC/Public Health Information Library)

there is not one kind of HIV, but many strains, increasing in number as HIV mutates. A related family of viruses called the Simian Immunodeficiency Viruses (SIVs) has been found in other primates.

Because of its method of transmission, AIDS attacks social networks in a particularly insidious manner. Various scientists and political groups have suggested over the years that AIDS is not caused by HIV, but is a natural result of lifestyle choices that result in high-risk behavior. Conspiracy theorists have proposed that AIDS emerged from the smallpox vaccination campaign or a contaminated batch of polio vaccine, or as the result of a deliberate plot by the U.S. Central Intelligence Agency.

Education programs promoting safer sex with condom use and the use of clean needles by drug users have helped calm the AIDS epidemic in developed nations. Although the vaccine that many researchers have sought has not emerged, drug therapies have emerged to slow the progress of the disease in an individual. These drugs have also proved successful in preventing transmission of HIV to a fetus by an infected mother. Used by themselves, azidothymidine (AZT) and other drugs select for drug-resistant strains of HIV as the virus mutates. Cocktails of drugs, combining several at a time, have been more successful in slowing the viral mutation rate in individual patients.

The World Health Organization (WHO) ignored the disease until 1986, then stepped into its natural role of trying to coordinate the struggle against AIDS/HIV. AIDS became a demographic disaster in sub-Saharan Africa in the 1990s, spreading primarily through heterosexual contact, ravaging adults in what should have been the productive prime of their lives. Orphans became a chronic problem. By the end of 2003, the Joint United Nations Programme on HIV/AIDS estimated that 20 million worldwide had died of AIDS, and more than 38 million people were infected with HIV. The majority of the cases were in sub-Saharan Africa, though rates in Russia, China, and Southeast Asia were all quickly rising.

See also Centers for Disease Control and Prevention; Medicine; Pharmacology; Smallpox Vaccination Campaign; World Health Organization

References

Gallo, Robert C. *Virus Hunting: Cancer, AIDS, and the Human Retrovirus. A Story of Scientific Discovery.* New York: Basic, 1991.

Garrett, Laurie. *The Coming Plague: Newly Emerging Diseases in a World out of Balance.* New York: Farrar, Straus and Giroux, 1994.

Schoub, Barry D. *AIDS and HIV in Perspective: A Guide to Understanding the Virus and Its Consequences.* Second edition. Cambridge: Cambridge University Press, 1999.

Shilts, Randy. *And the Band Played On: Politics, People and the AIDS Epidemic.* New York: St. Martin's, 1987.

Thomas, Patricia. *Big Shot: Passion, Politics, and the Struggle for an AIDS Vaccine.* New York: Public Affairs, 2001.

Agriculture

The rapid increase in world population during the twentieth century required more food. In the industrialized world, increased mechanization, herbicide use, and pesticide use have yielded increases in food production. At the same time, machinery and larger-scale operations have reduced the labor necessary to work a modern Western farm, while smaller subsistence operations around the world have rarely benefited from tractors or other large machines. By 1980, every American farmer fed around eighty people, and Australian farmers fed even more. By 1985, Western agriculture was thirty-six times more labor efficient than Third World agriculture (McNeill, 225). Except where government regulation has prevented it, the traditional family-owned farm has rapidly disappeared in the West in favor of larger commercial enterprises. Meat production—the raising of cattle, hogs, chickens, and turkeys—became industrialized, creating meat factories where heavy use of antibiotics keeps infections at bay. Scientific research has also contributed better methods for preserving foods during storage and transportation.

More land has been put under cultivation through the construction of large hydrological projects. In the first half of the twentieth century, such projects in the U.S. West created dams and waterworks to control flooding and open up new lands to irrigation. Since 1950, projects in Egypt, the Soviet Union, and India have emulated these American successes. Poor farming practices and erratic climate patterns have also dramatically increased the amount of land lost to desertification and increased silt runoff into rivers. Desertification has been so bad in Africa that, with accompanying population increases, Africa actually grew less food per person at the end of the century than the continent produced in 1960.

Developing Third World nations experienced even more dramatic increases in population than developed nations during the twentieth century. During the 1960s and 1970s, some prognosticators feared a coming worldwide shortage of food, especially within developing nations. Their salvation has come from increasing the amount of land under cultivation and developing new strains of staple crops. By the end of the century, however, most land capable of cultivation with current technologies was already being used.

An American plant specialist, Norman E. Borlaug (1914–), went to Mexico in 1944 to work for the Rockefeller Foundation on a project to create new strains of wheat. Borlaug's work and the efforts of the International Rice Research Institute in the Philippines created what became known as the Green Revolution. The new strains of wheat and rice produced higher yields and resisted diseases better, though they also required more water and depleted the soil more quickly, requiring fertilizers to help them grow. Borlaug did not believe that the Green Revolution had solved the population crisis; rather, the increased food production had bought time to find more permanent solutions to the problem. Largely as a result of the Green Revolution, worldwide grain production increased from 692 million tons in 1950 to 1.9 billion tons in 1992, outpacing the growth in population (http://www.actionbioscience.org/biotech/borlaug.html). Average production of food grew from less than two tons per hectare in 1900 to over four tons per hectare in the 1990s, on an annual basis.

It is estimated that the use of artificial fertilizers creates enough extra food to directly feed a quarter to a third of the world's current population. The cost of artificial fertilizers favors large farms raising cash crops over smaller subsistence farming, and the Third World has experienced in the last half-century a steady increase in larger corporate farming. The phenomenon forces subsistence farmers onto more marginal lands or into urban centers to serve as industrial workers or join the ranks of the impoverished in vast shantytowns that surround many Third World cities. The larger farms using artificial fertilizers also tend to grow crops that respond more readily to fertilizers and to increase their yield, thus two-thirds of all grain production worldwide now comes from just three plants: rice, wheat, and maize. Farmers have found that increasing the use of artificial fertilizers will not increase food production any more unless new crop strains are developed.

Other increases in crop production have come from the surge in dam building and irrigation projects. Total irrigated hectares worldwide increased from 94 million in 1950 to 255 million in 1995 (McNeill, 180). By the end of the century, about 30 percent of the world's food production came from irrigated fields. The most ambitious of all these projects, the Three Gorges Dam project on the Yangtze River, is still under construction, and will displace over 1 million people to make way for a large lake. The consequences of Three Gorges are expected to be similar to the results of other like projects: flood control, stable water levels for navigation, hydroelectric power, irrigation water, ecosystem changes, and increased salinity in the soil.

The Soviet biologist Trofim Lysenko (1898–1976) promoted a theory of non-Mendelian genetics that found favor in the Stalinist Soviet Union and distorted the science of biology and practice of agriculture in that country for decades. Some Soviet scientists who disagreed with Lysenko were arrested and died from either execution or starvation. In 1964, after changes in Soviet leadership, Lysenko was finally discredited, though it took more time for the effects of Lysenko's regime to disappear.

As large farms have turned to producing single crops, this monoculture has encouraged pests, but discouraged the pest predators that like more varied environments. This development has led to heavier use of pesticides, which has led to the selection of more resistant pests, which have required greater amounts of pesticides. The overuse of synthetic pesticides, an issue effectively raised by the publication in 1962 of Rachel Carson's *Silent Spring,* created the political environmental movement. Environmentalists have tried to understand the ecological impact of new crop strains, and many have begun to emphasize the negative ecological consequences of the monoculture nature of the Green Revolution. Borlaug has responded by arguing that the first step toward social justice and environmental progress is having enough food. In 1970, American corn grow-

Norman E. Borlaug, 87, awarded the Nobel Peace Prize for his contributions to the Green Revolution, visiting Sharanjit Singh, a progressive Indian farmer, in 2001 (Pallava Bagla/Corbis)

ers lost 15 percent of their harvest to a fungus outbreak, and they promptly diversified their crop with different varieties of corn the next year (McNeill, 224).

Recent efforts to improve crop yields have moved from crossbreeding to more direct genetic manipulation by using the techniques of biotechnology. These efforts are aimed at making the crops more resistant to insects and other pests. Many of these new strains have been developed by private corporations for profit, raising important issues about patents and commercial control. The environmental activists who have decried the lack of genetic diversity that comes with monoculture and overuse of the new strains, which mean loss of genetic diversity, fear that genetically engineered crops, dubbed Frankenfoods, will present health dangers. Unfortunately, in order to meet the growing need for food, more refined techniques from biotech-

nology will be necessary to engineer transgenic cereal crops that are pest-resistant, fungus-resistant, and virus-resistant; can grow in marginal soils, and can yield ever larger returns for each acre. In a talk given in 2000, Borlaug warned that though current technology and technology already under development could feed ten billion people, humanity must get its population growth problem under control.

See also Biology and the Life Sciences;
    Biotechnology; Carson, Rachel; Ecology;
    Environmental Movement; Foundations;
    Genetics; Lysenko, Trofim; Population Studies

References

Beeman, Randal S., and James A. Pritchard. *A Green and Permanent Land: Ecology and Agriculture in the Twentieth Century.* Lawrence: University Press of Kansas, 2001.

CAST: The Science Source for Food, Agricultural, and Environmental Issues. http://www.cast-science.org/ (accessed February 13, 2004).

Fowler, Cary, and Pat Mooney. *Shattering: Food, Politics, and the Loss of Genetic Diversity.* Tucson: University of Arizona Press, 1991.

McHughen, Alan. *Pandora's Picnic Basket: The Potential and Hazards of Genetically Modified Foods.* New York: Oxford University Press, 2000.

McNeill, John Robert. *Something New under the Sun: An Environmental History of the Twentieth-Century World.* New York: Norton, 2000.

Winston, Mark L. *Travels in the Genetically Modified Zone.* Cambridge: Harvard University Press, 2002.

AIDS

See Acquired Immunodeficiency Syndrome

Alvarez, Luis W. (1911–1988), and Walter Alvarez (1940–)

The father-son team of Luis Walter Alvarez and Walter Alvarez discovered high concentrations of iridium in a 65-million-year-old layer of clay, leading them to propose the theory that the dinosaurs were driven to extinction by the impact of an asteroid or comet. Luis was of Spanish descent, born in San Francisco, where his father was a physician. Luis earned all his degrees in physics at the University of Chicago: his B.S. in 1932, M.S. in 1934, and Ph.D. in 1936. After his education, he moved to the University of California at Berkeley, where he used an early cyclotron to discover several new isotopes of mercury, hydrogen, and helium. During World War II, Luis worked on the Manhattan Project; he invented the detonators on the first atomic bomb. Unlike some of the other scientists, he did not regret the building of the bomb or its use, arguing that using the bomb had actually saved lives by forcing an end to the war. After the war, Luis returned to Berkeley to work on the linear accelerator. His team developed a new way to photograph the tracks of transitory atomic particles in a cloud chamber, allowing them to discover additional fundamental particles, many of them existing only in short-lived resonance states. In 1968 he received the Nobel Prize in

Physics for this work. Later in life, he indulged in other interests, including an effort to use cosmic rays to see if there were any hidden chambers in a pyramid at Giza. He found that the pyramid was made of solid rock.

Walter Alvarez was born in Berkeley, California, to Luis and his wife, Geraldine Smithwick. Luis and Geraldine later divorced. Walter gained an enthusiasm for rocks and geology from his mother. He earned a B.A. in 1962 in geology from Carleton College, located in Northfield, Minnesota, and a Ph.D. in geology from Princeton University in 1967. After working in South America, Europe, and Libya, Walter joined the geology department at the University of California at Berkeley in 1977.

Walter and his colleagues first went to Gubbio, Italy, to examine the rocks of the area for evidence of magnetic polarity reversals. They found such evidence, reinforcing support for plate tectonics, but also noticed a one-centimeter layer of clay between two beds of limestone. The layer below contained microfossils of the Cretaceous era and the layer above contained microfossils of the Tertiary era. This stratigraphical boundary, called the K-T boundary, is associated with the extinction of the dinosaurs. Walter wanted to know why this clay layer showed such a dramatic break between two major geologic eras. How long had it taken for this layer of clay to form?

Walter worked on the problem of the clay with his father; they first tried to measure the amount of isotope beryllium-10 in the clay, but found that the half-life of the isotope was too short to give them good results. Luis suggested that they measure the iridium in the clay bed. Iridium is deposited on Earth from extraterrestrial sources through meteorites and micrometeorites. They expected to find a small amount of iridium if the clay bed had formed slowly and none if it had formed quickly; instead, they found ninety times more iridium than expected.

After ruling out other possible causes, the

father and son and their colleagues published their findings, with the conclusion that a large comet or asteroid had hit Earth at the time of the K-T boundary and deposited the iridium. This impact could also have altered the planet's climate enough to drive the dinosaurs into extinction. This proposal, relying on a catastrophe, met resistance from geologists inclined to look for gradualist explanations. Catastrophism reminded them too much of creationism and pseudoscientific theories, like the proposal by the Russian-born writer Immanuel Velikovsky (1895–1979) that events of the Old Testament were caused by the motions of the planets. Later discovery of high levels of iridium around the world, all at the K-T boundary, showed that the cause of the iridium was not a local event. Other scientists in the 1980s identified an impact crater in Chicxulub, Yucatán, Mexico, as the source of the K-T boundary iridium, and a Shuttle Radar Topography Mission in 2000 confirmed the existence of the crater. Geologists have become convinced that, although most geological change is gradual, occasional, catastrophic geologic events have occurred in the past.

See also Creationism; Geology; Hess, Harry;
   Nobel Prizes; Plate Tectonics; Shoemaker,
   Eugene; Shoemaker-Levy Comet Impact;
   Wilson, John Tuzo

References
Alvarez, Luis W. *Alvarez: Adventures of a Physicist.*
   New York: Basic, 1987.
Alvarez, Walter. *T. Rex and the Crater of Doom.*
   Princeton: Princeton University Press, 1997.

Anthropology

Anthropology became an established and respected academic discipline in the first half of the twentieth century. In the United States, the cultural relativism of Franz Boas (1858–1942) held sway; Boas emphasized the collection of details about native cultures and argued that no group or race was any better, in a moral sense, than any other group or race. This stance emerged as a reaction to the brutal treatment of native peoples around the world in the nineteenth century, which was justified by a sense of moral superiority, as well as a reaction to Social Darwinism, which justified social classes as a consequence of innate genetic endowment. In Britain, Alfred Reginald Radcliffe-Brown (1881–1955) and Bronislaw Malinowski (1884–1942) had created a form of social anthropology called functionalism. The two men disagreed about what functionalism was, but drew their ideas from biology, seeing human culture as an organism. In France, Émile Durkheim (1881–1955) saw human behavior as purely social. Sigmund Freud (1856–1939) was not an anthropologist, but he felt that his psychological theory of psychoanalysis was applicable to all problems of human behavior. Freud discussed native customs and behavior as instances of the Oedipus complex in several books. He argued that civilization was a necessary construct to restrain the primitive emotions of the id in people.

Claude Lévi-Strauss (1908–) published his *Elementary Structures of Kinship* in 1949 and his *Structural Anthropology* in 1961, which established him as a major thinker in French society. Translations of his books spread his influence in other nations. In these books and others, all written in challenging and complex prose, Lévi-Strauss tried to reduce human behavior and culture to their essentials and illustrate the structured relationships between those essentials. His greatest success came from describing the incest taboo and kinship systems as a foundation for culture. In religion, he argued that myths are attempts to resolve contradictions in human behavior, rather than to create justifications for human behavior, as others had argued. Lévi-Strauss sought to explain human behavior and culture in a way unbounded by the historical flow of time, to find truths that applied to all people at all times. His "structuralism" was not applied only to anthropology; it was just as influential in philosophy, literary criticism, and the study of religions.

A challenge to social anthropology came from Edward O. Wilson in 1975, with his influential *Sociobiology: A New Synthesis*. Wilson argued that human behavior is dictated by biology and natural selection, with culture and society providing only a veneer over the struggle for reproduction. Many anthropologists condemned sociobiology and its intellectual child, evolutionary psychology, as a new form of Social Darwinism. They feared that dominant industrial cultures would use sociobiological arguments to justify destroying native cultures. With the encroachment of industrial society on traditional native groups, many anthropologists had become advocates for the peoples they studied.

Margaret Mead (1901–1978), a disciple of Boas, made her reputation with a study of adolescent girls in Samoa, published in 1928 as *Coming of Age in Samoa*. A particularly vicious attack on her methodology and conclusions by the anthropologist Derek Freeman (1916–2001) in the late 1980s and early 1990s was partially inspired by the renewed emphasis on nature as opposed to nurture, a shift partially inspired by the rise of sociobiology. The rise of feminism in the 1960s and its sensitivity to issues of gender and social roles also influenced anthropology. Some feminists have argued that prehistoric cultures were egalitarian matriarchies. This argument has foundered on the fact that no matriarchies have been found in any of the hundreds of separate cultures documented by anthropologists in the last two centuries.

While cultural anthropology studies human behavior and social practices, physical anthropology studies the physical nature of humans. A major emphasis in this field has been the continuing effort to apply Charles Darwin's (1809–1882) theory of natural selection to human beings and find precursors to modern humans. The Piltdown Man, discovered in 1912, was conclusively shown to be a hoax in 1953. This finding meant that England was not the birthplace of humanity. The discoveries of Peking Man and Java Man pointed to East Asia as the birthplace of humanity and the best place to look for fossils of earlier hominids.

The paleoanthropologist Louis S. B. Leakey (1903–1972) argued that Africa was the birthplace of humanity and intended to prove it. Leakey and his wife, Mary Leakey (1913–1996), worked in northern Tanzania at Olduvai Gorge. In 1959, Mary found hundreds of fragments of a hominid skull in Olduvai Gorge within strata dated to 1.75 million years ago; these fragments she reconstructed into a primate that Louis called *Zinjanthropus boisei*. This primate was thought for a time to be a direct ancestor of modern humans, but further discoveries led even Louis to agree that it represented a separate line of hominids, and the skull was reclassified as *Australopithecus boisei*. Leakey also thought that the study of the great apes might lead to insights into the behavior of humanity's ancestors. He recruited Jane Goodall (1934–) to study chimpanzees, Dian Fossey (1932–1985) to study mountain gorillas, and Biruté Galdikas (1946–) to study orangutans in Borneo.

In 1978 at Laetoli, Mary Leakey found the fossilized footprints of three hominids dated at 3.6 million years old. The three are thought to be the footprints of a man, woman or smaller man, and a child, and the gait of the footprints clearly shows that these hominids walked upright, much sooner than anyone had suspected. An expedition led Richard Leakey (1944–), the son of Mary and Louis, to the shores of Lake Turkana, where he excavated fossils from perhaps 200 individual hominids, including the 1.6-million-year-old Turkana Boy, an almost complete skeleton of *Homo erectus*. The discoveries at Lake Turkana showed that at least three different species of hominids lived together in close proximity millions of years ago, though controversy remains over which line of hominids led to modern humans. In 1974, Donald Johanson (1943–) found the first nearly complete skeleton of *Australopithecus afarensis*, which was named "Lucy." Despite these extraordinary finds in the second half of the twentieth century, the meaning of these discoveries is still elusive, and the so-called missing link between apes and humans has not been found. The continuing profusion of extinct hominid species simply creates confusion, making more difficult the continuing efforts to construct a solid line of evolution to modern humans.

Numerous skeletons of Neanderthals have been found ever since the nineteenth century. Scientists argued over whether Neanderthals were a precursor to modern humans or a separate species. A consensus gradually grew that Neanderthals were a separate sapient species, with their own ability to make tools, art, and graves for their dead. Neanderthals must also have coexisted with modern humans during the last ice age. Recent finds of Neanderthal skeletons in the late 1990s have allowed scientists to extract mitochondrial DNA. The current conclusions are that the DNA demonstrates that Neanderthals were a separate species, but that crossbreeding with modern humans was possible and may even have occurred.

See also Archaeology; Feminism; Fossey, Dian; Goodall, Jane; Leakey Family; Lévi-Strauss, Claude; Sociobiology and Evolutionary Psychology; Wilson, Edward O.

References
Eriksen, Thomas Hylland, and Finn Sivert Nielsen. *A History of Anthropology.* London: Pluto, 2001.
Guldin, Gregory Eliyu. *The Saga of Anthropology in China: From Malinowski to Moscow to Mao.* Armonk, NY: M. E. Sharpe, 1994.
McGee, R. Jon, and Richard L. Warms. *Anthropological Theory: An Introductory History.* Second edition. Mountain View, CA: Mayfield, 2000.
Spencer, Frank. *A History of American Physical Anthropology, 1930–1980.* New York: Academic, 1982.
Tattersall, Ian. *The Fossil Trail: How We Know What We Think We Know about Human Evolution.* New York: Oxford University Press, 1995.
Tattersall, Ian, and Jeffrey H. Schwartz. *Extinct Humans.* Boulder, CO: Westview, 2000.

Apollo Project

On May 25, 1961, in his first State of the Union address, President John F. Kennedy urged Congress to commit the United States to landing a man on the Moon and returning him safely to Earth before the end of the decade. Kennedy was primarily motivated by cold war competition. The United States had been embarrassed by the Soviet Union launching *Sputnik I,* the first artificial satellite, in 1957, and putting the first man into space just a month earlier, on April 12, 1961. As a struggle of competing ideologies, the cold war conflict between the superpowers depended as much on prestige as on military power, and the United States wanted to regain its prestige as the preeminent scientific and technological power on the planet.

The Mercury project put the first American astronaut into space on May 5, 1961. Other Mercury flights followed, sending astronauts into orbit in small capsules. A flight to the Moon and landing there required a much more sophisticated effort. The National Aeronautics and Space Administration (NASA) decided on an elaborate Moon effort that required the world's most powerful booster, a crew of three, the ability to twice rendezvous in space, and a stay in space of over a week. The Gemini project followed, with two-person capsules and practice that included rendezvousing two spacecraft while in orbit. A massive three-stage rocket, the *Saturn V,* provided enough thrust to push a three-man command capsule and a lunar lander into Earth's orbit.

The Ranger, Surveyor, and Lunar Orbiter projects sent unmanned robotic craft to the Moon to scout out the way for the Apollo astronauts. Ranger probes smashed into the Moon while sending back television pictures; Surveyor probes actually landed; and the Lunar Orbiters used radar and photography to chart the surface of the Moon, searching for safe landing sites for the Apollo astronauts.

The Apollo project faced a tragic setback in 1967 when a test of the command capsule atop an unfueled rocket on the pad resulted in a flash fire. Fed by the pure oxygen atmosphere of the capsule and by poor design, the fire killed three astronauts. The project halted

Apollo 11 *astronaut Buzz Aldrin on the surface of the Moon at Tranquility Base, July 20, 1969 (Original image courtesy of Neil Armstrong / NASA / Corbis)*

for over a year while the capsule was re-designed to be safer and a nitrogen-oxygen atmosphere was introduced for use while the spacecraft was on the ground.

When the project resumed, the first flights were proof of concept efforts. The astronauts of *Apollo 7* and *Apollo 9* only orbited Earth, while the astronauts of *Apollo 8* and *Apollo 10* traveled to the Moon, orbited it, and re-turned to Earth. The astronauts of *Apollo 8* orbited the Moon on Christmas Eve, 1968, read passages from the biblical Book of Genesis on the radio, and took a famous picture of Earth rising over the lunar landscape. The Soviets launched their own spacecraft, *Zond 5* and *Zond 6,* in 1968, into orbits around the Moon, successfully returning them to Earth. These modified Soyuz capsules, though originally designed to carry cosmonauts, carried only turtles, mealworms, bacteria, plants, and seeds.

The astronauts of *Apollo 11* were Neil Armstrong (1930–), Edwin E. Aldrin Jr. (1930–), and Michael Collins (1930–). While Collins stayed behind in the orbiting command capsule, Armstrong and Aldrin descended to the Moon in the lunar lander. Using rocket thrust to slow their descent, they landed on the four legs of the lander on July 21, 1969. As Armstrong stepped from the last rung of the ladder to the Moon's surface, he said: "That's one small step for man, one giant leap for mankind." The record was later altered to reflect what he meant: "That's one small step for a man, one giant leap for mankind." Armstrong and Aldrin spent less than a day on the lunar surface before using the bottom half of their lunar lander as a launching pad and returning to orbit in the upper half of their lunar lander.

From 1969 to 1972, a total of twelve astronauts walked on the Moon, the only humans ever to do so. *Apollo 13* suffered the explosion of an oxygen tank aboard the command module while on the way to the Moon, forcing the astronauts to cancel their Moon landing and desperately improvise to get home safely.

In terms of science, the Apollo astronauts brought back 838 pounds (380 kilograms) of lunar rocks and spent a total of 296 hours exploring the lunar surface. Later missions even included a lunar rover to allow the astronauts to range across the lunar surface. The astronauts left scientific instruments on the Moon to monitor seismic activity and reflect back lasers from Earth to accurately measure the distance between Earth and the Moon. The Apollo project cost $25 billion, a considerable investment for that time; the project became the first big customer of integrated circuits and played a key role in accelerating the growth of the semiconductor industry. Public interest declined after the first landing, and three later scheduled landings were cancelled due to federal funding cuts.

The Soviet program to beat the Americans to the Moon suffered a crippling setback when a rocket exploded on the pad and killed dozens of key engineers and technical personnel. After *Apollo 11* reached the Moon's surface first, the Soviets announced that they were not interested in following suit and kept their own failed effort secret. Surplus Apollo hardware was used in the three Skylab missions of 1973–1974, during which American astronauts lived in space aboard a small space station, and in the Apollo-Soyuz mission of 1975, during which American astronauts and Soviet cosmonauts rendezvoused in orbit and shook hands inside a temporary tunnel connecting the two spacecraft.

See also Big Science; Cold War; Integrated Circuits; National Aeronautics and Space Administration; Shoemaker, Eugene; Space Exploration and Space Science

References

Armstrong, Neil, Michael Collins, and Edwin E. Aldrin Jr. *First on the Moon: A Voyage with Neil Armstrong, Michael Collins and Edwin E. Aldrin, Jr.* Boston: Little, Brown, 1970.

Chaikin, Andrew. *A Man on the Moon: The Voyages of the Apollo Astronauts.* New York: Viking, 1994.

Collins, Michael. *Carrying the Fire: An Astronaut's Journeys.* New York: Farrar, Straus and Giroux, 1974.

Compton, William David. *Where No Man Has Gone Before: A History of Apollo Lunar Exploration Missions.* Washington, DC: National Aeronautics and Space Administration, 1989.

Crouch, Tom D. *Aiming for the Stars: The Dreamers and Doers of the Space Age.* Washington, DC: Smithsonian Institution, 1999.

Shepard, Alan, and Deke Slayton. *Moon Shot: The Inside Story of America's Race to the Moon.* Atlanta: Turner, 1994.

Archaeology

Other branches of science changed the practice of archaeology during the twentieth century. In 1948, Willard F. Libby of the University of Chicago developed the technique of radioactive carbon dating. This allowed archaeologists to measure the amount of naturally occurring radioactive isotope carbon-14 in wood, ashes, or bones from archaeological sites and date them, based on the decay of the carbon-14. Ground-imaging radar, global positioning systems, remote sensing, computers, and trace-element analysis are some of the many other scientific and technological advances that offered new tools to archaeology.

In the 1960s, archaeologists embraced processional archaeology, also called New Archaeology. This movement started in America and rapidly found adherents in Britain and Scandinavia, spreading later to other developed countries. Processional archaeology draws more upon the discipline of anthropology, and those who practice it strive to become more than just collectors of artifacts and recorders of stratigraphic layers. They recognize that the context of an artifact is more important than the artifact itself. The new tools allow archaeologists to examine how individuals actually lived their lives in the past. In common with similar movements in other social sciences and the humanities, the New Archaeology seeks to understand the world of the common people, not just the elite. In the 1980s, a loosely defined post-processual archaeology movement developed in Britain, urging even more emphasis on individuals as active agents in their world and arguing even more strongly for context.

Archaeologists as a community increased in numbers and importance during the second half of the twentieth century, as the affluence of developed countries sustained their activities. Laws to promote preservation of archaeological sites were passed in many developed countries. Activism by descendants of native peoples also led to laws compelling museums and archaeologists to return artifacts, and especially human remains, for burial according to native customs. The Kennewick Man, discovered on the bank of the Columbia River in 1996, illustrated the tensions involved, in this case between scientific investigation of a 9,000-year-old skeleton and five Native American nations who demanded possession of the skeleton under the authority of the Native American Graves Protection and Repatriation Act, passed by the United States Congress in 1990.

Historical archaeology emerged as a sub-discipline of its own in the 1970s, concentrating on archaeological sites of recent origin. Environmental archaeology also emerged in the 1970s, drawing on the skills of geologists, botanists, and biologists to more thoroughly understand the environment of the past through seed remains, pollen samples, soil analysis, and other such techniques. The subdiscipline of salvage archaeology emerged to preserve sites or conduct quick excavations in the face of new construction. The most famous example of salvage archaeology occurred in 1960s, when an international effort moved important ancient Egyptian monuments to higher ground to avoid their being submerged by the rising waters behind the Aswan Dam on the Nile River.

The invention of scuba gear by Jacques Cousteau (1910–1997) allowed him to undertake an undersea archaeological excavation. The first professional archaeologist to take up scuba was George F. Bass of the University of Pennsylvania in 1960, pioneering many of the techniques for working underwater. Other archaeologists adopted the technology, recovering artifacts from ancient shipwrecks, and later excavating the sites of prehistoric lake settlements in Europe. Scuba also permitted the raising of shipwrecks, for example the Swedish warship *Vasa,* sunk in 1628 in Stockholm harbor, raised in 1961.

Many notable finds excited the interest of laypeople in archaeology. The royal tomb of the first emperor of China, Qin Shi Huang, was unearthed in 1974. Other chambers were later discovered, including a large vault containing 6,000 terra-cotta statues of warriors and horses. The life-sized army wore real clothes, boots, and armor, and bore weapons.

The bogs of northern Europe preserved bodies cast into them by ancient people. The

Terra-cotta soldiers from the mausoleum of China's first emperor in Xian, China, unearthed in the 1970s (Bettmann / Corbis)

Lindow Man, from an English bog, was scientifically excavated in 1984, revealing considerable detail. A receding glacier on the Italian-Austrian border in 1991 revealed Similuan Man (popularly called the "Iceman"), the 5,000-year-old body of a man remarkably preserved by the ice. The preservation was so good that police were initially summoned, on the assumption that the corpse came from a contemporary crime. In the Andes Mountains, children left as sacrifices were preserved by the high-altitude air. The first to be found, a young woman in 1995, was dubbed the Ice Maiden. The preservation of DNA, and even the contents of stomachs left from final meals, added to the importance of these finds.

See also Anthropology; Cousteau, Jacques; Libby, Willard F.

References

Cotterell, Arthur. *The First Emperor of China: The Greatest Archeological Find of Our Time.* New York: Holt, Rinehart, and Winston, 1981.

Fagan, Brian M. *Time Detectives: How Archaeologists Use Technology to Recapture the Past.* New York: Simon and Schuster, 1995.

Fowler, Benda. *Iceman: Uncovering the Life and Times of a Prehistoric Man Found in an Alpine Glacier.* Chicago: University of Chicago Press, 2001.

Silberman, Neil. *Between Past and Present: Archaeology, Ideology, and Nationalism in the Modern Middle East.* New York: Henry Holt, 1989.

Thomas, David Hurst. *Skull Wars: Kennewick Man, Archaeology, and the Battle for Native American Identity.* New York: Basic, 2000.

Trigger, Bruce G. *A History of Archaeological Thought.* Cambridge: Cambridge University Press, 1989.

Artificial Intelligence

Artificial intelligence research involves the effort to create computer hardware and software that behave as humans do and actually think. Early computer pioneers, such as the mathematicians Alan Turing (1912–1954) and John von Neumann (1903–1957), intentionally developed electronic computers as the first step toward the creation of genuine thinking machines. In 1950, Turing created the Turing Test, a way of testing an intelligent machine. If a person could interrogate an intelligent machine via a teletype and at the end of the conversation not tell whether or not there was another person or a machine at the other end of the teletype, then that machine was intelligent. Although contemporary researchers no longer commonly accept the Turing Test as an absolute criterion, it serves as a good indicator of the goals of these early pioneers.

The term *artificial intelligence* (AI) was first coined in a two-month summer workshop at Dartmouth College in 1956, organized by the mathematicians Marvin Minsky (1927–) and John McCarthy (1927–). The first two decades of AI research were dominated by researchers at the Massachusetts

Institute of Technology (MIT), Carnegie Tech (later renamed Carnegie Mellon University), Stanford, and International Business Machines (IBM). European efforts and Japanese efforts later became important. The second high-level computer language, called Lisp (from list processor), was created specifically in 1958 by McCarthy at MIT for AI research.

The history of AI has been characterized by a series of theories that showed initial promise when applied to limited cases, leading to optimistic declarations that intelligent machines were just around the corner, and disappointment as the theories failed when applied to more difficult problems. In the early 1980s, expert systems showed promise. An expert system is created as a series of rules, with an inference engine, covering a narrow field of expertise. The first successful commercial expert system configured orders for Digital Computer Corporation computer systems. The promise of expert systems created a commercial expansion of the field, only to end in disappointment by the end of the 1980s as the expense and limitations of expert systems became apparent.

AI researchers developed a close relationship with cognitive science, the study of how humans think. One result of this relationship was the creation of neural networks, implemented in either hardware or software; these networks simulate the way the human brain uses neurons to form patterns and change the relationships within those patterns. The concept of fuzzy logic also grew out of this relationship, allowing software to make decisions when input data remain uncertain and incomplete.

Robotics is related to AI, in that many of the difficult problems in each field are similar. Industrial robots that perform limited tasks have become common, though general-purpose robots that can correctly perceive the natural world, through sight or other sensory means, and react to that sensory data remain visions of the future. Significant work at MIT's AI Lab in the 1990s has developed a different bottom-up approach to robotics, using simple algorithms to create machines that move like insects.

Vision is a form of pattern recognition, and AI research eventually led to pattern recognition of optical characters, which became common by the 1990s. Effectively recognizing handwriting patterns also became possible in the 1990s. The solution of the related problem of voice recognition has resulted in simple commercial applications.

In 1997, an IBM supercomputer called Big Blue defeated Garry Kasparov (1963–), the world chess champion, in a chess tournament. Kasparov later complained that Big Blue had been programmed to specifically defeat him, though all grand masters train to defeat specific opponents. Many commercial computer games and video games have rudimentary AI algorithms to provide an artificial opponent for the human player.

See also Cognitive Science; Computers; Minsky, Marvin; Neumann, John von; Penrose, Roger

References
Crevier, Daniel. *AI: The Tumultuous History of the Search for Artificial Intelligence.* New York: Basic, 1993.
Hogan, James P. *Mind Matters: Exploring the World of Artificial Intelligence.* New York: Ballantine, 1997.
Hsu, Feng-hsiung. *Behind Deep Blue: Building the Computer That Defeated the World Chess Champion.* Princeton: Princeton University Press, 2002.
Kurzweil, Raymond. *In the Age of the Intelligent Machine.* Cambridge: MIT Press, 1990.
Minsky, Marvin. *The Society of Mind.* New York: Simon and Schuster, 1988.
Newborn, Monty. *Kasparov versus Deep Blue: Computer Chess Comes of Age.* New York: Springer Verlag, 1996.

Asimov, Isaac (1920–1992)

As one of the leading science fiction authors and science writers of the twentieth century, Isaac Asimov brought his knowledge of science and his fascination with the wonder of science to a mass audience. He was born in Petrovichi, Russia, and his family emigrated to New York City when he was only three years old. He graduated from high school at age fifteen and entered Columbia University. He received a B.S. in chemistry in 1939, and an M.A. in the same field in 1941. During World War II, he worked at the U.S. Naval Air Experimental Station in Philadelphia. After the war and a short tour in the army, he returned to Columbia, earning a doctorate in 1948. He taught biochemistry at Boston University until 1958, when he became a full-time writer. Father to two children by his first wife, Asimov died of Acquired Immunodeficiency Syndrome (AIDS) in 1992, the result of an HIV infection he contracted when undergoing triple-bypass heart surgery in 1983.

Asimov is considered one of the three grand old masters of the science fiction field (Arthur C. Clarke [1917–] and Robert A. Heinlein [1907–1988] being the other two). Asimov's 1941 short story, "Nightfall," is considered by many to be the finest science fiction short story ever written. The story describes a world with six suns whose complex orbits allow night to come only briefly every 2,000 years. The sight of the stars, heretofore hidden by daylight, drives many people insane, destroying their civilization. His series of Foundation novels, first published in the early 1950s and resumed in the 1980s, is considered by many critics to be the finest science fiction series ever written. His Robot series of short stories and novels developed a set of three robotic laws that is often quoted:

1. A robot may not injure a human being, or, through inaction, allow a human being to come to harm.
2. A robot must obey orders given it by human beings except where such orders would conflict with the First Law.
3. A robot must protect its own existence as long as such protection does not conflict with the First or Second Law. (*I, Robot,* 6)

His nonfiction writing, including regular essays on science for numerous magazines, put him in the position of being the leading science popularizer of his time. His straightforward writing style enabled him to explain

Isaac Asimov, science fiction author and prolific science populizer (Douglas Kirkland / Corbis)

complex scientific ideas while remaining accurate. Examples of his books include *The Intelligent Man's Guide to Science* (1960), *Great Ideas of Science* (1969), *The History of Physics* (1984), and *Asimov's Chronology of Science and Discovery* (1989).

His work reflected concerns with overpopulation, the environment, the role of science in people's lives, and revealed a skeptical, humanist outlook toward religion and claims of the paranormal. Considering the complex content of his books, Asimov is certainly the most prolific writer of all time, with some four hundred books, hundreds of short stories, and thousands of articles to his credit. His prodigious memory and compulsive work habits helped him generate millions of words.

See also Acquired Immunodeficiency Syndrome;
    Clarke, Arthur C.; Media and Popular Culture;
    Sagan, Carl; Science Fiction

References
Asimov, Isaac. *I, Asimov.* New York: Doubleday, 1994.
———. *I, Robot.* New York: Doubleday, 1950.
Asimov, Janet Jeppson, editor. *Isaac Asimov: It's Been a Good Life.* Amherst, NY: Prometheus, 2002.

Associations

Scientific academies and associations played a key role in the development of science prior to the twentieth century, providing a way for scientists to meet with each other and sponsoring publications in journals and books as a way to disseminate scientific knowledge. These roles have continued through the twentieth century and into the twenty-first. Scientific associations are usually found organized along national lines or within scientific disciplines. The Royal Society in the United Kingdom, founded in 1660, is the premier example of a national society; it has remained vigorous, publishing journals, funding research, and electing distinguished scientists to be fellows of the society.

Other prominent national societies have played important roles. The American Association for the Advancement of Science (AAAS), founded in 1848, grew out of an earlier association of geologists. The weekly journal of the AAAS, *Science,* founded in 1880, eventually became one of the most important scientific journals in the world. The Royal Swedish Academy of Sciences selects the winners of the scientific Nobel prizes. The Soviet Academy of Sciences played a much stronger role than is normally true of national societies, directing the practice of science in the Soviet Union through a centralized system of control not seen in other countries, except for similar national organizations in other communist countries.

The German Max Planck Society, founded in 1948 as a successor to the Kaiser Wilhelm Society for the Advancement of Science, received significant funding from the federal and state governments of West Germany and has continued to receive state support from the new united German government. The society inherited institutes from the Kaiser Wilhelm Society and founded more as Germany recovered from World War II. The institutes have devoted themselves to basic research, producing numerous Nobel prize winners. The society promotes research in its own institutes and at universities, and supports the career development of junior scientists.

Every scientific discipline has an association. One of the best ways for historians to see that a scientific subdiscipline has been successfully created is to look for the founding of the accompanying association. Examples of associations in America include the American Psychological Association, American Astronomical Society, American Medical Association, and the Geological Society of America. Comparable associations exist in other countries.

National associations often function as a method to promote interdisciplinary research, as do the broader international associations. The International Geophysical Year (IGY) of 1957–1958 was organized and run by the International Council of Scientific Unions (ICSU), an association founded in 1931 to foster international cooperation in science. Associations also engage in media relations, campaigns for funding, political activism, and promotion of science as a way to knowledge.

See also International Geophysical Year; Nobel Prizes; Soviet Academy of Sciences; Universities

References
American Association for the Advancement of Science. http://www.aaas.org/ (accessed February 13, 2004).
Dael, Wolfle. *Renewing a Scientific Society: The American Association for the Advancement of Science from World War II to 1970.* Washington, DC: American Association for the Advancement of Science, 1989.
Pyenson, Lewis, and Susan Sheets-Pyenson. *Servants of Nature: A History of Scientific Institutions, Enterprises, and Sensibilities.* New York: Norton, 1999.
Royal Society. http://www.royalsoc.ac.uk/ (accessed February 13, 2004).

Astronomy

The first half of the twentieth century revolutionized astronomy and cosmology. Giant telescopes in America, endowed mostly by private foundations, pushed out the limits of the known universe. The Milky Way became only one galaxy among many when other galaxies were discovered. In 1929 the American astronomer Edwin Powell Hubble (1889–1953) found that all galaxies are receding from each other, and that the faster a galaxy is receding, the farther away it is, and the more its spectra shift toward the longer wavelengths of the red end of the electromagnetic spectrum. Albert Einstein's (1879–1955) theory of relativity and the emergence of quantum mechanics offered a new foundation for understanding the stars. The years since 1950 have revealed that the cosmos is much stranger and much larger than expected, populated with neutron stars, quasars, pulsars, black holes, colliding galaxies, interstellar dust, and the elusive dark matter.

New techniques of telescope construction have made ever more powerful telescopes. The governments of the United States, Japan, Europe, and the Soviet Union became important sponsors of astronomical research, reflecting an emphasis on big science projects, though funds from private foundations have remained important. Radio, x-ray, and infrared astronomy emerged to contribute knowledge from their shares of the electromagnetic spectrum. The opening of the space age in 1957 has enabled various satellite telescopes and other satellites to observe the universe unhindered by the atmosphere. Computers have allowed astronomers to harness these major advances in observation by making more sensitive images, enhancing communication among scientists, and enabling the creation of sophisticated models of physical processes.

American and Soviet unmanned and manned space exploration completely changed human understanding of the solar system. From 1969 to 1972, the Apollo project sent twelve astronauts to the Moon, the only astronomical body ever visited by humans. American spacecraft built by the Jet Propulsion Laboratory of the National Aeronautics and Space Administration (NASA) visited every planet in the solar system except for Pluto. Soviet spacecraft visited Mars and Venus. European spacecraft, sponsored by the European Space Agency, visited comets and observed the Sun. Among the most impressive efforts were the twin American *Voyager* spacecraft, which explored the outer gas giants, finding erupting volcanoes on Jupiter's moon Io, and possible giant oceans covered with ice on other large moons of Jupiter and Saturn. The 1994 impact of the comet Shoemaker-Levy 9 demonstrated that the impact craters found on Earth, other planets, and various moons were not just a relic of the past, but indicated a present danger. The geologist Eugene Shoemaker (1928–1997) devoted his life to finding and understanding such dangers. Such an impact on Earth would be catastrophic for human civilization and maybe even for all higher life-forms on the planet.

In the 1940s, Fred Hoyle and others advanced steady state theories to explain the origin of the universe. According to this hypothesis, matter was continually being created as the universe expanded. The opposing theory proposed that the universe began in a single event, which Fred Hoyle ironically labeled the "Big Bang." Both theories sought to explain why the universe was expanding. The discovery of distant, intense sources of energy called quasars (quasi-stellar radio sources) in 1963 and the discovery in 1964 of a background microwave radiation temperature near the level the Big Bang theorists had predicted shifted scientific support to favor the Big Bang theory. Cosmology became an exciting field, as scientists used relativity theory and quantum mechanics to strive to understand how the universe was created in the first few minutes after the Big Bang. Advances in particle physics came to bear directly on cosmology through the study of the Big Bang.

Searching for a way to explain why heavier

The Saturn V *rocket lifts off on the way to the Moon on July 16, 1969, beginning the* Apollo 11 *mission. (National Aeronautics and Space Administration)*

elements were created in a steady-state universe, Hoyle and his associates showed that a nuclear process in stars called nucleosynthesis turned lighter chemical elements into heavier elements. Scientists developed a more sophisticated understanding of stellar evolution: how stars were created, the nature of their life cycle, and the process of their demise. The equations of relativity and quantum mechanics predicted that some stars might collapse into neutron stars, where matter was packed so tightly by gravity that atoms collapsed into their constituent parts. The graduate student Jocelyn Bell Burnell (1943–) discovered a pulsar in 1967, a rapidly rotating neutron star that broadcast rapid bursts of intense radio waves like a manic lighthouse. The discovery of pulsars inclined scientists to suspect that black holes might also exist. Black holes are formed when a star contains enough mass to collapse past the neutron star stage into a singularity, from which not even light can escape. John A. Wheeler (1911–), Subrahmanyan Chandrasekhar (1910–1995),

Stephen Hawking (1942–), and Roger Penrose (1931–) are important theorists on black holes. In 1971 an orbiting satellite observatory detected intense x-rays from a binary system in the constellation Cygnus. Many astronomers believe that a black hole (called Cygnus X-1) is in orbit around a blue supergiant some 6,000 light-years away in our own galaxy, sucking material away from its companion star and causing a lot of x-ray noise in the process. Black holes are now thought to exist at the center of galaxies, consuming stars and growing ever more massive as billions of years pass. In 1994, the Hubble space telescope found what many consider to be evidence of a large black hole in the heart of the M87 galaxy.

In 1995, Michel Mayor (1942–) and Didier Queloz (1966–), Swiss astronomers at the Geneva Observatory, found strong evidence through measuring Doppler shifts that a planet of at least half the mass of Jupiter orbited 51 Pegasi, a star similar to the Sun about forty-five light-years away. Since 1995, numerous other extrasolar planets have been discovered by observations of subtle Doppler shifts. The planets discovered so far are all the size of gas giants, but smaller than brown dwarfs.

The discovery of extrasolar planets has added impetus to the existing efforts in the search for extraterrestrial intelligence (SETI). In 1960, Frank D. Drake (1930–) listened for possible extraterrestrial radio transmissions from two nearby stars, Tau Ceti and Epsilon Eridani. Each star is similar to our own Sun. Drake did not find success, nor did other American and Soviet scientists in the following decades, but SETI efforts gained considerable momentum in the 1990s from the development of new technologies and the infusion of private funding. The astronomer Carl Sagan (1934–1996) often promoted efforts in SETI, as well as popularizing science and planetary astronomy.

Archaeoastronomy, the study of how past cultures understood astronomy, also grew during the later part of the twentieth century.

The orientation of Egyptian monuments and Stonehenge to various types of stellar axis had already been recognized, but now archaeoastronomers found similar alignments of prominent ancient buildings to stellar events, such as the summer and winter solstices and the fall and spring equinoxes, among Mesoamerican cultures, in Europe (in its megalithic monuments), among the Anasazi of the desert in the southwestern United States, and in China.

See also Apollo Project; Bell Burnell, Jocelyn; Big Bang Theory and Steady State Theory; Big Science; Black Holes; Chandrasekhar, Subrahmanyan; European Space Agency; Extrasolar Planets; Foundations; Hawking, Stephen; Hoyle, Fred; Jet Propulsion Laboratory; National Aeronautics and Space Administration; Penrose, Roger; Particle Physics; Physics; Pulsars; Quasars; Sagan, Carl; Satellites; Search for Extraterrestrial Intelligence; Shoemaker, Eugene; Shoemaker-Levy Comet Impact; Space Exploration and Space Science; Telescopes; Voyager Spacecraft; Wheeler, John A.

References

Aveni, Anthony. *World Archaeoastronomy.* Cambridge: Cambridge University Press, 1989.

Dick, Steven J. *The Biological Universe: The Twentieth-Century Extraterrestrial Life Debate and the Limits of Science.* Cambridge: Cambridge University Press, 1996.

Hirsh, Richard F. *Glimpsing an Invisible Universe: The Emergence of X-Ray Astronomy.* New York: Cambridge University Press, 1983.

Hoskin, Michael, editor. *The Cambridge Illustrated History of Astronomy.* New York: Cambridge University Press, 1997.

North, John. *The Norton History of Astronomy and Cosmology.* New York: Norton, 1995.

Overbye, Dennis. *Lonely Hearts of the Cosmos: The Story of the Scientific Quest for the Secret of the Universe.* New York: HarperCollins, 1991.

B

Bardeen, John (1908–1991)

John Bardeen shared two Nobel Prizes in Physics for his work on the invention of the transistor and research on superconductivity. Bardeen was born in Madison, Wisconsin, where his father was dean of the medical school. He attended an experimental elementary school that allowed him to skip grades and he graduated from high school at the age of fifteen. He earned a B.S. in 1928 and an M.S. in 1929 from the University of Wisconsin, both in electrical engineering. In 1936, he earned a Ph.D. in mathematical physics from Princeton University. During World War II, he served as a civilian physicist with the U.S. Naval Ordnance Laboratory.

After the war, Bardeen joined Bell Telephone Laboratories in Murray Hill, New Jersey, the premier industrial research laboratory in the world. He worked with William Shockley (1910–1989) and Walter H. Brattain (1902–1987) to invent the transistor in 1947, a replacement for vacuum tube technology. Transistors led to smaller and cheaper electronics, making possible the space program, more advanced computers, and integrated circuits. Bardeen shared the 1956 Nobel Prize in Physics with Shockley and Brattain.

In 1951, Bardeen joined the University of Illinois at Urbana-Champaign as a professor of electrical engineering and a professor of

John Bardeen, physicist and two-time Nobel Prize winner. (Bettmann/Corbis)

physics, where he returned to an earlier interest in superconductivity. Superconductivity had been discovered in 1911 when the

Dutch physicist Heike Kamerlingh Onnes (1853–1926) found that some metals lost all resistance to the flow of electrons at temperatures near absolute zero. Onnes received the Nobel Prize in Physics for this discovery only two years later. Bardeen worked on this problem, later joined in his efforts by two research assistants, Leon J. Cooper (1930–) and J. Robert Schrieffer (1931–). In 1957, Bardeen published with his two assistants the BCS (an acronym from the initials of their three last names) theory of superconductivity, offering a theoretical foundation for superconductivity. For this achievement, Bardeen shared the 1972 Nobel Prize in Physics with Cooper and Schrieffer. Their theory eventually led to the development of superconducting materials that operated at more normal temperatures, enabling the development of powerful electromagnets of smaller size that consumed less energy than previous electromagnets.

See also Computers; Integrated Circuits; Müller, K. Alex; Nobel Prizes; Physics

References

Hoddeson, Lillian. *True Genius: The Life and Science of John Bardeen: The Only Winner of Two Nobel Prizes in Physics.* Washington, DC: Joseph Henry, 2002.

"John Bardeen." Special Issue of *Physics Today* 45 (April 1992).

Barghoorn, Elso (1915–1984)

The paleobotanist Elso Barghoorn discovered microfossils, eventually pushing back the known age of life on Earth to 3.4 billion years. Barghoorn was born in New York City, raised in Dayton, Ohio, and went to graduate school at Harvard University in 1937. His interests centered on paleobotany, and four years later he earned a doctorate in botany. During World War II, Barghoorn served on Barro Colorado Island in the Panama Canal Zone, where he studied the filamentous fungi that rotted the clothing of soldiers in the tropical regions of the Pacific Ocean. After the war, he returned to Harvard and eventually became a professor of botany.

In 1953, a geologist from the University of Wisconsin, Stanley A. Tyler (1906–1963), brought photographs of microscopic chert that he had found in the Gunflint Formation in Ontario, Canada, to a conference at Harvard. There he was introduced to Barghoorn, and the paleobotanist decided that the photographs showed fossils of microscopic algae and fungi. Tyler and Barghoorn published their conclusions in the journal *Science* the following year.

Ever since Charles Darwin (1809–1882) convinced scientists that life had existed for a long period of time on Earth and evolved through a process of natural selection, scientists had been creating a tree of life. The tree began in the Cambrian period, about 550 million years ago, when an explosion of life left a profusion of fossils. The Precambrian era lacked any known fossils. Many thought that the boundary between the Cambrian and the Precambrian marked the point when single-celled life evolved into more complex forms of multicellular life. In the past, some geologists had identified microfossils in Precambrian deposits, but these bits of evidence were dismissed as geological artifacts. Tyler and Barghoorn's article began to change this perception.

In the Soviet Union, Boris Vasil'evich Timofeev (1916–1982) also identified Precambrian microfossils. The Soviet scientists at the Institute of Precambrian Geochronology in Leningrad developed a technique of dissolving rocks in acid, allowing the microfossils in them to be collected. This technique was prone to contamination, which, combined with cold war suspicions, prevented Timofeev's finds from being commonly accepted in the West. Nevertheless, other scientists, such as Preston Cloud (1912–1991) and the Australian Martin F. Glaessner (1906–1989), began to get involved in the exciting new field.

Tyler and Barghoorn continued to collect samples and photographs from the Gunflint Formation, though Tyler's death in 1963 left Barghoorn and his graduate students to carry on. Cloud was able to figure out where they had collected their samples in Ontario and worked on the problem on his own. When Cloud was ready to publish his research, Barghoorn hurried to complete his own paper, and both were published in *Science* in 1965, with the Barghoorn and Tyler paper appearing first. That paper prompted media interviews of Barghoorn and further interest in the field.

In subsequent years, Barghoorn remained an active leader in the field of microfossils, and the field is now well established. Eventually, microfossil finds from around the world pushed back the earliest known life on Earth to 3.5 billion years ago. Barghoorn also worked on lunar samples brought back by the Apollo astronauts, seeking in vain any evidence of microfossils. After his death, the saga of microfossils continued with the announcement in 1996 that a meteorite from Mars, recovered from the surface of Antarctica, contained microfossils. This evidence of extraterrestrial unicellular life is still disputed within the scientific community.

See also Cold War; Evolution; Geology
References
Knoll, Andrew H. "Elso Sterrenberg Barghoorn, Jr." *Proceedings of the American Philosophical Society* 135, no. 1 (1991): 87–90.
Schopf, J. William. *Cradle of Life: The Discovery of Earth's Earliest Fossils.* Princeton: Princeton University Press, 1999.

Barton, Derek (1918–1998)

Sir Derek Harold Richard Barton revolutionized organic chemistry with his development of conformational analysis. He was born in Gravesend, Kent, England; his father's carpentry business allowed Barton to attend a private school. He received a B.S. in chemistry from the Imperial College of Science and Technology in 1940, and a Ph.D. in organic chemistry two years later. Recruited into the war effort during World War II, he investigated the properties of invisible inks. With the war over, he returned to his alma

mater, where he taught inorganic chemistry and earned a second doctorate in chemistry.

In 1949, Barton accepted a visiting professorship at Harvard for a year, where he studied steroids. Building upon the little-known work of Odd Hassel (1897–1981), a German-trained Norwegian chemist, Barton created conformational analysis, showing how the shapes of organic carbon rings dictated their chemical and biological properties. Barton and Hassel shared the 1969 Nobel Prize in Chemistry for their work. Returning to the United Kingdom, Barton worked at the University of London and University of Glasgow, before settling down at Imperial College in 1957. He was knighted in 1972.

Barton did not rest as he grew older and always remained vigorously engaged in research. When a mandatory retirement age approached in England, he moved to France in 1978 to assume a position as director of the Institute for the Chemistry of Natural Substances in Gif-sur-Yvette, France. His first marriage resulted in a son and ended in divorce. His second marriage was to a French professor. When retirement again threatened, Barton moved to the United States in 1986, where Texas A&M University offered him a distinguished professorship and the opportunity to continue his research activities. The death of his second wife led to a third marriage in Texas before his own death in 1998.

See also Chemistry; Nobel Prizes
References
Barton, Derek H. R. *Some Recollections of Gap Jumping.* Washington, DC: American Chemical Society, 1991.
Nobel *e*-Museum. "Derek Barton—Biography." http://nobelprize.org/chemistry/ (accessed February 12, 2004).

Bell Burnell, Jocelyn (1943–)

Jocelyn Bell Burnell discovered pulsars in 1967 as a graduate student. Susan Jocelyn Bell was born in Belfast, Northern Ireland, where her father was an architect. One of his jobs included work at the Armagh Observatory, and

The astronomers Jocelyn Bell and Antony Hewish among the wires of the radio telescope, with which she discovered the first pulsar, in East Anglia, England, in 1967 (Hencoup Enterprises Ltd. / Photo Researchers, Inc.)

when she accompanied him to work, Bell became fascinated by astronomy through her conversations with observatory staff. Her early schooling was academically disappointing, but the Mount School, a private Quaker boarding school, allowed her to recover and learn to excel. Raised as a member of the Society of Friends (Quakers), Bell remained active in religious work her entire life.

In 1965 she earned a B.Sc. degree in physics from the University of Glasgow. As the only woman physics student, she at times experienced teasing and cruel comments. She learned not to blush and to laugh about the difficulty of being a pioneering woman in a field dominated by males. She entered Cambridge University in 1965 as a graduate student in astronomy. Her thesis advisor, Antony Hewish (1924–), had found the funds to build a radio telescope to study quasars. His graduate students built the 4.5-acre telescope themselves, placing instruments atop a thou-

sand nine-foot posts and stringing numerous cables from pole to pole. Bell became adept at the manual labor of working cable and electronics.

When the telescope began operation in July 1967, Hewish assigned Bell the job of going through the printouts of data. This demanding task required attention to detail and developing a sense of which signals came from terrestrial origins and which came from the stars. Within six months the telescope generated 3.5 miles of squiggles on paper. In October, Bell noticed a half-inch of "scruff" that intrigued her. Working backward, she found that the small signal always corresponded to a specific sidereal time, 23 hours and 56 minutes, indicating that it always came from the same spot in the sky. A more sensitive recorder installed in November allowed Bell and Hewish to look at the scruff in more detail. They found that the signal pulsed in intensity every 1.33 seconds. This was much too fast for a variable star, and its eerie regular intensity made some on the research team wonder if they had actually stumbled across a transmission from an extraterrestrial intelligence. A second signal, discovered in December in a different part of the sky, pulsed at a rate of 1.2 seconds. With the signals so far apart from each other, thoughts of extraterrestrial intelligence were laid to rest. Bell and the team had discovered a new phenomenon. When Hewish announced the discoveries in February, Bell had already found another two pulsars. The discovery electrified astronomers.

In 1968, Bell received her Ph.D. and married Martin Burnell. Her husband's employment as a government official demanded occasional relocations, forcing her to move with him. She left the intellectual mainstream of radio astronomy and joined the faculty of the University of Southampton. In 1973, she resigned from the university to take care of her newborn son. A year later, she joined the Mullard Space Science Laboratory at University College in London, where she worked in x-ray astronomy. In 1974, Hewish and Martin Ryle (1918–1984) shared the Nobel Prize in Physics for their work in radio astronomy. The noted astronomer Fred Hoyle (1915–2001) publicly objected to the injustice of not also awarding Bell Burnell the Nobel Prize. Hoyle pointed out that she had made the actual discovery of pulsars, a difficult feat, and that the prizes were supposed to be for discoveries. Bell Burnell herself disagreed, noting that she was just a graduate student at the time. She eventually received a variety of other prestigious awards in astronomy for her discovery of pulsars.

Bell Burnell worked at the Royal Observatory in Edinburgh, Scotland, from 1982 to 1991 and became a full professor of physics at the Open University in Milton Keynes in 1991. Her marriage ended in divorce in 1989.

See also Astronomy; Hoyle, Fred; Nobel Prizes; Pulsars; Quasars; Search for Extraterrestrial Intelligence; Telescopes

References

Greenstein, George. *Frozen Star: Of Pulsars, Black Holes, and the Fate of Stars.* New York: Freundlich, 1983.

McGrayne, Sharon Bertsch. "Jocelyn Bell Burnell," 359–379. *Nobel Prize Women in Science.* Second edition. Secaucus, NJ: Carol, 1998.

Woolgar, S. W. "Writing an Intellectual History of Scientific Achievement: The Use of Discovery Accounts." *Social Studies of Science* 6 (1976): 395–422.

Berners-Lee, Timothy (1955–)

Timothy Berners-Lee invented the World Wide Web, the technology that made the Internet accessible to the masses. Berners-Lee was born in London to parents who were both mathematicians and who had worked as programmers on one of the earliest computers, the Mark I at Manchester University. He graduated with honors and a bachelor's degree in physics from Oxford University in 1976. In 1980, he went to work at the Conseil Européen pour la Recherche Nucléaire (CERN), a nuclear research facility on the French-Swiss border, as a software developer.

The physics community at CERN used computers extensively, with data and documents scattered across a variety of different computer models, often created by different manufacturers. Communication between the different computer systems was difficult. A lifelong ambition to make computers easier to use encouraged Berners-Lee to create a system to allow easy access to information. He built his system on two existing technologies: computer networking and hypertext. Computer networks had been invented in the 1960s and were difficult to use, though a worldwide Internet—a network of networks—had emerged. Hypertext was developed in the 1960s by the development team at Stanford Research Institute led by the computer scientist Douglas Engelbart (1925–), based on the idea that documents should have hyperlinks in them connecting to other relevant documents.

Berners-Lee created a system that delivered hypertext over computer networks using a set of rules called the hypertext transfer protocol (HTTP). He simplified the technology of hypertext to create a display language that he called hypertext markup language (HTML). The final innovation was to create a method of uniquely identifying any particular document in the world. He used the term *universal resource identifier* (URI), which became *universal resource location* (URL). In March 1991, Berners-Lee gave copies of his new *WorldWideWeb* programs, a web server and text-based web browser, to colleagues at CERN. By that time, Internet connections at universities around the world were common, and the World Wide Web (WWW) caught on quickly, as other people readily converted the programs to different computer systems. A team of staff and students at the National Center for Supercomputing Applications at the University of Illinois at Urbana-Champaign released a graphical web browser in February 1993, making the WWW even easier to use. The WWW made it easy to transfer text, pictures, and multimedia content from computer to computer. Berners-Lee's original vision of the WWW included the ability for consumers to interactively modify the information they received, though this proved technically difficult and has never been fully implemented.

An Internet economy based on the WWW emerged in the mid-1990s, dramatically changing many categories of industries within a matter of only a few years. A major key to the success of the WWW was the generosity on the part of CERN and Berners-Lee in not claiming any financial royalties for the invention. Berners-Lee moved to the Massachusetts Institute of Technology in 1994, where he became director of the World Wide Web Consortium. This organization, under the guidance of Berners-Lee, has continued to coordinate the creation of new technical standards to enable the WWW to grow in ability and power.

See also Computers; Internet
References

Berners-Lee, Tim. *Weaving the Web: The Original Design and Ultimate Destiny of the World Wide Web by Its Inventor.* San Francisco: HarperSanFrancisco, 1999.

Gillies, James. *How the Web Was Born: The Story of the World Wide Web.* New York: Oxford University Press, 2000.

Berson, Solomon A. (1919–1972)

See Yalow, Rosalyn (1921–)

Big Bang Theory and Steady State Theory

During the 1940s, 1950s, and 1960s, two scientific theories of the origin of the universe competed for dominance. In 1929, Edwin Powell Hubble (1889–1953) discovered that all galaxies were receding from each other as part of an expanding universe. Hubble also found that the faster a galaxy was receding, the farther away it was, and the more its spectra shifted toward the longer wavelengths of the red end of the electromagnetic spectrum. The obvious conclusion was that the universe was expanding. Earlier theorists, notably the Russian physicist Alexander Friedmann in 1922, had already proposed that the universe began as a primordial egg or in some other form of initial creation, rather than that the universe had always existed.

In the 1940s, nuclear physicists began to examine the problem. They approached this central issue of cosmology by asking where the chemical elements came from. In 1942, George Gamow proposed that the elements found today and the proportions in which they are found came about as the result of the breakdown of the heavy elements that composed the initial universe. Gamow later reversed himself in 1946 and concluded that the universe initially consisted of light elements only, which later combined to form the heavier elements. In 1948, Gamow and his student Ralph Alpher wrote a paper on "The Origin of Chemical Elements," one of

Arno Penzias (left) and Bob Wilson standing before the Bell Labs horn antenna that they used to discover background radiation. The antenna provided the first strong evidence supporting the Big Bang theory. (Roger Ressmeyer/Corbis)

many contributions to the Big Bang theory that Gamow made and a key step to the modern understanding of nucleosynthesis, the name given to the way the nuclear processes within stars create heavier elements out of lighter elements.

Fred Hoyle (1915–2001), Hermann Bondi (1919–), and Thomas Gold (1920–) published an alternate theory, also in 1948, the steady state theory of cosmology. Hoyle became the leading proponent of the steady state theory and vigorously defended the idea that the universe had always existed, with matter being constantly created to fuel its expansion. The opposing theory proposed a moment of creation, a primordial egg that exploded to create the universe. Hoyle dubbed the opposing theory the Big Bang, and the name stuck. During the 1950s, debate over the two competing theories raged among astronomers, with Gamow and Hoyle representing the opposing sides.

The discovery in 1963 of quasars, intense energy sources on the outer edge of the known universe, lent credence to the Big Bang, since the steady state theory could not explain them, while Big Bang theorists argued that quasars were glimpses of the earlier state of the universe. In June 1964, Arno Penzias (1933–) and Robert Wilson (1936–), working at Bell Labs in New Jersey, took over an abandoned horn antenna originally created to communicate with an early Telstar communications satellite. They wanted to use the antenna for radio astronomy, and after modifying the antenna, pointed the cone to the sky as a test. The antenna happened to be set to the 7.35-centimeter wavelength and they expected to hear nothing, since our galaxy does not emit anything at that spectrum. They instead picked up a signal that corresponded to 3.5 kelvins, just above absolute zero. Ten months of tests showed that this microwave signal came from all directions in the sky.

Contact with a group working at Princeton alerted the bewildered duo of Penzias and Wilson that the Big Bang theory predicted a background radiation temperature near what they had found. This became the pivotal piece of evidence that the steady state theory had not predicted and could not explain. In 1978 Penzias and Wilson received the Nobel Prize in Physics for their accidental discovery.

By the end of the 1960s, especially bolstered by the evidence of background radiation in the universe, the Big Bang theory had gained preeminence, and the steady state theory became a relic of history. A NASA team of scientists announced in 1992 that the *Cosmic Background Explorer* (COBE) satellite had surveyed the sky and detected minor variations in the cosmic background radiation, exactly what Big Bang theorists had predicted. Further efforts in cosmology have focused on refining and detailing the theory, concentrating especially on the first few minutes after the birth of the universe. Speculation has also focused on the question of whether there is enough matter in the universe to eventually reverse the expansion of the universe so that the universe will start to contract, eventually leading to a "big crunch," when all matter comes together in a giant black hole. At that point perhaps another Big Bang will occur, beginning a new universe, perhaps part of a continuing cycle of Big Bangs and big crunches.

Among those whose religious beliefs point to the creation of the universe by God, the Big Bang has often seemed more compatible with their point of view than the steady state theory. Those who believe that God and the universe have always existed find the steady state theory more compatible. Other creationists have rejected both the Big Bang and steady state theories, creating pseudoscientific theories based on literal interpretations of scripture of how the universe was created.

See also Astronomy; Creationism; Gamow, George; Hawking, Stephen; Hoyle, Fred; National Aeronautics and Space Administration; Quasars; Satellites

References

Alpher, Ralph A., and Robert Herman. *Genesis of the Big Bang.* New York: Oxford University Press, 2001.

Fox, Karen G. *The Big Bang Theory: What It Is, Where It Came from, and Why It Works.* New York: Wiley, 2002.

Gribbon, John. *In Search of the Big Bang: Quantum Physics and Cosmology.* New York: Bantam, 1986.

Hawking, Stephen W. *A Brief History of Time: From the Big Bang to Black Holes.* New York: Bantam, 1988.

Kragh, Helge. *Cosmology and Controversy: The Historical Development of Two Theories of the Universe.* Princeton: Princeton University Press, 1996.

Mather, John C., and John Boslough. *The Very First Light: The True Inside Story of the Scientific Journey Back to the Dawn of the Universe.* New York: Basic, 1996.

Overbye, Dennis. *Lonely Hearts of the Cosmos: The Story of the Scientific Quest for the Secret of the Universe.* New York: HarperCollins, 1991.

Big Science

The term *Big Science* refers to scientific projects involving centralized control over large numbers of scientists and technical staff, supported by large amounts of funding, usually from governments. These projects often include the application of big technology, which is often characterized by expensive and physically large technology. Big science is also characterized by scientific projects, which have grown in size from small teams to large, widely distributed enterprises. Although much of the scientific and technological research in the Soviet Union was centrally directed by the Soviet Academy of Sciences or industrial ministries, only some of the projects reached a scale to be considered Big Science.

Ancient civilizations built impressive structures, such as the Egyptian pyramids, the Great Wall of China, and major canals to connect rivers to oceans, all requiring highly organized states that mobilized thousands of workers over a period of years. Although technological projects on that scale continue today, Big Science is a unique category. The

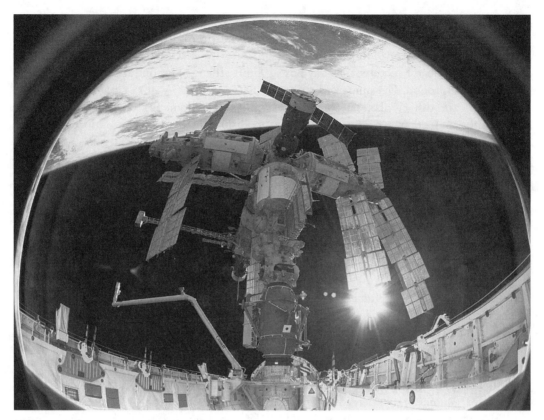

The space shuttle Atlantis *docks with the Russian space station Mir as they orbit the earth in tandem, November 15, 1995. (National Aeronautics and Space Administration)*

American Manhattan Project during World War II which developed the first atomic bombs in 1945, is the prototype of big science projects. This first big science project cost $2 billion and involved 100,000 workers. The American physics community, riding on the laurels of the Manhattan Project, turned to other applied science projects to develop the hydrogen bomb and the nuclear reactor. In the quest to understand ever smaller components of the atom, large particle accelerators were built, some of them many miles in length. The physicist Alvin M. Weinberg (1915–), the director of the Oak Ridge National Laboratory, popularized the term *Big Science* in 1961 to describe the new scale of this research.

Space exploration only became possible with the backing of the wealthiest governments. The Apollo project cost $25 billion dollars and succeeded in landing a human on the moon in 1969; a total of six successful landings brought back 838 pounds (380 kilograms) of lunar rocks. The Apollo project became the first big customer of integrated circuits and played a key role in accelerating the growth of the semiconductor industry. After the cancellation of further Apollo landings, the National Aeronautics and Space Administration (NASA) created the space shuttle, the most complex machine ever designed by humanity, first launched in 1981.

Big science is also characterized by building scientific instruments that single scientists could never build for their own individual use. The best examples of this kind of instrument are the early large telescopes. Telescopes on the ground cost a considerable amount of money and are usually built by consortiums of universities and governmental organizations.

The premier space-based telescope, the Hubble Space Telescope (HST), launched from a space shuttle in 1990, revolutionized astronomy, providing clearer pictures than any possible from the ground. The numerous spacecraft sent to other planets to return photographs and other sensor readings, including landings and probes on Venus, Mars, and Jupiter, have resulted from a long-term commitment of financing and resources by American, European, and Soviet (now Russian) space agencies. The International Space Station, begun in the 1990s, was the most expensive ongoing big science project as the millennium came to a close, though many critics have doubted its scientific usefulness, especially when its political usefulness seemed to be the main motivation.

The Human Genome Project marked the coming of Big Science to the biological sciences. Begun in 1990, the project announced in 2003 that it had mapped 99.99 percent of the human genome (100 percent was not reachable because people are genetically different from each other), finishing the project in thirteen years instead of the originally projected fifteen years and coming in under budget at $2.7 billion. The political leaders of the six major nations involved signed the statement announcing final success of the international project. The biological sciences have built on the success of the Human Genome Project to start other large projects, including the Protein Structure Initiative, the Single Nucleotide Polymorphism Consortium, and the International HapMap Project.

Inspired by the dramatic success of the Manhattan Project, Ronald Reagan (1911–2004), the president of the United States, launched in 1983 the development and construction of the Strategic Defense Initiative (SDI), a space-borne system to shoot down incoming intercontinental ballistic missiles. After the expenditure of billions of dollars, the dream proved to be beyond the reach of current science and technology; it must be counted as a failure of big science ambitions, even though research continues on a reduced scale with limited goals—shooting down a few missiles rather than thousands. Another failure came with the ambitious attempt to built the Superconducting Supercollider (SSC) in Texas, with a circumference of 52 miles (87 kilometers) and plans to produce collisions of protons and antiprotons at 40 TeV (trillions of electron-volts). Construction began in 1989, but disputes within the physics community about where limited government funding should go, escalating costs, and federal politics caused the SSC to founder on federal budget cuts in 1993 after being partially built. A major rationale for the SSC, national prestige, no longer mattered as much after the cold war ended in 1991. The Human Genome Project continued to receive funding because of promised advances in medical research.

Other big science projects included the International Geophysical Year of 1957–1958, the smallpox vaccination campaign that succeeded in eradicating that dread disease in 1979, various oceanographic research initiatives, and the continuing research presence in Antarctica. Big science has become big business, as government-sponsored projects have funneled funding into universities, aerospace companies, national laboratories, and international laboratories like the Conseil Européen pour la Recherche Nucléaire (CERN). Advances in genetic engineering have led to successful bioengineering corporations. Though created by scientific developments, biotechnology corporations usually are not examples of Big Science, lacking the large number of people and the same scale of funding. Big science projects receive considerable public attention and media coverage, leading to a distorted view of science by many laypeople. Critics of Big Science often point out that the big projects vacuum up so much of the available funding that smaller scientific efforts have difficulty finding funding.

See also Apollo Project; Cold War; Human Genome Project; Hydrogen Bomb; International Geophysical Year; Media and Popular Culture; National Aeronautics and

Space Administration; National Science Foundation; Nuclear Physics; Oceanography; Particle Accelerators; Smallpox Vaccination Campaign; Soviet Academy of Sciences; Soviet Union; Space Exploration and Space Science; Telescopes

References

Galison, Peter, and Bruce Hevly, editors. *Big Science: The Growth of Large-Scale Research.* Stanford, CA: Stanford University Press, 1992.

Kevles, Daniel. "Big Science and Big Politics in the United States: Reflections on the Death of the SSC and the Life of the Human Genome Project." *Historical Studies in the Physical and Biological Sciences* 27 (1997): 269–297.

———. *The Physicists: The History of a Scientific Community in Modern America.* Revised edition. Cambridge: Harvard University Press, 1995.

Price, Derek J. de Solla. *Little Science, Big Science . . . and Beyond.* New York: Columbia University Press, 1963.

Rhodes, Richard. *The Making of the Atomic Bomb.* New York: Simon and Schuster, 1986.

Weinberg, Alvin M. *Reflections on Big Science.* Cambridge: MIT Press, 1967.

Bioethics

Bioethics is the study of how to apply moral and ethical principles in the practice of medicine and the life sciences. After World War II, the victorious Allies convened a war crimes tribunal at Nuremberg, Germany. Judges from among the Allies tried top-ranking Nazis for war crimes and punished the convicted by execution or imprisonment. The trials brought out the complicity of many German physicians and medical researchers in the Nazi murder of Jews, other ethnic minorities, and the mentally or physically handicapped. Investigators discovered that medical researchers had also performed horrific experiments on prisoners, including studies involving fatal exposure to cold water or low pressure, deliberately burning patients, infecting them with diseases, and performing experimental surgical procedures. Japanese medical doctors and scientists in the infamous Unit 731 also engaged in similar aberrant research. Although many commentators considered these activities to be an aberra-

tion, the faith of many commentators in the goodness of the medical profession was shaken. Other cases of medical malfeasance also came to light in subsequent years.

From 1932 to 1972, to take one notorious example, the United States Public Health Service conducted a study during which physicians withheld treatment for syphilis from 400 African-American men in Alabama in order to document and study the effects of syphilis as the disease ran its fatal course. The Tuskegee Institute, founded and staffed by African Americans, cooperated in the study. An effective cure for syphilis was not in fact available at the beginning of the study, but even after penicillin became widely available after World War II, it was not given to the subjects. The study ended only when exposed by a newspaper article. Episodes like the Tuskegee experiment led to an ethical movement where researchers needed to obtain informed consent of all human subjects in any research project.

The first institute to study bioethics was founded in 1969 in the United States. Other institutes, organizations, scholarly journals, and academic conferences followed. Medical schools now often include bioethics in their curriculum. The Human Genome Project set aside up to 5 percent of its funding to study ethical, legal, and social issues involved in the project. One of the objections to the project was that the knowledge might lead to eugenics, allowing people to tailor the DNA of their unborn children to conform to modern images of attractiveness, athleticism, and intellectual ability. Cultures that value docility in women might strive to find gene sequences that predispose women to that trait.

Bioethical discussions have also focused on abortion, in vitro fertilization, animal experimentation, organ transplants, cloning, stem cell research, biotechnology, and suicide. Abortion gradually became legal in the United States and Great Britain in the late 1960s and early 1970s. Abortion had been legal earlier in other Western countries, became legal later in other countries, and remains illegal in some countries. In vitro fertilization, first successfully demonstrated in humans in 1978, raised many ethical questions, such as what to do with frozen embryos that are no longer needed. If the parents divorce, who retains control over frozen embryos?

Medical science has long used animals as test subjects in lieu of humans. In the last several decades, animal activists have become increasingly vocal in their opposition to this testing. Medical researchers have successfully argued that they need animal testing, but the activism has led to improved conditions within laboratories. Psychologists have also used animals, not only in the classic rat mazes, but in more sophisticated research. In the 1940s and 1950s, Harry F. Harlow (1905–1981) used affection deprivation studies on rhesus monkeys to prove that affection and love are primary psychological needs. Harlow's research methods, condemned nowadays as inhumane, would not be approved by current ethics review boards, but his results are now part of the accepted wisdom in contemporary psychology.

The first successful organ transplant occurred in 1954 in Boston when a patient received the kidney of his identical twin brother and lived for another eight years. Organ transplants raise a host of ethical questions, especially considering the shortage of donated organs. Where do we get the organs, who gets to receive them, and can we increase the organ supply by using animal organs? In most Western countries, the distribution of organs is controlled by a strictly egalitarian system, though stories have circulated that China has harvested organs from executed prisoners for distribution to those with money or political connections. Most people do not die in a way that allows many of their organs to be harvested. The question also arises, when is a person dead? The legal systems of various legal jurisdictions and religious groups give different definitions of traditional death and brain death (which can be difficult to diagnose).

With the successful cloning of Dolly the sheep in 1996, cloning became a real issue in bioethics. Should humans be cloned? Many scientists have spoken out in opposition to cloning on ethical and practical grounds; other scientists have welcomed cloning and other genetic engineering advances as ways for humans to take conscious control of human evolution. Stem cell research and biotechnology both promise medical miracles. The debate over undifferentiated stem cells, which come from fetuses or umbilical cords, is intertwined with the debate on abortion. Successful stem cell therapies could create an enormous demand for stem cells that would be difficult to meet in an ethical manner.

Issues of death and dying are important in bioethics. The psychiatrist Elisabeth Kübler-Ross (1926–2004) found that terminally ill patients went through a series of five emotional stages as they dealt with their impending death. Family members of terminally ill patients also go through the same stages, as do the bereaved. Kübler-Ross also promoted the hospice movement to create places near hospitals to care for the physical and emotional needs of terminally ill patients. What is a good life and what is a good death are important questions, and here religious perspectives and the perspectives of bioethics can combine to find answers. The extremely controversial practice of physician-assisted suicide also reflects attempts to decide what is a good death.

See also Biotechnology; Cloning; Ethics; Human Genome Project; In Vitro Fertilization; Kübler-Ross, Elisabeth; Medicine; Organ Transplants; Religion

References

Jones, James H. *Bad Blood: The Tuskegee Syphilis Experiment.* New York: Free Press, 1981.

McGee, Glenn, editor. *Pragmatic Bioethics.* Second edition. Cambridge: MIT Press, 2003.

Paul, Ellen Frankel, and Jeffrey Paul. *Why Animal Experimentation Matters: The Use of Animals in Medical Research.* New Brunswick, NJ: Social Philosophy and Policy Foundation, 2001.

Pence, Gregory E. *Brave New Bioethics.* Lanham, MD: Rowman and Littlefield, 2002.

Veatch, Robert M. *The Basics of Bioethics.* New York: Prentice Hall, 2000.

Biology and the Life Sciences

The practice of biology and other life sciences in the last half of the twentieth century has been dominated by the theoretical insights of the theory of natural selection, advances in genetics, and the development of new technologies. For centuries, the issue of vitalism had tormented the life sciences. How is it that living things seemed to be animated and different from inanimate things? The scientific revolution of the sixteenth and seventeenth centuries explained the inanimate world in mechanical terms, but life seemed to be an exception, relying on some unknown "vital principle" that could not be understood in mechanical terms. Advances in ontogeny that explained the development of living organisms, the discovery of the structure of DNA, and a greater understanding of the inner working of cells via cytology finally put vitalism to rest in the twentieth century.

In the 1930s and 1940s, the Russian-born zoologist Theodosius Dobzhansky (1900–1975), the ornithologist Ernst Mayr (1904–), and others reinvigorated Charles Darwin's (1809–1882) theory of natural selection and its emphasis on random mutations, creating what became known as the modern synthesis by combining the theory of natural selection with Mendelian genetics. The paleontologist Stephen Jay Gould (1941–2002) and a colleague promoted the theory of punctuated equilibria, according to which evolution did not always occur in gradual steps, taking long periods of time; at times evolutionary change happened in bursts. A consequence of the modern synthesis is that many scientists have argued that the taxonomic system of naming living things should be changed to reflect their evolutionary history.

The microbiologist Lynn Margulis (1938–) proposed an additional twist to evolution in 1967 with her development of serial endosymbiosis theory (SET), first published in 1967. Margulis argued that eukaryotic cells, which contain nuclei, evolved when non-nucleated bacteria fused together billions of years ago. She also demonstrated that microbes could acquire induced characteristics, and that organelles inside cells were really former bacteria that had fused with other cells. These theories, together with her revival of the importance of symbiosis, came to be accepted. Margulis also supported the Gaia hypothesis, developed by the scientific polymath James Lovelock (1919–) in the late 1960s, which proposed that Earth was like a living organism, maintaining surface and atmospheric conditions so that the chemistry and temperature would sustain life.

The biologist Edward O. Wilson (1929–), the world's leading expert on ants, showed that ants communicate via pheromones; he also developed important contributions to the theory of natural selection. His controversial creation of sociobiology in the 1970s argued that many social behaviors in insects, animals, and humans have a biological basis and are selected for as part of evolution. The English zoologist Richard Dawkins (1941–) expanded on sociobiology and promoted the idea of evolutionary psychology. He also argued that evolution occurred on the level of genes, rather than the level of individuals or species.

Scientists also made important contributions to the study of animal behavior within the framework of natural selection. The zoologist Konrad Lorenz (1903–1989), popularly known for the way ducklings and goslings imprinted on him as their surrogate parent, helped found and promote the science of modern ethnology. Primatologists such as Jane Goodall (1934–) and Dian Fossey (1932–1985) showed that many primate behaviors were not that different from human behaviors.

The discovery of the structure of DNA by the zoologist James D. Watson (1928–) and the biologist Francis Crick (1916–2004) in 1953 revolutionized the study of genetics and led to a shift in focus toward understanding life on the molecular level rather than only on the larger scale preferred by traditional biologists and zoologists. Molecular biology became a dominant field, not only drawing the

Macrophotograph of a female spider, Araneus quadratus, *on the stem of a plant. The spider measures about 18 mm across. (Dr. Jeremy Burgess / Photo Researchers, Inc.)*

most funding and best talent, but intellectually dominating the other subfields within biology. For molecular biologists, the secrets of biology were found in genes, proteins, and enzymes, not on the larger scale of cells, plants, or animals. A similar divide is found between physicists, who often are looking at the smallest possible scale, and chemists, who apply the atomic-level knowledge gained by physicists, but also work on a larger scale to develop practical applications. An exception to this trend, the geneticist Barbara McClintock (1902–1992), known as a traditional biologist, looked at the whole organism with a holistic sense of intuition, rather than from the narrow view that molecular biologists brought to their work. McClintock showed in 1951 that transposable genes existed that could move from chromosome to chromosome.

The increasing understanding of genetics led to the development of genetic engineering in the early 1970s by the biochemists Herbert Boyer (1936–) and Stanley Cohen (1935–). Genetic engineers often used the bacteria *Escherichia coli (E. coli)* and the mustard weed, *Arabidopsis thaliana,* for manipulation in their work. A biotechnology industry developed from genetic engineering, leading many to believe that just as the twentieth century was dominated by physics, the twenty-first century will be dominated by biology.

Oceanography provided new vistas for biologists. One of the more extraordinary discoveries came in 1977: deep-sea hydrothermal vents spewing hot water mixed with minerals from geologically active areas. Scientists were astonished at the unique ecology surrounding the vents. New species of tubeworms, clams, crabs, shrimp, mussels, and other kinds of life have been catalogued. Although the food chain of life in the rest of the ocean is dependent on the basic process of photosynthesis, converting sunlight into energy, the bacteria around the vents use chemosynthesis to create energy, converting the sulfides into organic carbon. The more complex life-forms form symbiotic relationships with the chemosynthetic bacteria and survive off the energy they create. A microorganism discovered in 1982 at a vent in the Pacific Ocean, *Methanococcos jannaschii,* is now considered to be part of a new domain of life consisting of microorganisms called archaea. The two previously recognized domains of life consisted of bacteria and eukaryotes. Archaea are characterized by the lack of a cell nucleus and a smaller number of genes than the organisms in the other domains of life.

The publication in 1962 of *Silent Spring* by the biologist Rachel Carson (1907–1964), documenting the dangers of synthetic pesticides, is often credited with launching the modern environmental movement. In the 1960s, many scientists in the life sciences became concerned about the expanding human population. The entomologist Paul R. Ehrlich (1932–) devoted his life to campaigning for a reduction in the population growth rate and an eventual reduction of human population. This theme became important to the emerging environmental movement, combined with concerns about pollution, species extinction, and the loss of unspoiled wilderness. The only reason that the present human population has avoided starvation is that the Green Revolution of the 1950s through the 1970s combined the use of nitrogenous fertilizers and new strains of staple crops to dramatically increase food production in Third World nations. The environmental movement also drew on the burgeoning insights of ecology, which accentuated the interconnectedness of all beings in the web of life.

For centuries, the botanical riches of tropical rainforests have made them a source of biodiversity, including new medicines. Medical anthropologists, representatives of pharmaceutical companies, and other scientists have sought the sources of drugs used by indigenous peoples and catalogued other substances that might be turned into useful drugs. Environmentalists have emphasized the usefulness of this biodiversity as an additional argument in favor of preserving the rainforests and stopping the extinction of species.

Botany also experienced many advances. During the first half of the century, botanists had discovered that plants secrete hormones, and during the second half, they came to understand how these hormones work. The hormone ethylene, for instance, signals the plant to ripen its fruit, change the size of individual cells, or develop more roots. In the 1950s, botanists discovered the chemical phytochrome, which is light sensitive and signals the plant how to respond to the length of the day. The American biochemist Melvin Calvin (1911–1997) at the University of California at Berkeley used the radioactive isotope carbon-14 to trace the process of photosynthesis in the green alga *Chlorella.* He received the 1961 Nobel Prize in Chemistry for this breakthrough. Deep Green, the Green Plant Phylogeny Research Coordination Group, a project started at the University of California at Berkeley in 1994, coordinated the work of many scientists, examining the phylogenetic makeup of numerous plants in an attempt to reconstruct the evolutionary relationships among all green plants. The project found that the evolutionary history revealed in those genes was much more complex than initially thought. Botanists had always thought that green plants initially emerged from seawater plants, but Deep Green showed that primitive freshwater plants are the original ancestral stock from which all known green plants now on land are descended.

See also Agriculture; Biotechnology; Boyer, Herbert; Carson, Rachel; Cohen, Stanley N.; Crick, Francis; Deep-Sea Hydrothermal Vents; Dobzhansky, Theodosius; Ecology; Ehrlich, Paul R.; Environmental Movement; Evolution; Fossey, Dian; Genetics; Gilbert, Walter; Goodall, Jane; Gould, Stephen Jay; Green Revolution; Lorenz, Konrad; Mayr, Ernst; McClintock, Barbara; Microbiology; National Geographic Society; Oceanography; Population Studies; Psychology; Sociobiology and Evolutionary Psychology; Symbiosis; Watson, James D.; Wilson, Edward O.

References

Joyce, Christopher. *Earthly Goods: Medicine-Hunting in the Rainforest.* Philadelphia: Little, Brown, 1994.

Judson, Horace Freeland. *The Eighth Day of Creation: Makers of the Revolution in Biology.* Expanded edition. Woodbury, NY: Cold Spring Harbor Laboratory, 1996.

Magner, Lois N. *A History of the Life Sciences.* Second edition. New York: Marcel Dekker, 1994.

Mayr, Ernst. *The Growth of Biological Thought: Diversity, Evolution, and Inheritance.* Cambridge: Harvard University Press, 1982.

Smocovitis, Vassiliki Betty. *Unifying Biology: The Evolutionary Synthesis and Evolutionary Biology.* Princeton: Princeton University Press, 1996.

Biotechnology

Civilizations have used biotechnology to ferment food and beverages for thousands of years, for example in the making of cheeses and bread. By the end of the twentieth century, biotechnology based on genetic engineering, manipulating genes to create new bacteria or plants, had become around a $50-billion-a-year industry in just the United States. Most early genetic engineering focused on medical research, though developing new strains of crops for agriculture also soon became important. The discovery of the structure of DNA by the zoologist James D. Watson (1928–) and the biologist Francis Crick (1916–2004) in 1953 eventually led the biochemists Herbert Boyer (1936–) and Stanley Cohen (1935–) to develop the techniques of genetic engineering in the early 1970s. Recombinant DNA enables scientists to combine DNA molecules from two or more sources together inside cells or a test tube. After being inserted into a host organism, such as *Escherichia coli (E. coli),* thus modifying the organism, the new creation is then able to reproduce.

Recognizing the power of this new technology, scientists initially agreed to a year-long moratorium on further research until research guidelines could be developed to minimize the risk of a genetically engineered organism accidentally escaping into the wild. The National Institutes of Health (NIH) announced safety guidelines for recombinant DNA technology in 1976, and as time went by without a major mishap, these controls were relaxed. The ethical standard switched from proving that a given experiment would not be harmful to proving that a given experiment would be harmful.

In 1975, Boyer and a venture capitalist formed Genentech, one of the first genetic engineering firms. The company's first products were variations of *E. coli* that produced insulin and human growth hormone. Genentech patented these new forms of life, though many critics questioned whether life itself should be patented. Walter Gilbert (1932–), who pioneered gene-sequencing techniques, formed with his partners the biotechnology company Biogen in 1978. Numerous other biotechnology firms followed, as venture capitalists saw biotechnology as possessing potential comparable to that of integrated circuits and computers. The scientist as an entrepreneur became a common figure in the biotechnology industry.

The nascent biotechnology industry argued that without patents they could not protect their financial investment from other companies simply copying their technology and selling it as their own. General Electric developed an oil-eating bacterium for possible use in cleaning up oil spills and applied for a patent, which led to a controversial landmark 1980 decision by the U.S. Supreme Court, *Diamond v. Chakrabarty,* that permitted life created in a laboratory to be patented. In 1988, the U.S. Patent Office took the next step by granting a patent on a transgenic mouse developed at Harvard University. The mouse had been altered to be susceptible to breast cancer. Other mice have been altered to be susceptible to various diseases so that they might be used to test new medicines and vaccines. In 1997, the U.S. Patent Office began to grant patents on expressed sequence tags (EST), short sequences of human DNA, causing a chorus of protest by critics that often the patent applicants could not even describe what the sequence actually did. These patent rights made the current form of the biotechnology industry possible.

The development of the polymerase chain reaction (PCR) technique by the American biochemist Kary Mullis (1944–) in 1985 made easier the creation of large amounts of genetic samples for testing, making the process of genetic engineering much more efficient. Advances in computer technology in the form of control equipment and database technology also helped make biotechnology and large-scale sequencing projects possible. The Human Genome Project, launched in 1990 and completed in 2003, provided a completed sequence of 3.1 billion pairs, making up 35,000 to 40,000 genes. In order to gain experience and develop better gene-sequencing technologies, the project also sequenced the genomes of simpler organisms, such as *Escherichia coli* and the genome of the mouse. Genetic engineering also made possible cloning, the creation of a genetic duplicate of a previous organism. Forensic analysis using DNA fingerprinting, first proposed in 1984, by the end of the century became an important tool in solving crimes in which bodily fluids or physical samples are recovered from the crime scene. Scientists also began to use variations of DNA fingerprinting and analysis of mitochondrial DNA on larger populations in order to understand historical demographic movements of ethnic groups.

In the 1990s, experiments in gene transfer therapy began, in which a gene is introduced into a patient (often via a virus) because the patient either lacks that gene or that gene does not function properly. Gene therapy may prove effective in treating diseases such as cystic fibrosis and Huntington's disease that have a strong genetic component. Ultimately, gene therapy could be used to permanently alter a patient so that the patient's body then continues to create any protein or enzyme that the body previously lacked.

By the end of the twentieth century, biotechnology had proved itself to be an important new science and technology, with promises that it may be as important as any previous technology ever invented by humans. The history of humanity has seen biological evolution through natural selection superseded by cultural evolution, and with biotechnology, the possibility of direct biological evolution through deliberate genetic engineering of the human genome is on the horizon. The ethical disputes surrounding this potential development promise to fuel one of the biggest controversies of the twenty-first century.

See also Bioethics; Boyer, Herbert; Cloning; Cohen, Stanley N.; Computers; Crick, Francis; Ethics; Genetics; Gilbert, Walter; Human Genome Project; Microbiology; National Institutes of Health; Watson, James D.

References

Aldridge, Susan. *The Thread of Life: The Story of Genes and Genetic Engineering.* New York: Cambridge University Press, 1996.

Bud, Robert. *The Uses of Life: A History of Biotechnology.* New York: Cambridge University Press, 1993.

Magnus, David, Arthur Caplan, and Glenn McGee, editors. *Who Owns Life?* Amherst, NY: Prometheus, 2002.

Oliver, Richard W. *The Coming Biotech Age: The Business of Bio-Materials.* New York: McGraw-Hill, 1999.

Shannon, Thomas A. *Genetic Engineering: A Documentary History.* Westport, CT: Greenwood, 1999.

Sterckx, Sigrid, editor. *Biotechnology, Patents and Morality.* Second edition. Aldershot, UK: Ashgate, 2000.

Birth Control Pill

The development of an effective birth control pill has had such a dramatic social effect that it is usually just called "the pill." In 1928, two American physicians, George Corner (1889–1981) and Willard M. Allen (1904–), discovered progesterone, a hormone that prompted the uterus to prepare for egg implantation. Other hormones were later discovered that regulated the human menstrual cycle, and scientists realized that an effective

Margaret Sanger, American advocate of birth control (Bettman / Corbis)

chemical contraceptive might be developed to fool a woman's body into assuming that she was already pregnant. The two main problems facing such a proposal were the legal limitations in America on birth control and the difficulty of obtaining sufficient natural supplies of the hormones.

Margaret Sanger (1879–1966), an American nurse and activist who coined the phrase "birth control" in 1914, campaigned for years to educate the American public about birth control techniques and provide contraceptive devices. At that time, the laws of many states in the United States made even birth control education illegal. A series of lawsuits eroded the effects of these laws, and Sanger's Birth Control League became Planned Parenthood. The botanist Marie Stopes (1880–1958) became the leading birth control activist in Britain, fighting a controversial fight similar to Sanger's.

The heiress Katherine Dexter McCormick (1875–1967) asked Sanger in 1951 what the birth control movement most needed. Sanger wanted a cheap, effective method of birth control. They contracted with the physiologist Gregory Pincus (1903–1967), who directed the Worcester Foundation for Experimental Biology, to undertake this task. Pincus found that a Boston gynecologist, in an attempt to regulate the menstrual cycles of his patients to help them with infertility problems, had been experimenting with expensive injections of progesterone and found that it was an effective contraceptive. Having found a solution, Pincus then needed a good source of cheap progesterone.

Though Pincus did not know it at the time, the chemical problems were almost solved. The American chemist Russell E. Marker (1902–1995) discovered in the 1940s a way to extract the steroid sapogenin from a Mexican desert plant and synthesize it into progestogen, a substitute for progesterone. He refused to patent his process, allowing others to freely use the technique.

In 1949, Syntex, a pharmaceutical company, hired the Austrian-born chemist Carl Djerassi (1929–) to work on synthetic steroids. Djerassi quickly found a compound that substituted for progesterone and could be administered orally. He applied for a patent on a progestin compound on November 22, 1951. At the same time, the Polish-born chemist Frank B. Colton (1923–), working at the G. D. Searle pharmaceutical company, found a similar substance and applied for a patent a year and a half after Djerassi did. Searle called its product Enovid.

Pincus conducted a secret trial, using the progestin products from Syntex and Searle, on fifty women in Massachusetts and was encouraged by the results. Larger clinical trials followed in Puerto Rico, Haiti, and Mexico City, where local laws were not so restrictive. They found that certain batches of Searle's product produced fewer side effects. Oddly enough the reason was that those batches had been contaminated with small amounts of estrogen. The contraceptive failure rate from the trial was under 2 percent.

In 1957, Searle received approval from the Food and Drug Administration (FDA) to market Enovid (containing small amounts of estrogen) for women with repeated miscarriages and menstrual disorders. By 1959, about half a million American women were using the product, though not nearly that many suffered from the "approved" problems. Searle requested approval for using the drug as an oral contraceptive, and in May 1960 the FDA approved the first drug prescribed for healthy women. The pill launched a revolution in birth control and sexual expression.

By 1969, researchers had proved that early complaints about side effects of the pill, such as blood clots, stroke, and heart attacks, were justified. They found a direct relationship between the amount of estrogen in the pill and the problems. In the early 1970s a progestin-only "mini-pill" was introduced. It worked by prompting changes in the cervix and uterus that inhibited the sperm from uniting with the egg. The mini-pill was less effective at contraception than the progestin-estrogen pills and did not become as popular. Innovations in the 1970s reduced the amount of estrogen in the first type of pills, keeping them as effective, but reducing the risks. In 1982, the biphasic pill was introduced, followed by the triphasic pill in 1984. These low-dose pills changed the ratio of progestin to estrogen from pill to pill during the normal 21-day cycle that pills are taken.

The feminist movement welcomed the pill and its implications for allowing women to control their own fertility. As medical problems emerged with the pill, some factions within the feminist movement condemned the pill as being foisted on women by an uncaring medical profession and drug companies. Other feminists condemned drug companies for not also developing chemical or hormonal contraceptives for men, leaving the onus of contraceptive responsibility on women only.

See also Feminism; Foundations; Medicine; Pharmacology; Population Studies

References

Briant, Keith. *Marie Stopes: A Biography.* London: Hogarth, 1962.

Clarke, Adele E. *Disciplining Reproduction: Modernity, American Life Sciences, and 'the Problem of Sex.'* Berkeley and Los Angeles: University of California Press, 1998.

Djerassi, Carl. *This Man's Pill: Reflections on the 50th Birthday of the Pill.* New York: Oxford University Press, 2001.

Gray, Madeline. *Margaret Sanger: A Biography of the Champion of Birth Control.* New York: Putnam, 1979.

Marks, Lara V. *Sexual Chemistry: A History of the Contraceptive Pill.* New Haven: Yale University Press, 2001.

Tone, Andrea. *Devices and Desires: A History of Contraceptives in America.* New York: Hill and Wang, 2001.

Watkins, Elizabeth Siegel. *On the Pill: A Social History of Oral Contraceptives, 1950–1970.* Baltimore: Johns Hopkins University Press, 1998.

Black Holes

Black holes are a form of mathematical singularities, collapsed stars whose concentrated mass produces such intense gravity that even light cannot escape. Although Albert Einstein (1879–1955) admitted that the equations of his theory of general relativity could be worked to create a singularity, he thought it physically impossible that such a thing could exist. The German astronomer Karl Schwarzschild (1873–1916) predicted that collapsed stars would not emit radiation, and he tragically died in World War I only months after sending a letter about his calculations to Einstein. Schwarzschild's ideas were considered a mere mathematical quirk. Then two papers published in 1939, both coauthored by Robert Oppenheimer, used the theories of relativity and quantum mechanics to examine what would happen if massive stars collapsed at the end of their life cycle. The equations predicted that the mass would collapse so far down that gravity would overcome the nuclear forces separating subatomic particles and a singularity would be formed.

In a remarkable example of a physicist reinventing himself later in life, John A. Wheeler (1911–) turned from nuclear physics and building bombs to relativistic theory. He focused his attention on singularities and promoted research in this area in the 1950s and 1960s. In 1967, Wheeler gave a talk in New York City on singularities, which he called gravitationally completely collapsed objects. The phrase was a mouthful, and an anonymous member of the audience suggested the term *black hole*. Wheeler used the term in a talk later that year, which was published in 1968, and he is now credited with coining the term. Soviet scientists used the term *frozen star* until the term *black hole* became common.

The Soviet scientists Yakov Boris Zeldovich (1914–1987) and O. H. Guseyov proposed in 1965 that black holes could be located if they were part of a double system, in which the black hole sucked material off its companion star, spitting out x-rays during the process. Roger Penrose (1931–) and Stephen Hawking (1942–) began to work on the problem of black holes and in 1970 published a theory that argued that, although black holes could not be observed, relativistic radiation from near the event horizon of black holes should be detectable. The event horizon is the boundary of black holes, the boundary inside which incoming mass can no longer escape the intense gravity of the black hole. The Schwarzschild radius is used to describe the radius of the event horizon of a black hole. Drawing on the work of Subrahmanyan Chandrasekhar (1910–1995) and others, theoretical physicists calculated the stellar mass necessary for a star to eventually collapse into a black hole.

Although theory predicted the existence of black holes, observational evidence was nonexistent. The discovery of pulsars in 1967 offered hope, when scientists concluded that pulsars were really spinning neutron stars. The same chain of mathematical reasoning that led to the idea of neutron stars also led to that of black holes. In 1971, an orbiting satellite observatory detected intense x-rays emanating from a binary system in the constellation Cygnus. Many astronomers believe that this system, called Cygnus X-1, consists of a black hole in orbit around a blue supergiant some 6,000 light-years away in our own galaxy, sucking material away from its companion star and causing a lot of x-ray noise in the process.

The study of singularities not only is about the eventual fate of large stars, but also finds application in the theory of the Big Bang. The Big Bang is thought to be a massive singularity that exploded to create the universe. In the 1970s, Hawking proposed that in the immediate aftermath of the Big Bang, numerous miniature black holes were formed, each no larger than a proton. Combining quantum mechanics with relativity theory showed Hawking, to his astonishment, that these miniature black holes would emit radiation. Eventually they would explode, scattering energy and particles. This revolutionary idea, running counter to all current theory on black holes, became known as Hawking Radiation. According to his theory, these miniature black holes were an important part of the sequence of events following the Big Bang, but no miniature black holes now remain.

Scientists also posited that black holes exist at the center of galaxies, consuming stars and growing ever more massive as billions of years pass. In 1994, the Hubble Space Telescope found what many consider to be evidence of a large black hole in the heart of the M87 galaxy. Astronomers have also found evidence to support the idea that our own local galaxy, the Milky Way, has its own large black hole at its center.

Quasars have gradually become identified as one of the proofs that black holes exist. The current thinking about quasars is that they are distant objects that give hints about the first couple of billion years after the Big Bang. The great energy output of each quasar is the result of energy being emitted by a large spinning black hole found at the center of a galaxy, where stars are literally being consumed by the ravenous black hole. Matter shooting from the black hole forms jets of energy and debris thousands of light-years long, and the matter is moving so fast that relativistic effects are created, such as time slowing relative to the rest of the universe.

> *See also* Astronomy; Chandrasekhar, Subrahmanyan; Hawking, Stephen; Oppenheimer, J. Robert; Penrose, Roger; Quasars; Satellites; Telescopes; Wheeler, John A.

> **References**
> Hawking, Stephen W. *A Brief History of Time: From the Big Bang to Black Holes.* New York: Bantam, 1988.
> Wheeler, John Archibald, with Kenneth Ford. *Geons, Black Holes and Quantum Foam: A Life of Physics.* New York: Norton, 1998.

Boyer, Herbert (1936–)

The biochemists Herbert Boyer and Stanley Cohen (1935–) developed the techniques that genetic engineering is based upon. Herbert Wayne Boyer was born in Pittsburgh, Pennsylvania, where his father worked on the railroad and as a coal miner. As a high school football player whose coach also taught science, Cohen soon developed an interest in science. In 1954 he enrolled in pre-medical studies at Saint Vincent College in Latrobe, Pennsylvania, intent on becoming a doctor. The discovery of the structure of DNA by James D. Watson (1928–) and Francis Crick (1916–2004) a year earlier had excited him, and he named his two cats Watson and Crick. Boyer graduated four years later with a B.S. in biology and chemistry. He attended the University of Pittsburgh for his doctorate and graduated in 1963, followed by three years of postdoctoral work at Yale. Boyer married in 1959 and fathered two children.

In 1966, Boyer became an assistant professor at the University of California at San Francisco. Boyer and his students worked on restriction enzymes found in *Escherichia coli (E. coli)*. These bacteria live in the human digestive system, helping us digest our food. Geneticists favor *E. coli* because it is simple in structure and reproduces quickly in culture. Boyer learned how to use enzymes to cut away a selected strand from the DNA molecule.

At a 1972 conference in Hawaii, Boyer met Stanley N. Cohen, a biochemist at Stanford University. Cohen, extending the work of another Stanford biochemist, Paul Berg (1926–), was also working with *E. coli* and had created a way to remove plasmids from cells and reinsert them into other cells. Plasmids are small rings of genetic material separate from the main DNA in a cell. Boyer and Cohen worked together to combine their techniques to cut away strands of DNA from one plasmid and insert it into another plasmid. These new plasmids were then injected into *E. coli* bacteria to create a new organism, and genetic engineering was born.

In 1975, the venture capitalist Robert Swanson approached Boyer about forming a company to exploit this new technique. Genentech became one of the first genetic engineering firms, and Boyer served as a vice president. Genentech quickly created variations of *E. coli* that produced insulin and human growth hormone. They patented these new forms of life, and in a landmark 1980 decision, the U.S. Supreme Court agreed that life created in a lab could be patented. Patent rights allowed Genentech to protect its investment from other companies. Patenting a life-form was very controversial; the Supreme Court's decision made the current form of the genetic engineering industry possible.

Boyer became a full professor in 1976, the same year that Genentech was formed, and has remained in academia, though he made millions of dollars from Genentech. He has continued to do research and has received numerous awards, including the National Medal of Science in 1990.

See also Biotechnology; Cohen, Stanley N.;
    Crick, Francis; Genetics; Watson, James D.
References
Genentech. http://www.gene.com/ (accessed
    February 12, 2004).
Hall, Stephen S. *Invisible Frontiers: The Race to
    Synthesize a Human Gene.* New York: Atlantic
    Monthly, 1987.

Broecker, Wallace S. (1931–)

A geologist and oceanographer who pioneered the study of ocean circulation cycles, Wallace S. Broecker has also been one of those scientists who have sounded a warning about global warming. Wallace S. Broecker (Wally) was born in Chicago into a family of fundamentalist Christians, where his father ran a gas station. He studied at Wheaton College from 1949 to 1952 before transferring to Columbia University and completing an

A.B. a year later. He received his master's degree in 1956 and his doctorate in geology in 1958. He remained at Columbia as a faculty member at the Lamont-Doherty Earth Observatory, and in 1977 he became the Newberry Professor of Earth and Environmental Sciences at Columbia. He married in 1952, fathered six children, and now describes himself as areligious.

Broecker's early work concentrated on using chemical and radioactive tracers to track water circulation patterns in the world's oceans. The Geochemical Ocean Sections Study (GEOSECS), sponsored by the Scripps Institute of Oceanography, collected data from the Atlantic, Pacific, and Indian Oceans between 1972 and 1977. Broecker partially led the GEOSECS effort and published *Tracers of the Sea* in 1982 based on his conclusions from GEOSECS. Broecker's most important contribution came from his study of the thermohaline circulation system in the northern Atlantic, which circulates heat from the tropics up to the water off of northwest Europe, making that part of the world much warmer than would normally be the case. This effect, which operates in some ways like a conveyer belt, is now called Broecker's Conveyor Belt.

Broecker was also interested in the geology of ice ages and paleoclimatology. His 1966 paper with his student Jan Van Donk used analysis of deep-sea cores to show that Milankovitch's ice age theory was correct. The Serbian scientist Milutin Milankovitch (1879–1958) posited in 1920 that the ice ages were caused by astronomical variations in Earth's orbit. Broecker went on to explain the Younger Dryas period, a time just after the end of the last ice age when a drop in temperature almost restarted the ice age, as the result of an interruption in the ocean heat conveyor system.

Based on his understanding of oceanography, Broecker became an outspoken iconoclast on the dangers of global warming and climate change. His study of ice age cycles convinced him that Earth can switch quite suddenly from one mode to another and that global warming could inadvertently trigger another ice age. His 1985 book, *How to Build a Habitable Planet,* is pessimistic about humanity's effect on the global climate. Among his many awards, Broecker received the National Medal of Science in 1996.

See also Environmental Movement; Geology; Global Warming; Oceanography

References
Calvin, William H. "The Great Climate Flip-Flop." *Atlantic Monthly* 281, no. 1 (January 1998): 47–64.
Stevens, William K. "Scientist at Work: Wallace S. Broecker, Iconoclastic Guru of the Climate Debate." *New York Times,* March 17, 1998, F1.

C

Carson, Rachel (1907–1964)

The biologist Rachel Carson combined her understanding of science with her writing talent to produce books that transformed how people viewed the natural world and alerted the world to the dangers of pesticides. More than any other writer, Carson shaped the emergence of modern environmentalism out of an older conservation tradition. Born in Springdale, Pennsylvania, Carson enjoyed the outdoors and loved books. In 1925, she earned a partial scholarship to the Pennsylvania College for Women. Her family sacrificed so that she could attend college, and she majored in English, with ambitions to become a writer. A required course in biology turned her interests to science. She earned another partial scholarship, took on more debt, and transferred to Johns Hopkins University, graduating magna cum laude in 1929. Three years later she earned a master's degree in zoology from Johns Hopkins. She wanted to work on a doctorate, but finances forced her to turn to teaching jobs.

The death of her father in 1935 and the death of her older sister a year later left Carson with a mother and two nieces to care for. She found a job with the U.S. Bureau of Fisheries (later to become the U.S. Fish and Wildlife Service), writing radio scripts for them, and eventually rose to editor in chief of all publications for the service. On her own time she wrote popular articles on natural history for newspapers and magazines. A book published in 1941, *Under the Sea-Wind,* sold poorly. Her next book, *The Sea around Us,* published in 1951, won the National Book Award, was translated into many languages, and brought her fame as a writer who could bring the magic of nature and science to the masses. She earned enough from the book to resign from government service and devote herself full-time to writing. A 1955 book, *The Edge of the Sea,* further established her reputation.

Her final book, *Silent Spring,* published in 1962, was a well-documented polemic on the dangers of synthetic pesticides, which she labeled "elixirs of death." These pesticides were chemical toxins that enabled high crop yields by killing small insects and other pests. Among the pesticides she described was DDT (dichloro-diphenyl-trichloroethane), developed in 1942 by Paul Müller of Switzerland, who received a Nobel Prize in Physiology or Medicine in 1948 for this breakthrough. DDT was toxic in minute amounts, and Carson described the research that showed that exposure to DDT poisoned agricultural workers and was persistent enough to pass through the food chain. Eggs from hens fed alfalfa from fields sprayed by DDT

Rachel Louise Carson, biologist and environmental activist, 1952 (Bettmann/Corbis)

See also Agriculture; Environmental Movement
References
Lear, Linda. *Rachel Carson: Witness for Nature.* New York: Henry Holt, 1997.
Rachel Carson.org. http://www.rachelcarson.org/ (accessed February 12, 2004).

Centers for Disease Control and Prevention

Founded in 1946 as part of the United States Public Health Service, the Communicable Disease Center (CDC) first used a building from the defunct Office of Malaria Control in War Areas in Atlanta, Georgia. The CDC initially worked to combat malaria and other insect-borne communicable diseases. As its funding, staff, and responsibilities expanded, the government changed the name of the CDC to the Center for Disease Control in 1970. A decade later, with more centers being added to its organization, the name changed again to Centers for Disease Control. The final name change came in 1992, to Centers for Disease Control and Prevention, emphasizing the role of proactive disease prevention, though the organization did not change its well-known initials. The CDC is now an agency of the federal Department of Health and Human Services and employs 8,500 scientists, physicians, and staff around the United States and the world.

In 1961, the CDC took over publishing the Morbidity and Mortality Weekly Report (MMWR), a weekly compilation of deaths and disease causes from all over the United States. The MMWR is often the tool that alerts the CDC to new disease outbreaks, and it played this role when acquired immunodeficiency syndrome (AIDS) initially surfaced in 1981. The CDC also publishes statistics on smoking, violence, sexually transmitted diseases, heart diseases, and other causes of illness and mortality besides communicable diseases. Programs on occupational safety, injury prevention, and environmental health also fall under the expanded CDC purview. An older federal agency, the National Institutes of Health (NIH), overlaps with the

showed concentrations of the chemical. DDT passed into milk when cows ate hay sprayed by the pesticide, and DDT also devastated bird populations. Carson argued that modern agriculture should turn to alternatives, such as biological controls and the release of sterile insect males.

The chemical and agricultural industries attacked Carson and her book, spreading falsehoods about her character and her scientific work. They derided her scientific credentials (only a master's degree), mocked her status as a popular writer, and slyly insinuated that her gender disqualified her from serious scientific work. Desperately ill with cancer, Carson pressed for pesticide legislation on the national and state levels. After her death, DDT and some other pesticides were banned from the United States and other Western nations, and many bird populations that had been declining began to come back. *Silent Spring* has remained a prophetic classic.

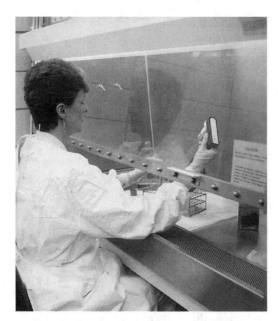

AIDS researcher at the Centers for Disease Control and Prevention (CDC/Public Health Information Library)

CDC in its responsibilities, at times causing duplication of efforts and bad feelings between the two organizations, though in general the NIH concentrates on basic research and the CDC works on recognizing and controlling communicable diseases.

The physician Donald A. Henderson worked at the CDC from 1955 to 1966 before leaving to join the World Health Organization (WHO), where from 1966 to 1979 he led the smallpox vaccination campaign that successfully eliminated smallpox in the wild. The CDC maintains one of the few known samples of the smallpox virus in its freezers in a maximum-containment laboratory.

An outbreak of a new type of pneumonia at a Philadelphia convention of the American Legion in 1976 led to twenty-nine dead and many more sick. The CDC successfully isolated the cause, *Legionella pneumophilia,* or Legionnaire's Disease, a year later. Studies by the CDC showed that contaminated water spread the disease, even through air conditioning systems, and had caused earlier outbreaks that had remained a mystery up until then. The year 1976 also saw a major misstep by the CDC. When soldiers at Fort Dix, New Jersey, became ill with influenza, and one died, the blood work found a variant of influenza called the swine flu. Alarmed that a repeat of the 1918–1919 influenza pandemic might occur, the CDC urged the federal government to initiate preventive vaccinations. The pandemic failed to materialize, embarrassing the president, Gerald R. Ford (1913–), the federal government, and the CDC.

The CDC created the Epidemic Intelligence Service (EIS) in 1951 to respond to epidemic outbreaks around the world. EIS has been instrumental in responding to outbreaks of new exotic diseases, including the Ebola, Marburg, and Lassa fevers in Africa. The CDC and EIS provide support for problems outside of the United States because of the recognition that the age of airline travel has made every corner of the world so interconnected that disease outbreaks can no longer be considered only a local matter.

See also Acquired Immunodeficiency Syndrome;
  Medicine; National Institutes of Health;
  Smallpox Vaccination Campaign; World Health
  Organization

References

Centers for Disease Control and Prevention. http://www.cdc.gov/ (accessed February 13, 2004).

Etheridge, Elizabeth W. *Sentinel for Health: A History of the Centers for Disease Control.* Berkeley and Los Angeles: University of California Press, 1992.

Garrett, Laurie. *The Coming Plague: Newly Emerging Diseases in a World out of Balance.* New York: Farrar, Straus and Giroux, 1994.

Mullen, Fitzhugh. *Plagues and Politics: The Story of the United States Public Health Service.* New York: Basic, 1989.

Chandrasekhar, Subrahmanyan (1910–1995)

Subrahmanyan Chandrasekhar made important contributions to astrophysics that laid the theoretical basis for black holes, quasars, neutron stars, and pulsars. Subrahmanyan Chandrasekhar, called Chandra by all who knew him, was born in Lahore, a city in the British colony of India (later to become Pakistan), to a high-caste family. His father was a government official and musicologist, and his mother was a literary scholar. His uncle, Sir Chandrasekhar Venkata Raman (1888–1970), who won the Nobel Prize in Physics in 1930, inspired Chandra to become a scientist. Chandra earned a degree in physics from Presidency College at the University of Madras in 1930, having already written his first two published scientific papers (at the age of eighteen). A scholarship allowed him to go to England to study at Trinity College at Cambridge University. On the voyage he read *The Internal Constitution of Stars,* by the eminent Cambridge astronomer Sir Arthur Stanley Eddington (1882–1944). Eddington argued that all stars went through a life cycle that ended with them becoming white dwarfs.

Applying the theory of special relativity and quantum statistics to the problem, Chandra concluded that Eddington's theory only applied to stars that contained up to 1.44 times the mass of the Sun. If a star contained more mass, the collapse of the star after a supernova would mean that the force of gravity would compress the atoms of the star into subatomic particles through relativistic degeneracy. This value of 1.44 times the mass of the Sun later became known as the Chandrasekhar Limit. Chandra continued to work on his theory at Cambridge and discussed his work extensively with Eddington. When Chandra presented his theory in 1933 to the Royal Astronomical Society, Eddington arranged to follow Chandra with his own surprise presentation. He ridiculed relativistic degeneracy and Chandra's ideas. Chandra was emotionally shattered by this assault. He earned his Ph.D. in 1933 and remained at Cambridge for the next four years, continuing to do research in order to build support for his theory. Although many eminent physicists privately told him that he was on the right track, the fame and authority of Eddington prevented them from offering more public support.

Chandra returned to India to marry in 1936, and the next year took his bride with him to America. His dispute with Eddington made an academic appointment difficult in Britain, so he found a position at the University of Chicago. He worked peripherally on the Manhattan Project, became a U.S. citizen in 1953, and turned out a long stream of scientific research, including work on radiative transfer, plasma physics, and a mathematical model of spinning black holes. Chandra did not hold a grudge against Eddington, later eulogizing him, but he did make a deliberate effort to be open to the ideas of his own students. Though his most influential work was done in the 1930s—following the tradition that most physicists make their major contribution while in their twenties or thirties—the significance of his work was not generally recognized for another two decades. The 1983 Nobel Prize in Physics was jointly awarded to Chandra and William A. Fowler (1911–1995), giving Chandra credit for work done half a century earlier. Fowler had also worked on the problem of stellar evolution, independently of Chandra.

See also Astronomy; Black Holes; Nobel Prizes; Pulsars; Quasars

References
Wali, Kameshwar C. *Chandra: A Biography of S. Chandrasekhar.* Chicago: University of Chicago Press, 1991.
————. *S. Chandrasekhar: The Man behind the Legend.* London: Imperial College Press, 1997.

Chaos Theory

Chaos theory, sometimes called the science of complexity, is a new mathematical understanding of physical phenomena represented by systems of nonlinear equations. Classical Newtonian physics predicted that the physical world behaved in a mechanical, deterministic manner, best described by linear equations. Nonlinear equations contain infinite quantities and can often only be solved approximately. The great French mathematician and astronomer Jules Henri Poincaré (1854–

1912) realized at the turn of the century that Newton's equations were not always so straightforward and that small differences in initial conditions of nonlinear equations could have dramatic effects when calculating the orbits of planets. The development of computers, allowing approximate solutions to large systems of equations, enabled scientists to examine nonlinear equations more closely.

The meteorologist Edward N. Lorenz (1917–) introduced the term *butterfly effect* to describe his own insight that a small change in initial conditions can have a dramatic effect on later events, or as he titled a 1972 paper, "Predictability: Does the Flap of a Butterfly's Wings in Brazil Set off a Tornado in Texas?" Because of the nature of chaos, he argued, even the best knowledge of current weather conditions would not enable meteorologists to predict weather too far into the future. Lorenz used computer modeling to illustrate his ideas and developed a set of well-known equations used in chaos theory. Computers became the essential tool of chaos theorists.

The mathematician Benoit Mandelbrot (1924–) coined the word *fractals* to describe physical shapes that have the same jagged shape no matter what scale you choose to use when looking at them. He noticed that his computer-generated color images resembled shapes in nature, such as the curve of coastlines, the patterns in turbulent liquids, snowflakes, blood vessels, and other natural systems that reflect chaotic behavior. In 1979, he developed his Mandelbrot Set to support his theory of fractals, and his 1982 book, *The Fractal Geometry of Nature,* with its gorgeous computer-generated pictures of fractals, helped scientists understand that fractals were a fundamental element of emerging chaos theory.

The physicist and Nobel laureate Murray Gell-Mann (1929–) and others founded the Santa Fe Institute in New Mexico in 1984 to promote the study of chaos theory and other implications of complexity. Chaos theory is probably misnamed, since the use of the word *chaos* implies that there are no constraints; in reality, chaotic systems confine themselves to a limited range of behavior. The implications of chaos theory are still in the early stages, but clearly the Newtonian understanding of physical phenomena is being significantly revised. Chaos theory has been applied to almost every area of science, including ecology, thermodynamics, chemistry, astrophysics, biology, economics, cognitive science, medicine, and even political geography.

See also Computers; Gell-Mann, Murray; Lorenz, Edward N.; Mandelbrot, Benoit

References

Briggs, John, and F. David Peat. *Turbulent Mirror: An Illustrated Guide to Chaos Theory and the Science of Wholeness.* New York: Harper and Row, 1989.

Gell-Mann, Murray. *The Quark and the Jaguar: Adventures in the Simple and the Complex.* San Francisco: W. H. Freeman, 1994.

Gleick, James. *Chaos: Making a New Science.* New York: Penguin, 1987.

Lorenz, Edward N. *The Essence of Chaos.* Seattle: University of Washington Press, 1993.

Pagels, Heinz R. *The Dreams of Reason: The Computer and the Rise of the Sciences of Complexity.* New York: Simon and Schuster, 1988.

Chemistry

Beginning in the early nineteenth century, chemistry has flourished as a successful science, not only explaining the chemical processes of the world, but contributing to technological advances. After World War II, chemistry continued to change dramatically, and many chemistry graduates found employment in a burgeoning set of related fields: biochemistry, genetics, pharmacology, chemical engineering, environmental science, geochemistry, and chemical physics. Biochemistry and molecular biology have grown so fast since the 1970s that college instructors have difficulty finding textbooks current with the field.

New tools allowed chemists to peer more deeply into the physical world. The transmission electron microscope (TEM), developed by the German physicist Ernst Ruska (1906–1988) in 1931, works like a traditional light telescope, but uses electrons instead of photons for viewing objects smaller than traditional light-based microscopes could image. The first electron micrograph of an intact cell was published in 1945. The scanning electron microscope (SEM), first developed in 1942 at the RCA Laboratories in the United States, did not become a commercial product until the 1960s; it resolved smaller objects than the TEM by scanning a beam of electrons across an object and using magnetic lenses to focus the resulting image. The scanning tunneling microscope (STM), invented in 1979 by the German physicist Gerd Binnig (1947–) and the Swiss physicist Heinrich Rohrer (1933–) at the International Business Machines (IBM) Zürich Research Laboratory, can image down to the atomic level. Ruska, Binnig, and Rohrer shared the 1986 Nobel Prize in Physics.

The new field of x-ray crystallography emerged after World War I, where x-rays reflected through crystals resulted in photographs. The American chemist Linus Pauling (1901–1994) used x-ray diffraction photography in the 1930s to analyze the crystal structure of inorganic molecules and to revolutionize the understanding of chemical bonds. The momentous discovery of the structure of the deoxyribonucleic acid (DNA) molecule by the zoologist James D. Watson (1928–) and the biologist Francis Crick (1916–2004) in 1953 revolutionized the study of genetics. They based their discovery on x-ray diffraction photographs taken of the DNA molecule by the chemist Rosiland Franklin (1920–1958). Dorothy Crowfoot Hodgkin (1910–1994) determined the chemical structure and shape of penicillin, vitamin B_{12}, and insulin by using x-ray crystallography. The chemical formula for B_{12} that she announced in 1957 was $C_{63}H_{88}N_{14}O_{14}PCo$, and the structure of the insulin molecule that she revealed in 1969 contained 777 atoms. Computer programs helped her in her work on insulin molecules,

performing complex calculations on measurements taken from x-ray photographs.

Spectroscopy, the analysis of light to determine the chemical composition of the light's source or of the place the light was reflected from, was first discovered by the Swiss Johann Balmer (1825–1898) in 1885. It became an important tool for chemists. In 1950, two English chemists, Ronald G. W. Norrish (1897–1978) and George Porter (1920–2002), developed the technique of flash photolysis: they used flashes of light each only a fraction of a second long to apply spectroscopic analysis to extremely fast chemical reactions. Porter later used lasers to generate flashes of light only a nanosecond long.

Polymers, macromolecules made up of smaller molecules (called monomers), were first discovered in the late nineteenth century. DNA itself is a macromolecule. As the twentieth century progressed, chemists understood polymers better and got better at manufacturing these wonder materials of modern technological civilization, with over a hundred different polymers and plastics in use at the end of the century. Organic polymers have been used in medicine and surgery, and some scientists hold out hope of using polymers to create artificial organs.

The English chemist Derek H. R. Barton (1918–1998) revolutionized organic chemistry with his development of conformational analysis in the early 1950s, demonstrating how the shapes of organic carbon rings dictated their chemical and biological properties. The crystalline forms of carbon are important compounds, including diamonds and graphite, with many industrial uses. The 1996 Nobel Prize in Chemistry went to three chemists, the Americans Robert F. Curl Jr. (1933–) and Richard E. Smalley (1943–) and the Englishman Sir Harold W. Kroto (1939–), for their discovery of carbon balls, fullerenes, which are formed in symmetric shapes similar to the geodesic design of the innovative thinker Buckminster Fuller (1895–1983). The buckminsterfullerene, composed of sixty carbon molecules, is known as the buckyball.

Some other notable advances in chemistry have included the discovery of dozens of new elements, all predicted by the periodic table, usually created in cyclotrons or particle accelerators, since element 43 (technetium) and all elements with an atomic number greater than 83 are found only as radioactive isotopes. The chemist Willard F. Libby (1908–1980) developed the process of carbon-14 dating in the late 1940s, one of the most important tools ever given archaeologists. In 1962, the British-born Neil Bartlett (1932–) synthesized the first noble gas, xenon hexafluoroplatinate, $XePtF_6$. More noble gases followed, used as powerful oxidizing agents in laboratories. In the 1970s, when an anticancer medicine was found in Pacific yew tree bark, chemists created a method of synthesizing the drug (paclitaxel) so that yew trees would not have to be harvested.

One of the great success stories of chemistry in the first half of the century became a nightmare in the second half. First invented in 1928, chlorofluorocarbons (CFCs) became a wonder chemical, manufactured for use in air conditioners and aerosol cans, and the basis of a multibillion-dollar industry by the 1970s. Drawing on the work of Dutch-born chemist Paul Crutzen (1933–), the Mexican chemist Mario Molina (1943–) and the American chemist F. Sherwood Rowland (1927–), working at the University of California at Irvine, discovered in 1974 that CFC gases accelerated the decay of the ozone layer. Further research confirmed continuing damage to the ozone layer, with possible catastrophic results for human civilization. The Montreal Protocol international agreements of 1987, 1990, and 1992 committed signatories to phasing out the use of CFCs.

See also Archaeology; Barton, Derek; Cold Fusion; Crick, Francis; Crutzen, Paul; Franklin, Rosiland; Fuller, Buckminster; Genetics; Hodgkin, Dorothy Crowfoot; Lasers; Libby, Willard F.; Medicine; Molina, Mario; Nobel Prizes; Ozone Layer and Chlorofluorocarbons; Particle Accelerators; Pauling, Linus; Pharmacology; Rowland, F. Sherwood; Watson, James D.

References

Aldersey-Williams, Hugh. *The Most Beautiful Molecule: The Discovery of the Buckyball.* New York: Wiley, 1997.

Chemical Heritage Foundation. http://www.chemheritage.org/ (accessed February 13, 2004).

Fruton, Joseph. *Proteins, Enzymes, Genes: The Interplay of Chemistry and Biology.* New Haven: Yale University Press, 1999.

Kohler, Robert. *From Medical Chemistry to Biochemistry: The Making of a Biomedical Discipline.* New York: Cambridge University Press, 1982.

Levere, Trevor H. *Transforming Matter: A History of Chemistry from Alchemy to the Buckyball.* Baltimore: Johns Hopkins University Press, 2001.

Reinhardt, Carsten, editor. *Chemical Science in the 20th Century: Bridging Boundaries.* Weinheim, Germany: Wiley-Vch, 2001.

Chomsky, Noam (1928–)

Noam Chomsky revolutionized the study of linguistics by arguing that language and grammar arose from innate traits in the human brain. Avram Noam Chomsky was born in Philadelphia, where his father was a noted Hebrew scholar who had emigrated from the Ukraine fifteen years earlier. His mother was also a Hebrew scholar, as well as a writer of children's books. In 1945 Chomsky entered the University of Pennsylvania, where he became a protégé of the linguist Zellig Harris (1932–2002). He earned a bachelor's degree in linguistics in 1949 and remained at the University of Pennsylvania to receive his master's degree in 1951 and his doctorate in 1955. He then joined the faculty of the Massachusetts Institute of Technology and has remained at MIT, with occasional semesters at other institutions as a visiting professor. Chomsky married a fellow linguist in 1949, and they have three children.

Drawing on some of the ideas of Harris, Chomsky published *Syntactic Structures* in 1957. At that time, ideas about the origin of language were shaped by behaviorist notions, such as those offered by the noted psychologist B. F. Skinner (1904–1990). The behaviorists argued that newborn babies had a blank mind, a tabula rasa, and that children acquired language through learning and mimicry. Chomsky argued that human beings were in fact born with the innate ability to understand the generative grammars that form the basis of all human languages. Children use this innate ability to learn the languages that they are exposed to. Chomsky cemented his linguistic theory in 1965 with *Aspects of the Theory of Syntax* and in 1975 with *The Logical Structure of Linguistic Theory*. Research in cognitive science later confirmed Chomsky's ideas. The impact of Chomsky on linguistics is comparable to the impact of Charles Darwin (1809–1882) on evolution and biology or Albert Einstein (1879–1955) on physics. Chomsky's devastating critique of Skinner's behaviorism helped diminish that theory's influence on psychology, and Chomsky's ideas have important implications for other areas of psychology as well as for cognitive science, anthropology, neurology, and sociology.

Although Chomsky has remained engaged in his linguistic studies, he has also gained considerable attention for his extensive political publications. Energized by the antiwar movement that opposed American involvement in Vietnam, Chomsky became politically active in 1964. His essays, books, and lectures have focused on the self-serving nature of American foreign policy. His political orientation consists of a curious blend of anarchism, socialism, and libertarianism. He displays a visceral distrust of authority and institutions that claim authority. In the United States, he usually aligns himself with leftists. His writings have become influential outside of the United States, and he remains far better known outside his country than within it.

See also Cognitive Science; Psychology; Skinner, B. F.

References

Barksy, Robert F. *Noam Chomsky: A Life of Dissent.* Cambridge: MIT Press, 1997.

Haley, Michael C., and Ronald F. Lunsford. *Noam Chomsky.* New York: Twayne, 1994.

Clarke, Arthur C. (1917–)

One of the leading science fiction authors and science writers of the twentieth century, Sir Arthur C. Clarke also conceived of the idea of communications satellites. Arthur Charles Clarke was born in Minehead, Somerset, England, where his father was a postal worker. Interested in science from a young age, Clarke built his own telescope at age thirteen and drew detailed maps of the Moon. During World War II, he served in the Royal Air Force as a radar technician and began publishing science fiction, short stories, and technical papers. In a 1945 article for *Wireless World,* he proposed the idea of satellites in geosynchronous orbit remaining stationary above Earth and providing relays for radio transmissions. Commentators often referred to him as the "father of the communications satellite" after this became a common technology twenty years later, though it was not his article that inspired the actual building of communications satellites. After the war, Clarke attended King's College in London, and in 1948 he graduated with a B.Sc. degree in physics and mathematics. Clarke continued to publish articles and books on the possibility of space travel while working as an assistant editor for the journal *Science Abstracts.*

Clarke's second book, *Exploration of Space* (1951), sold well enough to encourage Clarke to become a full-time writer. With this book and later books, Clarke became a world-famous science writer, concentrating on technology and the future of space exploration. After his only marriage proved to be a failure, Clarke moved to Ceylon (later called Sri Lanka), where he made his permanent home and became an avid scuba diver and writer on undersea topics.

Along with Isaac Asimov (1920–1992) and Robert A. Heinlein (1907–1988), Clarke is considered one of the three grand old masters of the science fiction field. His 1953 novel, *Childhood's End,* is among his best, describing a future in which an alien species arrives to raise up humanity from its

Arthur C. Clarke, science fiction author and science writer, 1976 (Bettmann/Corbis)

childhood. He collaborated with the famous filmmaker Stanley Kubrick on the landmark 1968 film *2001: A Space Odyssey,* considered by many to be one of the finest science fiction movies of all time. He served as a television commentator during the first Moon landing, a vindication of the faith and vision, influenced by science fiction, that he and other futurists shared. In 2000, Clarke was knighted.

In 1962, Clarke formulated three laws, which became well-known:

> When a distinguished but elderly scientist states that something is possible he is almost certainly right. When he states that something is impossible, he is very probably wrong.
> The only way of discovering the limits of the possible is to venture a little way past them into the impossible.
> Any sufficiently advanced technology is indistinguishable from magic. (*Profiles of the Future* 14, 21)

See also Asimov, Isaac; Media and Popular Culture; Sagan, Carl; Satellites; Science Fiction

References
Clarke, Arthur C. *Greetings, Carbon-Based Bipeds! Collected Essays, 1934–1998.* New York: St. Martin's, 1999.
———. *Profiles of the Future: An Inquiry into the Limits of the Possible.* New York: Harper & Row, 1973.
McAleer, Neil. *Arthur C. Clarke: The Authorized Biography.* Chicago: Contemporary, 1992.

Cloning

Cloning is the process of creating a genetic duplicate of a previous organism. Cloning is common in nature as a method of reproduction. As scientists came to understand the role of the nucleus in a cell and the role of genetic material, the idea emerged of removing the nucleus from an egg cell and replacing it with the nucleus from another cell. In 1952, the American embryologist Robert Briggs (1911–1983) and a colleague tried to do this with eggs from the North American leopard frog. Frog eggs were used because they were relatively large and easy to manipulate. They were only partially successful, and later work showed that the embryonic age of a cell was very important to success. After eggs are fertilized, they start a rapid timetable of division and differentiation, as different cells become precursors to each type of cell found in an animal's body.

In 1970, repeated experiments with frog eggs produced animals that reached the tadpole stage before they died. By 1984, the technique had been improved enough to lead to the successful cloning of a sheep. By 1994, calves had been cloned, by a process transferring a nucleus into an egg that had reached an amazing seven divisions. This last success came after it was accidentally found that if an egg was starved of nutrients, it restarted its embryonic timetable.

In 1996, at Roslin Institute, a government research facility near Edinburgh, Scotland, a team led by Ian Wilmut (1944–) and funded by a pharmaceutical corporation decided to take adult cells (not from an egg as in previous efforts) from a sheep, starve them of

Dolly the sheep, the first animal cloned from DNA taken from an adult animal (Najlah Feanny / Corbis Saba)

nutrients, and inject them into sheep eggs. An electrical pulse started the cell dividing. Out of 434 attempts, one succeeded, and on July 5, 1996, Dolly the sheep was born. Her existence was announced February 22, 1997. This extraordinary event captured the world's imagination.

With this successful cloning from an adult somatic cell, people immediately asked themselves whether humans should be cloned, and if so, when. This idea had originally been explored in the realm of science fiction, some of it insightful and some of it seriously lacking in scientific understanding. Many people are concerned about the ethics of human cloning, wondering how two identical people would be able to achieve distinct identities. This fear is caused by a belief in genetic determinism. It ignores the question, is it our genes that make us unique individuals or are we more than our genes?

Many scientists have spoken out in opposition to cloning on ethical and practical grounds. The sociobiologist Edward O. Wilson (1929–), however, welcomed cloning and other genetic engineering advances as ways for humans to take conscious control of human evolution. There have been spurious claims of cloning humans, but so far no publicly sanctioned experiments.

Other animals have been cloned from adult cells since 1996, and for unknown reasons, serious medical problems are often common in cloned animals. Suffering from a lung infection and chronic arthritis, Dolly was euthanized at the age of six in 2003. Efforts have also been made to use the DNA from frozen mammoths and elephant eggs to allow a female elephant to carry a mammoth to term. So far this effort has met with no success.

See also Bioethics; Medicine; Science Fiction;
  Wilson, Edward O.

References

Brannigan, Michael C., editor. *Ethical Issues in Human Cloning: Cross-Disciplinary Perspectives.* New York: Seven Bridges, 2001.

Kolata, Gina. *Clone: The Road to Dolly, and the Path Ahead.* New York: William Morrow, 1998.

MacKinnon, Barbara, editor. *Human Cloning: Science, Ethics, and Public Policy.* Urbana: University of Illinois Press, 2000.

Nussbaum, Martha, and Cass Sunstein, editors. *Clones and Clones: Facts and Fantasies about Human Cloning.* New York: Norton, 1998.

Stone, Richard. *Mammoth: The Resurrection of an Ice Age Giant.* Cambridge, MA: Perseus, 2001.

Wilmut, Ian, Keith Campbell, and Colin Tudge. *The Second Creation: Dolly and the Age of Biological Control.* New York: Farrar, Straus and Giroux, 2000.

Cognitive Science

Cognitive science is the study of cognition, the process by which humans think. The invention of the electronic digital computer founded the field of cognitive science. The American mathematician Marvin Minsky (1927–), a pioneer in the field of artificial intelligence, was among many who saw electronic computers as the first step toward the creation of genuine thinking machines. Subsequent efforts in artificial intelligence have fallen far short of expectations. To believe that the development of better computers will lead to artificial intelligence requires a mechanistic theory of the mind, according to which there is no essential difference between machines and humans.

During the 1960s and 1970s, the idea that computers offered a model of how the human mind actually worked dominated the field. During the 1980s, when neurologists used new tools to understand the brain, cognitive science moved away from the computer as a brain model and included research into how the brain actually worked. Neural networks, which had first been experimented with in the early 1950s, made a resurgence, as computer programs tried to solve problems just as the human brain did, through association rather than sequential execution.

By its nature, the study of cognitive science is an interdisciplinary activity, though some sixty academic departments or academic centers in cognitive science have been founded at different universities. Anthropology, cognitive psychology, linguistics, computer science, artificial intelligence, neurology, mathematics, logic theory, and the philosophical study of the mind are all elements of cognitive science. The Cognitive Science Society was founded at a meeting at the University of California at San Diego in 1979, and in 1986 that university founded a department in cognitive science. Researchers in many different disciplines added their expertise in the effort to understand how the human brain really worked. The American linguist Noam Chomsky (1928–) revolutionized the field of linguistics in the late 1950s by arguing that language and grammar arose from innate traits in the human brain. The British mathematician Roger Penrose (1931–), after making significant contributions to the theory of black holes, turned to artificial intelligence. His studies in the 1980s and 1990s led him to argue that the human brain can carry out processes that no computer can do. This conclusion ran contrary to the general consensus in the field.

The problem of consciousness plagues cognitive science. How is it that people are self-aware? Some have argued that the problem does not really exist because our perception of consciousness is actually an illusion. Neurologists have not found a way to associate consciousness with any particular neural pattern or part of the brain. After decades of assuming that the problem would disappear, in the 1980s a new subdiscipline of cognitive neuroscience emerged, which made the assumption that consciousness remained a central problem to be examined.

The Harvard-based educator Howard Gardner (1943–) drew on neuroscience to argue that intelligence is not a single quantity in people, but that individuals possess multiple types of intelligence. Gardner identified seven types of intelligence: linguistic, musical, logical-mathematical, spatial, bodily-kines-thetic, interpersonal, and intrapersonal. He found these different intelligences by studying how brain damage isolated different abilities, how the brain worked in idiot savants and other exceptional individuals, how intelligence might have evolved through natural selection, and what psychologists had found through experiments and psychometrics. Gardner and other researchers have also identified four possible additional intelligences: naturalist, spiritual, existential, and moral. Individuals are not dominated by a single type of intelligence, but possess each type of intelligence present to a different degree. Gardner's ideas have had a significant effect on education; the theory shows teachers how to modify their curriculum to take advantage of the different learning styles associated with each type of intelligence.

See also Artificial Intelligence; Chomsky; Noam; Computers; Minsky, Marvin; Neuroscience; Penrose, Roger; Psychology

References

Bechtel, William, and George Graham, editors. *A Companion to Cognitive Science*. Oxford: Blackwell, 1998.

Dupuy, Jean-Pierre. *The Mechanization of the Mind: On the Origins of Cognitive Science*. Translated by M. B. DeBevoise. Princeton: Princeton University Press, 2000.

Gardner, Howard. *Frames of Mind: The Theory of Multiple Intelligences*. Second edition. New York: Basic, 1993.

————. *Intelligence Reframed: Multiple Intelligences for the 21st Century*. New York: Basic, 1999.

Penrose, Roger. *Shadows of the Mind: A Search for the Missing Science of Consciousness*. New York: Oxford University Press, 1994.

Pinker, Steven. *How the Mind Works*. New York: Norton, 1997.

Cohen, Stanley N. (1935–)

The biochemists Stanley N. Cohen and Herbert Boyer (1936–) developed the techniques that genetic engineering are based upon. Stanley N. Cohen was born in Perth Amboy, New Jersey, and received his bachelor's degree from Rutgers University and a doctorate in biochemistry from the University of Pennsylvania in 1960. After serving in various academic medical positions in New York City; Ann Arbor, Michigan; Bethesda, Maryland; and Durham, North Carolina, Cohen moved west to Stanford University.

Cohen and his students extended the work of another Stanford biochemist, Paul Berg (1926–), by creating a way to remove plasmids from cells and reinsert them into other cells. Plasmids are small rings of genetic material separate from the main DNA in a cell. Cohen was using *Escherichia coli (E. coli)* bacteria as the basis of his work, which are the bacteria that live in the human digestive system, helping us digest our food. *E. coli* are favored by geneticists because they are simple in structure and reproduce quickly in culture.

At a 1972 conference in Hawaii, Cohen met Herbert Boyer, a biochemist at the University of California at San Francisco. Boyer was working on restriction enzymes found in *E. coli*. He had learned how to use the enzymes to cut away a selected strand from the DNA molecule. Cohen and Boyer worked together to combine their techniques to cut away strands of DNA from one plasmid and insert it into another plasmid. These new plasmids were then injected into *E. coli* bacteria and created a new organism. Genetic engineering was born. Cohen called the new organisms chimeras, but the technical term *clone* is more commonly used, though not in the broader sense that the layperson uses the term *clone*.

Boyer has teamed up with a venture capitalist to form a company called Genentech to exploit this new technology. Cohen has continued working at Stanford, increasing our knowledge about plasmid biology. He has received many awards, including the National Medal of Science in 1988. Cohen married in 1961 and has two children.

See also Biotechnology; Boyer, Herbert; Crick, Francis; Genetics; Watson, James D.

References

Hall, Stephen S. *Invisible Frontiers: The Race to Synthesize a Human Gene*. New York: Atlantic Monthly, 1987.

Cold Fusion

On March 23, 1989, the chemists Stanley Pons (1943–) and Martin Fleischmann (1927–) held a press conference with officials at the University of Utah where they announced that a simple laboratory experiment had created excess heat that could only be explained by a process of fusion occurring at room temperatures (thus the term *cold fusion*). Fleischmann was a respected senior electrochemist from England and a Fellow of the Royal Society. Normally the results of such an experiment would be presented at a scientific meeting or published in a scientific journal. Pons, Fleischmann, and university officials chose such an unorthodox method because another scientist, the physicist Steven E. Jones (1949–) at the nearby Brigham Young University, was also engaged in the same research. Whoever claimed priority on such a breakthrough technology stood to earn billions of dollars from a source of energy that promised to be environmentally friendly and cheap.

A cold fusion experiment (Leif Skoogfors / Corbis)

Even before successfully building the atomic bomb, scientists created early fission reactors with the potential to generate electricity. After World War II, intensive investment resulted in military and civilian fission-based nuclear reactors, used aboard ships and to generate electricity for cities. These reactors generated radioactive waste, though other impacts by the reactors on the environment were minor compared with equivalent coal-fired electrical generation plants. After the hydrogen bomb proved that fusion could be harnessed to create an explosion, scientists sought to harness fusion in reactors.

Fusion is the process that powers the Sun and the stars; it uses deuterium, a form of hydrogen, as its fuel. The promise of cheap electricity, created with reactors much safer than fission reactors and relying on a fuel easily extracted from ordinary water, excited policy makers. Over the following decades, scientists and engineers expended hundreds of millions of dollars pursuing hot fusion. They found fusion difficult to control, and fusion reactors required large amounts of power to raise the temperatures the millions of degrees sufficient to actually create fusion. The dream of using fusion has not yet been realized.

The experiment of Pons and Fleischmann consisted of a tabletop apparatus in which deuterium was electrolyzed by two electrodes made of palladium and platinum. The experiment generated more energy than was put into it. From the beginning, many scientists, especially physicists, doubted that such a simple experiment could yield such dramatic results. Other laboratories around the world attempted to duplicate the experiment; some met with partial success, and many failed. A laboratory in Italy announced they had detected excess neutrons in their experiment, an indication of fusion. The media went into a frenzy, and governments quickly announced funding for cold fusion research.

In a meeting of the American Physical Society in Baltimore on May 1, 1989, physicists condemned the entire idea of cold fusion. From a sociological standpoint, Pons and Fleischmann had three strikes against them. They were chemists claiming to succeed where generations of physicists working on hot fusion had failed; they had not followed the normal process of scientific publication; and their simple experiment proved not to be simple to reproduce. Cold fusion became a classic cautionary tale teaching the importance of repeating experimental results enough times to be certain and of following the accepted procedures of publication.

The story did not end there. Pons and Fleischmann left Utah for Nice, France, where the Japanese corporation Toyota funded their research. Scientists at the United States Office of Naval Research conducted over two hundred experiments in cold fusion during the 1990s and achieved inconsistent results. They did find hints that the composition of trace elements in the palladium electrode was very important and that microscopic cracks stopped the experiment from producing excess energy. When they published their results, their work was ignored. The Japanese government has funded a New Hydrogen Energy program, and research has also continued in China and Italy. Nevertheless, research on cold fusion has become pariah science, conducted on the fringes of respectable science.

See also Chemistry; Hydrogen Bomb; Nuclear
    Physics; Physics; Social Constructionism;
    Universities

References
Beaudette, Charles G. *Excess Heat: Why Cold Fusion
    Research Prevailed.* Second edition. South
    Bristol, ME: Oak Grove, 2002.
Daviss, Bennett. "Reasonable Doubt." *New Scientist*
    177, no. 2388 (March 29, 2003): 36–44.
Goodstein, David. "Pariah Science: Whatever
    Happened to Cold Fusion?" *American Scholar* 63
    (1994): 527–541.
Huizenga, John R. *Cold Fusion: The Scientific Fiasco
    of the Century.* Rochester, NY: University of
    Rochester Press, 1992.
"Low Energy Nuclear Reactions—Chemically
    Assisted Nuclear Reactions." http://www
    .lenr-canr.org/ (accessed February 13, 2004).
Simon, Bart. *Undead Science: Science Studies and the
    Afterlife of Cold Fusion.* Rutgers, NJ: Rutgers
    University Press, 2002.
Taubes, Gary. *Bad Science: The Short Life and Weird
    Times of Cold Fusion.* New York: Random
    House, 1993.

Cold War

The greatest war in history, World War II, was fought on battlefields and in the laboratory. Never had a war been so dependent on research in science and technology. Among the inventions that poured forth, many of them created on the basis of scientific research, were radar, computers, jet airplanes, short-range missiles, and the atomic bomb. A cold war developed in the late 1940s, a global ideological struggle between communism and its centrally controlled markets, led by the Soviet Union, and democracy and free markets, led by the United States and its Western European allies. Both antagonists realized how important science and technology were to their efforts, and funding levels for research in science and technology continued at unprecedented levels during the cold war, part of the race for newer weapons and prestige.

The Soviet Union exploded its first atomic bomb in 1949. The United States responded by developing a more powerful nuclear device, relying on fusion rather than fission, and exploded its first hydrogen bomb in 1952. The Soviets followed suit in 1953. The race for nuclear weapons, and developing aircraft and ballistic missiles to deliver them, resulted in a standoff, based on the doctrine of mutual assured destruction (MAD): both sides were reluctant to use their nuclear weapons for fear that a retaliatory strike would turn victory into defeat. An anti-ballistic missile (ABM) treaty was even signed in 1972 to prohibit the development of defensive missiles to shoot down incoming nuclear-tipped missiles, because of concerns that such a defensive system might undermine MAD. Frenzied open-air nuclear weapons testing during the 1950s led to scientists' becoming concerned about the accumulation of radiation in the atmosphere. Among others, the U.S. Nobel laureate Linus Pauling (1901–1994) campaigned for the United States and Soviet Union to at least stop open-air testing in the interest of public health. His efforts helped lead to the 1963 Nuclear Test Ban Treaty, and Pauling received the 1962 Nobel Peace Prize.

Soviet premier Leonid Brezhnev shakes hands with U.S. president Richard Nixon following talks on the Anti-Ballistic Missile treaty in 1972. (National Archives)

Actual combat encounters between the two superpowers proved to be rare, such as when reconnaissance aircraft were shot down, and the fact was usually quickly concealed in order not to escalate the situation. Other nations instead served as proxies, fighting ideologically based surrogate wars. Besides the economic and ideological arenas, the two superpowers competed by showing their scientific and technological prowess. The launching of *Sputnik I* in 1957 by the Soviet Union as part of the International Geophysical Year (IGY) initiated a "space race" between the two superpowers, as they sought international prestige by achieving firsts. The Soviets placed the first animal (1957), first man (1961), and first woman (1963) into orbit, but fell behind as the Apollo project successfully landed an American astronaut on the moon in 1969. Besides the public relations advantages, the space race contributed to stabilizing the cold war; spy satellites could now ascertain the strength

and deployment of opposing forces. Spy satellites also made cheating on arms controls treaties more difficult.

The governments of the United States, the Soviet Union, and their allies recruited their best and brightest to serve in defense-related research and development. Scientists and engineers developed more advanced computers, computer networks, the Internet, better medicines, better alloys, industrial ceramics, and technologies with no civilian use, like the neutron bomb. Britain, France, and China also poured funding into defense-related research, developing their own nuclear weapons and trying to maintain their own status as second-tier powers.

In an effort to break the nuclear stalemate and concentrate on defensive weapons rather than offensive weapons, Ronald Reagan (1911–2004), the ardently anticommunist president of the United States, proposed in 1983 the development and construction of the Strategic Defense Initiative (SDI), an orbiting system of satellites to shoot down incoming intercontinental ballistic missiles. Later dubbed Star Wars, after the popular movie, SDI implementation required the United States to withdraw from the ABM treaty, though the technology never advanced enough to require that step. As the project developed through the mid-1980s, the original ambition of a near-perfect shield formed by orbiting satellites armed with antimissile missiles and beam weapons quickly became more limited in conception, once it became apparent that the Soviet Union could build more intercontinental ballistic missiles much faster and cheaper than any possible SDI system that could be deployed. Although research continued on a much reduced scale, SDI disappeared as a strategic vision until the second Bush administration revived it in 2001, withdrawing from the ABM treaty, and hoping to develop a limited system to shoot down possible missiles from North Korea.

Though the cold war strained the economies of the United States and its allies, it eventually bankrupted the Soviet Union. An estimated quarter of the entire Soviet economy came to be devoted to their military-industrial complex, and the inefficiencies inherent in their centralized economy eventually caused an economic and political collapse. In November 1989, a popular uprising breached the Berlin Wall, and the Eastern European countries in the Soviet political orbit fell away. In 1991, the Soviet Union itself dissolved into its constituent republics, officially ending the cold war.

See also Apollo Project; Computers; Hydrogen Bomb; International Geophysical Year; Internet; Nuclear Physics; Pauling, Linus; Satellites; War

References
Baucom, Donald R. *The Origins of SDI, 1944–1983.* Lawrence: University Press of Kansas, 1992.
Fitzgerald, Frances. *Way out There in the Blue: Reagan, Star Wars and the End of the Cold War.* New York: Simon and Schuster, 2000.
Leslie, Stuart W. *The Cold War and American Science: The Military-Industrial-Academic Complex at MIT and Stanford.* New York: Columbia University Press, 1994.
Schwartz, Richard Alan. *The Cold War Reference Guide.* Jefferson, NC: McFarland, 1997.
Wang, Jessica. *American Science in an Age of Anxiety: Scientists, Anticommunism, and the Cold War.* Chapel Hill: University of North Carolina Press, 1999.

Computers

During World War II, various defense projects in America and Britain developed high-speed electronic calculators, among which were the first digital computers. Earlier computers were either mechanical or analog-based. Digital computers represent data as discrete pieces of information, or bits. These new computers were based on vacuum-tube technologies and experienced frequent failures. Engineers extending the frontiers of their craft, rather than scientists, built these computers.

The first truly digital computer that was more than a prototype was the Colossus, built by the British during World War II to help decrypt German radio traffic. The British government kept their success a secret for more than two decades, so the first acknowledged digital computer was an American invention. Completed in 1945, the ENIAC (Electronic Numerical Integrator and Computer) was a high-speed electronic calculator designed to create artillery ballistic tables, built by J. Presper Eckert (1919–1995) and John Mauchly (1907–1980) for the U.S. Army at the University of Pennsylvania Moore School of Electrical Engineering. While building ENIAC, Eckert and Mauchly developed for their next computer the concept of a stored program, where data and program code resided together in memory. This concept allowed computers to be programmed dynamically so that the actual electronics did not have to change with every program.

The noted mathematician John von Neumann (1903–1957) expanded on the concept

Computer operators program ENIAC, the first successful electronic digital computer, by plugging and unplugging cables and adjusting switches. (Corbis)

of stored programs and laid the theoretical foundations of all modern computers in a 1945 report and later work. His ideas came to be known as the "von Neumann Architecture." The center of the architecture is the repeating fetch-decode-execute cycle: instructions are fetched from memory and then decoded and executed in a processor. The execution of the instruction changes data that are also in memory. Eckert and Mauchly deserve equal credit with von Neumann for their innovations, though von Neumann's elaboration of their initial ideas and his considerable prestige lent credibility to the budding movement to build digital computers. Eckert and Mauchly went on to build the first commercial computer, the UNIVAC (Universal Automatic Computer), in 1951.

The Soviet Union built their first computer in 1950 at the Kiev Institute of Electric Engineering under the direction of S. A. Lebedev; it ran fifty instructions a second and used a memory of thirty-one 16-bit words. After moving to Moscow, Lebedev proceeded to supervise the development of the BESM (the acronym for "large electronic computer" in Russian) series of computers, which used magnetic drums and magnetic tapes for storage.

The transistor was invented in 1947 at the Bell Telephone Laboratories in Murray Hill, New Jersey, by John Bardeen (1908–1991), William Shockley (1910–1989), and Walter H. Brattain (1902–1987). By the late 1950s, the transistor became a useful commercial product, creating the second generation of computer hardware; it rapidly replaced vacuum tubes in computers and other electronic devices because transistors were much smaller, generated less heat, and were more reliable. Defense-related projects and space-related projects undertaken in the United States as part of the cold war became a major driver for computer-related innovation in both hardware and software.

The development of integrated circuits in the mid-1960s concentrated transistors on silicon wafers, inaugurating a third generation of computer hardware. In the 1970s, engineers packed enough logic gates onto an integrated circuit to create a microprocessor, or a computer on a chip. Microprocessors led to personal computers and cheap, handheld calculators, banishing slide rules to museums. The current fourth generation of hardware is based on microprocessors and ever more sophisticated integrated circuit chips.

In the 1960s a new kind of computer, the supercomputer, emerged. Expensive and oriented toward scientific computing, it was used in new centers in Europe, the United States, and Japan to offer scientists the opportunity to model physical systems. Since the late 1970s, most supercomputers have been parallel computers: many processors combine to work on different parts of a problem in parallel, rather than in sequence.

The 1950s brought the first higher-level programming languages. FORTRAN (FORmula TRANslator) was invented in 1957 as an easy way to write mathematically oriented scientific programs. The 1960s brought timesharing systems and the development of operating systems. In the 1970s, structured programming became prevalent, leading to programs that were easier to develop and less likely to have errors.

In the 1950s and 1960s, the academic discipline of computer science emerged, a unique combination of engineering, applied mathematics, and technical craft. Whereas other sciences have a natural part of the world that is the object of their study, computer scientists study and expand technological creations. Purists argue that computer science, if it is to be worthy of the name of science, should be only the study of algorithms, which are mathematically proven recipes for solving programming problems. Computer science included, from the 1950s, research into artificial intelligence, trying to create a combination of computer hardware and computer software that behaves as the human brain does. Research into artificial intelligence has led to advances in cognitive research.

Numerous computer scientists have dedicated their research toward the goal of making computers a tool extending the human mind, using the computer's calculating ability and ability to organize information. Douglas Engelbart (1925–) contributed the idea of hypertext in the 1960s as a way to navigate the interconnectedness of information. Engelbart and his research team also invented the windowed interface and the mouse. These ideas were extended by a team at the Xerox Palo Alto Research Center (PARC) in the 1970s to create local area networks, desktop publishing, laser printers, and the idea of a paperless office. The Xerox technologies ironically led to even greater consumption of paper. Xerox also created the graphical user interface, based on the combination of the mouse and bit-mapped graphics displays. The development of object-oriented programming helped program these new graphical environments and allowed programmers to model their programs more closely on real-world analogies.

In the 1960s, early networks endeavored to link computers together for communications. The Internet began with four nodes in 1969 as a creation of the cold war and became a worldwide network of networks, forming a single whole. Tim Berners-Lee (1955–) invented the World Wide Web in 1991, a technology that made the Internet accessible to the masses. E-mail, file transfer, and other network technologies have allowed scientists to communicate more quickly and access supercomputing resources from a distance.

Computers have created the technology of robotics, which allows scientists to send instruments deep into space, deep into the ocean, and into the hot craters of volcanoes. In the 1990s nanotechnology, the creation of microscopically small devices, emerged, with the potential to launch another technological revolution.

Computers are the great technological and scientific innovation of the second half of the twentieth century. Computers have made many scientific advances possible, from models of subatomic interactions to models of the birth and death of the universe. Computers have made possible space travel, advanced aircraft design, DNA sequencing, chaos theory modeling, and numerical analysis, as well as the creation of atmospheric models, molecular dynamics, and models of proteins. As computers have become ubiquitous in daily life, the dominant trends in computing are ever more processing power and greater ease of use.

Much as the ever more widespread use of the clock in the later Middle Ages changed the way people perceived time, and the refinement and spread of the steam engine in the eighteenth century stimulated scientists to think of the laws of nature in terms of machines, the success of the computer in the late twentieth century prompted scientists to think of the operation of the basic laws of the universe as being similar to the operation of a computer. The new physics of information has come to think of matter and natural laws as bits of information. For example, one can argue that not only do black holes consume energy and matter, they also destroy information. The computer, made possible by advances in science, has transformed the practice of science and the actual content of science.

See also Artificial Intelligence; Bardeen, John; Berners-Lee, Timothy; Chaos Theory; Cognitive Science; Cold War; Engelbart, Douglas; Integrated Circuits; Internet; Mathematics; Nanotechnology; Neumann, John von

References

Aspray, William. *John von Neumann and the Origins of Modern Computing.* Cambridge: MIT Press, 1990.

Ceruzzi, Paul E. *A History of Modern Computing.* Cambridge: MIT Press, 1998.

Garfinkel, Simson. *Architects of the Information Society: Thirty-Five Years of the Laboratory for Computer Science at MIT.* Cambridge: MIT Press, 1999.

Kaufmann, William J. *Supercomputing and the Transformation of Science.* New York: Scientific American Library, 1993.

McCartney, Scott. *ENIAC: The Triumphs and Tragedies of the World's First Computer.* New York: Walker, 1999.

Williams, Michael R. *A History of Computing Technology.* Second edition. Los Alamitos, CA: IEEE Computer Society, 1997.

Cousteau, Jacques (1910–1997)

Jacques-Yves Cousteau achieved fame for his technical inventions and indefatigable energy in exploring the world's oceans and promoting the sound use of the ocean's ecology. Though they lived in Paris, his parents returned to their home village of Saint-André-de-Cubzac for Cousteau's birth in 1910. His father worked as a personal assistant to an American millionaire, and the family moved often, including spending a year in New York City when Cousteau was ten years old. As a child he built a functional model of a marine crane and a small battery-powered automobile. Though considered a sickly child, Cousteau learned to swim and was drawn to the ocean. An earlier disinclination toward schoolwork changed as he grew older, and he graduated from College Stanislas in Paris in 1930, then entered the École Navale, France's naval academy. He served as a naval officer from 1933 to 1956 and retired as a *capitaine de corvette.*

Though he learned to pilot aircraft, Cousteau remained oriented toward the sea. In 1936, he was given a pair of goggles that allowed him to see underwater. He began working on a device to allow a human to breathe underwater. After France's defeat by Germany in 1940, Cousteau continued to serve with the Vichy navy and secretly joined the French underground resistance. For his espionage activities, he received the Légion d'Honneur and the Croix de Guerre with Palm after the war. During the war, Cousteau also continued working on his "Aqua-Lung," which the Germans allowed because they considered him harmless.

Cousteau succeeded in 1943 by using the automatic gas-feeder valve invented by Émile Gagnon. After the war, the French Air Liquids Company started to sell commercial versions of the Aqua-Lung, which came to be called a scuba (a self-contained underwater breathing apparatus).

In 1946, Cousteau founded the Undersea Research Group for the French navy and began extensive efforts to use his Aqua-Lung in oceanographic research. In 1951, Cousteau converted a British minesweeper into a research ship named *Calypso*. He lived aboard this ship and traveled the world on expeditions, raising money from his own private ventures, from public agencies such as the French Academy of Sciences, and from research foundations, such as the National Geographic Society. In 1957, Cousteau became director of the Oceanographic Museum of Monaco, a position he held until 1988. Also in 1957, he became head of the Conshelf Saturation Dive Program, a lengthy effort experimenting with long-term undersea human habitation.

During World War II, Cousteau made his first two underwater films. After the war he developed a new type of process that allowed better television filming underwater. He began to turn out film documentaries and then television documentaries based on the travels of the *Calypso*. The popular films won three Academy Awards (1957, 1959, and 1964) and numerous other film awards.

Cousteau married in 1937, and his two sons worked with him in his research. He wrote many books, often with coauthors, that sold millions of copies worldwide and were translated into many languages. Among the more prominent books were *The Silent World* (1953), *The Living Sea* (1963), and *The Ocean World of Jacques Cousteau* (1972), a series. In his books and films, Cousteau acted as a relentless educator who brought undersea exploration to the masses and campaigned for understanding and preserving the ocean environment.

See also Environmental Movement; National Geographic Society; Oceanography; Undersea Exploration

References
Cousteau Society. http://www.cousteau.org/ (accessed February 12, 2004).
Madsen, Axel. *Cousteau: A Biography.* New York: Beaufort, 1986.
Munson, Richard. *Cousteau: The Captain and His World.* New York: William Morrow, 1989.

Creationism

Creationism is the belief that a deity created the universe and life within the universe; the oddly named discipline of creationist science is focused primarily on a literal interpretation of the Judeo-Christian Book of Genesis. The theory of natural selection developed by Charles Darwin (1809–1882) created a crisis for biblical literalists on two points. Darwin and other geologically minded scientists argued for a long age of Earth, more than the six thousand years traditionally based on a literal interpretation of the book of Genesis. Darwin and his successors also taught that humans were not unique, not directly created by God, but just an evolved form of primate, sharing ancestry with such other primates as apes, chimpanzees, and monkeys. Darwinism and other aspects of modern thought prompted a conservative backlash and the revitalization of fundamentalist and evangelical forms of Christianity at the beginning of the twentieth century (in the Islamic world, modernity also prompted a similar phenomenon). George McReady Price (1870–1963), an American Seventh-day Adventist and a geologist, wrote numerous books putting forth his own geological theory that the flood described in the Bible had laid down all the strata and fossils during the year that water covered the entire planet.

The strength of the arguments by evolutionists came from fossils and geological strata. The doctrine of gradualism became dominant in geology, in part as a reaction to the creation story of Genesis. When the physicist Luis W.

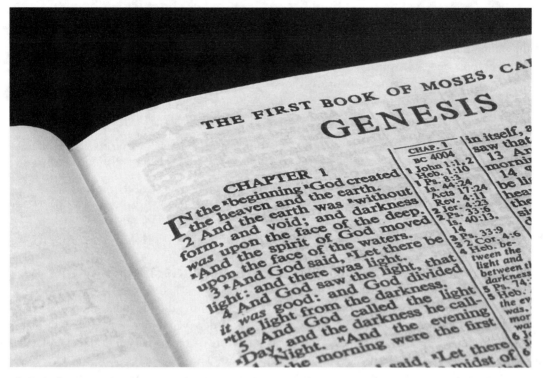

The book of Genesis in the Bible (Stan Rohrer/iStockPhoto.com)

Alvarez (1911–1988) and geologist Walter Alvarez (1940–) found that an asteroid had catastrophically hit Earth 65 million years ago, geologists were reluctant to abandon their ingrained gradualist assumptions. Catastrophism reminded geologists too much of creationism and pseudoscientific theories such as the proposal of the Russian-born writer Immanuel Velikovsky (1895–1979) that certain events described in the Old Testament were caused by the motions of the planets.

In the United States prior to World War II, Darwinism remained under serious attack from state legislatures and local school boards, though at the same time some people used the idea of Social Darwinism to justify harsh economic and social policies. Social Darwinists believed that poverty and other social ills came from poor genes, not flawed social structures. After the war that science had so obviously helped win, Darwinism enjoyed a revival, as a new evolutionary synthesis reemphasized the

principle of natural selection among evolutionists. Darwinism also came to be taught in the schools to a greater extent.

In a 1961 book, *The Genesis Flood,* a Texas hydraulic engineer, Henry M. Morris (1918–), and a theologian, John C. Whitcomb Jr. (1924–), revived Price's theories. As strict creationists, they argued that the universe was divinely created less than ten thousand years ago, that the Fall of Adam activated the second law of thermodynamics, that the theory of biological evolution was wrong, and that the flood explained all seemingly contradictory evidence. Morris and Whitcomb helped found the Creation Research Society in Ann Arbor, Michigan, in 1963. Other societies followed, and scientists holding doctorates participated. The societies have sponsored scientific research, published peer-reviewed journals, and undertaken educational outreach to promote what they term scientific creationism.

Beginning in the 1970s, creationist political activists within the United States have fought for state laws to mandate that biological evolution and scientific creationism be taught together as alternative theories; they have sought curriculum change on the local level. Creationists have created textbooks with biblical references removed, but all the same arguments remaining. They have found a measure of success, especially with local school boards, though a Supreme Court ruling in 1986 ruled that laws requiring equal treatment were unconstitutional on the basis of separation of church and state. Strict creationism has not been confined to the United States, with marginal creationist movements found in most countries that have significant Christian populations.

When the theory of the Big Bang began to gain prominence in the late 1940s, theologians were intrigued, since it seemed to be a variation on the idea of creation out of nothingness, as described in Genesis. Pope Pius XII (1876–1958) spoke positively in 1951 of the idea of a primeval atom. Fred Hoyle (1915–2001), creator of the steady state theory of the universe, complained bitterly that one reason people supported the Big Bang was that it resonated with their desire for a divine creation. Other theologians found the steady state theory, which required the continuous creation of matter to fuel the expansion of the universe, to be more congenial to their theological point of view.

Mainstream scientists reacted strongly against creationism, partly in forceful popular writings by Carl Sagan (1934–1996) and Stephen Jay Gould (1941–2002). Some creationists have abandoned a strict literal interpretation of Genesis, arguing that the days in the biblical account are really geological ages, or even abandoning Genesis entirely and arguing for a form of divinely guided evolution via natural selection.

See also Alvarez, Luis W., and Walter Alvarez; Big Bang Theory and Steady State Theory; Gould, Stephen Jay; Hoyle, Fred; Philosophy of Science; Religion; Sagan, Carl

References

Eve, Raymond A., and Francis B. Harrold. *The Creationist Movement in Modern America.* Boston: Twayne, 1990.

Hayward, James L. *The Creation/Evolution Controversy: An Annotated Bibliography.* Pasadena, CA: Salem, 1998.

Morris, Henry M., and Gary E. Parker. *What Is Creation Science?* El Cajon, CA: Master, 1987.

Numbers, Ronald L. *The Creationists: The Evolution of Scientific Creationism.* New York: Knopf, 1992.

Toumey, Christopher P. *God's Own Scientists: Creationists in a Secular World.* New Brunswick, NJ: Rutgers University Press, 1994.

Crick, Francis (1916–2004)

Francis Crick and James D. Watson (1928–) discovered the double helix molecular structure of deoxyribonucleic acid (DNA) that is the basis of modern genetics and molecular biology. Francis Harry Compton Crick was born outside of Northampton, in Northamptonshire, England, where his father manufactured boots and shoes. When the family business failed after World War I, the family moved to London, and Crick went to a private school at Mill Hill. In 1934 he entered University College in London, earning a B.S. in physics in 1937. He remained at University College for graduate studies, researching the viscosity of water at high temperatures. With the coming of World War II, Crick left school to conduct research on naval mines at the British Admiralty Research Laboratory.

Crick remained with the navy for two years after the war, and during that time he read Erwin Schrödinger's book *What Is Life? The Physical Aspect of the Living Cell,* which argued that the phenomenon of life could be reduced to chemical and physical factors. As an atheist, Crick disliked the notion of vitalism in biology, which is the belief that an unknown "vital principle" animated living beings that could not be explained in just physical terms, and so Schrödinger's ideas intrigued him. He chose to change his emphasis from physics to biology and in 1947 began working at the Strangeways Research Laboratory at Cambridge University on a Medical Research Council Studentship scholarship. After teaching himself biology and biochemistry, Crick moved to the Medical Research Council Unit at the Cavendish Laboratories to study under the chemist Max F. Perutz (1914–2002), a future Nobel laureate in chemistry in 1962. Scientists at that time knew that DNA formed the genes that passed on hereditary information from generation to generation. Scientists even understood many of the chemical characteristics of DNA, but they did not understand how it worked. The shape of the molecule eluded them.

Crick worked on this problem by drawing on the x-ray diffraction pictures of DNA taken by the chemist Rosiland Franklin (1920–1958) of the University of London, an associate of Crick's friend, the physicist Maurice H. F. Wilkins (1916–). In 1951, James D. Watson, a twenty-three-year-old wunderkind from America, came to Cambridge to do postdoctoral study at the Cavendish Laboratories. Crick and Watson developed a close relationship and collaborated on determining the chemical shape and structure of DNA. In February 1953, Watson realized that a double helix would match the shape that they sought. They created a model of their double helix out of beads, wire, and cardboard, and published their discovery in a letter to the journal *Nature* in April. In their landmark letter, they also observed that the double helix, with its zipper-like nature, offered a mechanism for gene replication. Crick, Watson, and Wilkins received the 1962 Nobel Prize in Physiology or Medicine. Franklin had died by that time, and the Nobel is not awarded posthumously.

Crick received his doctorate from Cambridge in 1953 for research unrelated to DNA and remained at Cambridge working on how DNA replication occurred. Important insights were provided by the physicist George Gamow (1904–1968) and others. Crick made contributions to explaining how ribonucleic acid (RNA) was used to transfer instructions from DNA to the cell to create proteins. In 1977, Crick left Cambridge to become the Kieckhefer Distinguished Research Professor at the Salk Institute of Biology in La Jolla, California. This move allowed him to change his research focus to neurobiology. Much of his research then concentrated on the nature of dreams, which he speculated were a way for the mind to discard experiences from the previous day that were not necessary to remember and would clutter up the neural connections of the mind.

In his 1981 book *Life Itself: Its Origin and Nature,* Crick drew attention to the fact that all life on Earth uses DNA, with the exception of mitochondria. He speculated that all life came from a single source of microorganisms from another planet. This theory of panspermia is similar to ideas that the noted astronomer Fred Hoyle advanced at about the same time. Crick's two marriages produced three children.

See also Biotechnology; Franklin, Rosiland; Gamow, George; Genetics; Hoyle, Fred; Neuroscience; Watson, James D.

References
Crick, Francis. *What Mad Pursuit: A Personal View of Scientific Discovery.* New York: Basic, 1988.
Judson, Horace Freeland. *The Eighth Day of Creation: Makers of the Revolution in Biology.* Expanded edition. Woodbury, NY: Cold Spring Harbor Laboratory Press, 1996.

Crutzen, Paul (1933–)

The chemist Paul Crutzen discovered that nitrogen oxide chemicals accelerated the decay of the ozone layer in Earth's atmosphere. Paul J. Crutzen was born in Amsterdam, the Netherlands. His father was a waiter. When he took his university entrance exams in 1951, sickness prevented him from achieving a score that qualified him for a university stipend. Crutzen instead attended Middelbare Technische School, middle technical school, for three years to become a civil engineer, and he worked as an engineer for the Bridge Construction Bureau of Amsterdam until 1958. Desiring an academic career, he moved to Sweden to work as a computer programmer in the Department of Meteorology at Stockholm University. He took courses in addition to his work and in 1963 qualified for the *filosofie kandidat,* which is the Swedish equivalent to a master's degree. Crutzen earned his doctorate in meteorology from Stockholm University in 1973.

Crutzen became interested in the photochemistry of atmospheric ozone after helping an American scientist in 1965 develop a model of oxygen allotrope distributions in the atmosphere. The ozone layer is created by the presence of an isotope of oxygen (O_3) in the stratosphere, which reduces the amount of ultraviolet radiation emitted by the Sun that reaches the planet's surface. Ozone is created naturally in the stratosphere, and without its protection, too many ultraviolet rays could lead to DNA mutations, cancer, and blindness. In 1970, while doing research at the Department of Atmospheric Physics at the Clarendon Laboratory at Oxford University in England, Crutzen published an article in the *Quarterly Journal of the Royal Meteorological Society* that showed that nitrogen oxides formed naturally by soil bacteria can rise up to the stratosphere. Once in the stratosphere, the nitrogen oxide molecules are broken apart by sunlight and react with ozone, accelerating the chemical processes that lead to the natural decay of ozone molecules.

Crutzen's research was not generally accepted until the Mexican chemist Mario Molina (1943–) and the American chemist F. Sherwood Rowland (1927–) discovered in 1974 that chlorofluorocarbon (CFC) gases, manufactured for use in aerosol cans and air conditioners, also accelerate the decay of the ozone layer. This discovery, and the realization that continued use of CFCs would deplete the ozone layer, with catastrophic results for human civilization, led to the Montreal Protocol international agreements of 1987, 1990, and 1992 to phase out the use of these gases. Crutzen, Molina, and Rowland received the 1995 Nobel Prize in Chemistry.

Crutzen moved to the National Center for Atmospheric Research in Boulder, Colorado, in the 1970s, where he investigated the effects of smoke from burning forests in Brazil on the atmosphere. This led to research with a colleague, John Birks, on what might happen from the fires started by a nuclear war. They concluded in the early 1980s that a significant amount of sunlight would be blocked, plunging a post–nuclear holocaust Earth into a "nuclear winter." Crutzen moved to the Max Planck Institute for Chemistry in Germany in 1980, and he also later became associated with the Scripps Institution of Oceanography. Crutzen married in 1958 and fathered two daughters.

See also Chemistry; Environmental Movement; Meteorology; Molina, Mario; Nobel Prizes; Rowland, F. Sherwood

References

Nobel *e*-Museum. "Paul J. Crutzen–Autobiography." http://nobelprize.org/chemistry/ (accessed February 12, 2004).

Cybernetics

See Ecology

D

Dawkins, Richard (1941–)

A leading popularizer of evolutionary theory, Richard Dawkins also helped found evolutionary psychology. He was born Clinton Richard Dawkins in Nairobi, Kenya, where his father worked as a colonial administrator. After spending the first eight years of his life in Africa, he moved with his family to a dairy farm near Chipping Norton, in Oxfordshire, England. Dawkins entered Oxford University in 1959, earning a bachelor's degree in 1962 and a doctorate in 1966, both in zoology. After two years at the University of California at Berkeley (1967–1969), he returned to Oxford as a faculty member and has remained there. Married three times, he fathered a daughter.

His success came not from his scientific work, but from his ability to write compelling arguments for his interpretation of Charles Darwin's theory of natural selection. His first book in 1976, *The Selfish Gene,* argued that natural selection occurs on the level of genes, rather than the level of individuals or species. He argued that individuals are just temporary carriers of genes and that physiology and behavior can be explained by the need for genes to propagate. The human brain introduced a complex new variable into natural selection, with the formation of cultural and intellectual ideas. Dawkins called these products of the mind memes, and argued that they propagate and are selected for in exactly the same way that genes are selected for.

How does complexity in organisms arise? Dawkins turned to computer models to show how seeming complexity can arise from simple interactions and the pressures of selection. In his third book, *The Blind Watchmaker: Why the Evidence of Evolution Reveals a Universe without Design* (1986), Dawkins described computer programs that he had written to mimic the process of natural selection. His programs simulate the development of more complex forms of life.

Not all scientists have accepted Dawkins's view of evolution, though Dawkins's evolutionary psychology and its kin, sociobiology, became powerful intellectual currents within science in the 1970s. Besides his books, Dawkins often appears on talk shows and gives lectures. In 1995, he became the first Charles Simonyi Professor of Public Understanding of Science, a position more suited to the role he had selected for himself. Dawkins has expounded his point of view with the fervor and conviction of a preacher. A committed atheist, he has also become well-known for his attacks on religion, which he views as an archaic meme suited to prescientific times.

See also Creationism; Evolution; Religion;
   Sociobiology and Evolutionary Psychology
References
Dawkins, Richard. *The Extended Phenotype: The Long
   Reach of the Gene.* Revised edition. New York:
   Oxford University Press, 1999.
"The World of Richard Dawkins." http://www.
   world-of-dawkins.com/ (accessed February
   12, 2004).

Deep-Core Drilling

During the International Geophysical Year
(IGY) of 1957–1958, glaciologists received
funding and support to more thoroughly explore the Arctic and Antarctica. Drills were
developed to draw out ice cores from locations in Greenland and Antarctica. The layers
of snow compacted into ice on the permanent ice caps created a stratigraphic record
similar to that provided by tree rings. Sophisticated analysis of ice in these cores revealed
annual temperatures and annual precipita-
tion, and preserved airborne particles from
plants, volcanic ash, and other chemicals.
Small bubbles in the ice also preserved samples of atmospheric gases from the past.

The Cold Regions Research and Engineering Laboratory (CRREL) of the U.S.
Army Corps of Engineers, located in
Hanover, New Hampshire, led the American
effort after the IGY, developing relationships
with Swiss and Danish geophysicists. The
Swiss glaciologist Henri Bader of Rutgers
University gained a reputation for his instrumental role in these early efforts. In 1966, a
CRREL project drilled all the way from the
surface of Greenland's ice cap to bedrock.
The Danes developed a lighter, more advanced drill to use in the combined Danish-
Swiss-American Greenland Ice Sheet Project
(GISP), reaching bedrock in 1981. These
early efforts took place close to the edge of
the ice cap, where working conditions were
easier, but the depth to drill to bedrock was
shorter.

The U.S. GISP2 was inaugurated in 1988,
accompanied by its European counterpart,
the *Greenland Ice Core Project* (GRIP).

These two efforts relocated to the center of
the ice cap, where conditions were harsher
and drilling more difficult, but the scientific
rewards greater. Aircraft from the New York
Air National Guard provided transportation
for scientists and ice cores. The Europeans
drilled from 1989 to 1992. The Americans
drilled for five years, and in 1993 GISP2 recovered the deepest ice core drilled in
Greenland: 3,053 meters deep. Greenland
ice cores reach back to 110,000 years ago;
cores from Antarctica reach back 450,000
years. Ice cores have also been recovered
from the Andes Mountains in South America,
Siberia, North America, and mountains in
Africa.

In order to store the ice cores from GISP2,
the National Science Foundation and U.S.
Geological Survey created a premier storage
facility for ice cores from around the world.
Fifty-five thousand cubic feet of freezer space
is kept at −33 degrees Fahrenheit (−36 degrees Celsius) at the National Ice Core Laboratory in Denver, Colorado. With the rise of
concerns over global warming, ice cores have
provided the best evidence of past climate
conditions, showing periods of heating and
cooling and proving that the climate is a dynamic force in history.

See also Geology; Global Warming; International
   Geophysical Year; Meteorology; National
   Science Foundation
References
Alley, Richard B. *The Two-Mile Time Machine: Ice
   Cores, Abrupt Climate Change, and Our Future.*
   Princeton: Princeton University Press, 2000.
Mayewski, Paul Andrew, and Frank White. *The Ice
   Chronicles: The Quest to Understand Global Climate
   Change.* Hanover, NH: University Press of New
   England, 2002.

Deep-Sea Hydrothermal Vents

The acceptance of the theory of plate tectonics in the 1960s led geologists to realize that
deep rifts and mountain ranges in the oceans
were actually the boundaries of tectonic
plates, either spreading apart into a rift or
pushing together into mountains. They pre-

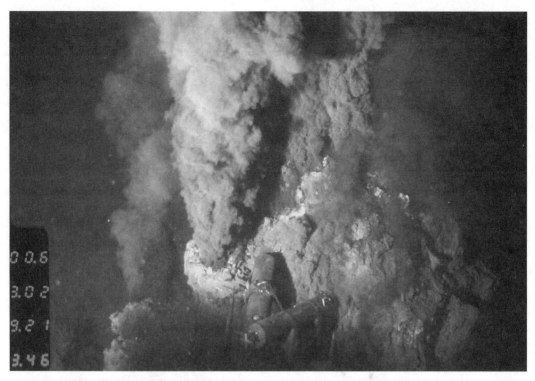

A black smoker in the Gulf of California's Guaymas Basin (Ralph White / Corbis)

dicted that volcanic activity at these rifts and mountains would lead to the venting of hot water from deeper in Earth's crust.

In 1977, the deep-sea submersible *Alvin* found deep-sea hydrothermal vents at about 2,700 meters in the Galapagos Rift near the Galapagos Islands. Vents in rifts, where the tectonic plates are spreading, tend to be spectacular, spewing what looks like white or black smoke. Sensors found the smoke, minerals mixed in with water, to be as hot as 400 degrees Celsius. These vents at times formed chimneys of accreted minerals, and researchers soon dubbed the vents and chimneys "black smokers," some as tall as 20 meters. Researchers were astonished by the chemistry of the black smokers, which were rich in sulfides, lead, cobalt, and other metals. Other vents that spewed white smoke, called "white smokers," were formed from gypsum and zinc, rather than sulfides. As the number of known deep-sea hydrothermal vents has grown, scientists have calculated

that up to a quarter of Earth's total heat input comes from these vents.

An even more important discovery by the *Alvin* was the unique ecology surrounding the vents. Large tubeworms as much as 2 meters long waved in the turbulence created by the hot water, alongside giant clams a third of a meter wide and new species of crabs, shrimp, mussels, and other types of worms. The food chain of most other life on Earth is dependent on the basic process of photosynthesis, the conversion of sunlight into energy. The fauna of the deep-sea vents are too deep for sunlight to reach. Species of bacteria around the vents use chemosynthesis to create energy, converting the sulfides into organic carbon. The more complex lifeforms form symbiotic relationships with the chemosynthesis bacteria and survive off them.

A microorganism called *Methanococcos jannaschii,* discovered in 1982 at a vent in the Pacific Ocean, came to be considered part of a new domain of life, made up of microorganisms named Archaea. The two previously recognized domains of life were bacteria and eukaryotes. Archaea are characterized by the lack of a cell nucleus and a smaller number of genes than other domains of life. Some scientists have proposed that Earth life originated in the rich chemical mix of deep-sea hydrothermal vents up to four billion years ago. They consider Archaea to be the most primitive domain and closest in kind to the first organisms to arise on Earth.

See also Biology and the Life Sciences; Ecology; Evolution; Geology; National Geographic Society; Oceanography; Plate Tectonics; Undersea Exploration; Volcanoes

References

Ballard, Robert D., with Will Hively. *The Eternal Darkness: A Personal History of Deep-Sea Exploration.* Princeton: Princeton University Press, 2000.

Van Dover, Cindy Lee. *The Ecology of Deep-Sea Hydrothermal Vents.* Princeton: Princeton University Press, 2000.

————. *The Octopus's Garden: Hydrothermal Vents and Other Mysteries of the Deep Sea.* Reading, MA: Addison-Wesley, 1996.

Dobzhansky, Theodosius (1900–1975)

Theodosius Dobzhansky pioneered the synthesis of Charles Darwin's theory of natural selection and the science of Mendelian genetics. Dobzhansky was born Feodosy Grigorevich Dobzhansky in Nemirov, Ukraine (then part of Russia), where his father was a mathematics teacher. He read Charles Darwin's *On the Origin of Species* as a teenager and majored in biology at the University of Kiev. Graduating in 1921, Dobzhansky taught in Kiev and at the University of Leningrad. Dobzhansky married in 1924 and fathered a daughter. In 1927, Dobzhansky came to the United States as a Fellow of the International Education Board of the Rockefeller Foundation, and studied with the Nobel laureate Thomas Hunt Morgan (1866–1945) at Columbia University. He impressed Morgan and moved with him to the California Institute of Technology, where Dobzhansky became an assistant professor of genetics. In 1936, he became a full professor, and in 1937 he published a great synthesis of Charles Darwin's theory of natural selection and Mendelian genetics under the title *Genetics and the Origin of Species.* Many commentators consider this to be the most important book on evolution written in the twentieth century. Dobzhansky became a U.S. citizen in 1937. In 1940, Dobzhansky returned to Columbia University as a professor of zoology. He later relocated to the Rockefeller Institute in 1962 and the University of California at Davis in 1971.

Dobzhansky's major area of study involved experiments with populations of the vinegar fly *Drosophila pseudoobscura,* in which the frequency of generations made the process of natural selection through genes more visible. He demonstrated that the frequency of certain genes among the population dramatically varied even during the course of a single year, as seasons put selection pressure on the population. Besides his own contributions to population and evolutionary genetics, Dobzhansky tirelessly and successfully promoted the perspective of using mathematical models of populations within genetics.

Later in his life, Dobzhansky wrote prolifically on his own and with other authors on the philosophical implications of evolution. His classic textbooks educated generations of scientists. His 1962 book, *Mankind Evolving,* condemned those who took biology as destiny, especially to justify racist positions; yet he also objected to those who argued that human behavior was not constrained by genetics. He believed that scientists had an obligation to understand genetics and natural selection, and to inform the public of this understanding so that humanity might learn to control the process of natural selection intelligently.

See also Evolution; Genetics

References

Ceccarelli, Leah. *Shaping Science with Rhetoric: The Cases of Dobzhansky, Schrödinger, and Wilson.* Chicago: University of Chicago Press, 2001.

Dobzhansky, Theodosius. *Mankind Evolving: The Evolution of the Human Species.* New Haven: Yale University Press, 1962.

Louis Levine, editor. *Genetics of Natural Populations: The Continuing Importance of Theodosius Dobzhansky.* New York: Columbia University Press, 1995.

Dyson, Freeman (1923–)

An important physicist, Freeman Dyson also became well-known for his speculative writings on nuclear weapons, extraterrestrial civilizations, genetics, and computers. Born in Crowthorne, Berkshire, England, Freeman John Dyson had as mother a lawyer and as father a music teacher who was later knighted. As a child, Dyson showed an aptitude for mathematics and consumed science fiction novels by Jules Verne, H. G. Wells, and Olaf Stapledon. He taught himself calculus from a mail-order textbook in 1938. In 1941, he enrolled at Cambridge University to study mathematics. In 1943, he joined the war effort as a civilian analyst in operations research for the Royal Air Force Bomber Command. He was horrified by the death rate among the bomber crews and the number of civilians killed on the ground. In later writings and conversations, he condemned the bureaucratic mentality of Bomber Command officers and their inability to innovate effectively.

After the war, Dyson returned to Cambridge and completed his B.A. in mathematics in 1945. A 1947 fellowship brought him across the Atlantic Ocean to Cornell University, where he studied physics with prominent physicists. During a bus trip through the western United States, Dyson had an insight that allowed him to show that the quantum electrodynamic (QED) theories of Richard P. Feynman (1918–1988), Julian Schwinger (1918–1994), and Sin-itiro Tomonaga (1906–1979) were mathematically equivalent, leading to the 1965 Nobel Prize in Physics for those three men. Dyson also studied at the Institute for Advanced Study in Princeton, New Jersey. After a brief time at Cornell as a

Freeman Dyson, physicist, 1985 (Douglas Kirkland/Corbis)

physics professor (1951–1953), Dyson returned to the Institute for Advanced Study at Princeton, where he remained. His first marriage in 1950 resulted in a son and a daughter. The marriage ended in divorce, and his second marriage yielded four daughters.

In 1958, Dyson took a temporary leave from the institute to work in La Jolla, California, on the Orion Project. Orion designed an ambitious spacecraft, with a large plate at its end, that was pushed through space by repeated atomic explosions. A small model using high explosives successfully demonstrated the soundness of the concept, though funding for the project languished, and the U.S. Air Force officially cancelled the effort in 1965. Dyson later speculated and wrote about the possibility of extraterrestrial civilizations. He posited that a truly advanced technological civilization would want to trap as much of their star's energy as possible and thus build a shell of planetoids around their star. This concept became known as the Dyson sphere. Dyson advocated the colonization of space by human beings in order to reintroduce social diversity and avoid the consequences of being trapped on only one planet.

Dyson also briefly worked on a low-fallout bomb, which later became the neutron bomb. He initially opposed the proposed nuclear test ban treaty in 1960, but later changed his mind and supported it wholeheartedly. He also lobbied effectively to create the U.S. Arms Control and Disarmament Agency. Dyson later became involved with the debate over recombinant DNA research in the 1970s and has speculated about the consequences of genetic technologies. Dyson published his first book at the age of fifty-five, and his talented writing on nuclear weapons, extraterrestrial civilizations, genetics, and computers has earned him a following. His work has resulted in many awards and honorary degrees.

See also Feynman, Richard; Nuclear Physics; Particle Physics; Physics; Science Fiction; Search for Extraterrestrial Intelligence

References

Dyson, Freeman. *Disturbing the Universe.* New York: Harper and Row, 1979.
———. *Weapons and Hope.* New York: Harper and Row, 1984.

E

Earthquakes

In 1923, an earthquake in Tokyo, Japan, killed 68,000 people. As might be expected, the Japanese government reacted by promoting scientific study of earthquakes and efforts to engineer more earthquake-resistant buildings. Because of delays in translating these efforts into English, Japanese expertise only slowly came to influence the rest of the world. American earthquake studies and earthquake engineering were spurred by the 1906 San Francisco earthquake. Charles Richter (1900–1985) developed his scale for measuring earthquake magnitude at the California Institute of Technology, though the Japanese scientist Kiyoo Wadati had already developed a similar system. The Japanese geophysicist Keiiti Aki (1930–), who worked mostly at American universities, is perhaps the most important seismologist of the last half of the twentieth century. Aki developed in 1966 a different measurement of earthquake size, the seismic moment, but the Richter scale remained the common measure for the public.

Geologists and seismologists knew that earthquakes were more frequent in some areas, such as along the Pacific Coast of America, Japan, and China, but did not understand why. During the 1960s, geology experienced a revolution, as geologists became convinced that continents really did drift and a 1970 article first used the term *plate tectonics*. The theory of plate tectonics proved to be a powerful explanatory mechanism that transformed geology from a science of particulars into a mature science with a body of integrated knowledge. As plates grid together, they create mountains from uplift, and where plates pull apart, they create rift valleys and opportunities for magma to well up from deeper in the Earth. Volcanoes and greater earthquake activity exist on the edges of continental plates. The rim of the Pacific Ocean is called the Ring of Fire because of numerous active volcanoes, and it has many seismic faults that cause numerous earthquakes.

During the cold war, the superpowers set up sensitive seismic monitoring stations to detect the waves from nuclear explosions. The technology became so good that international nuclear test bans were possible because nuclear tests could not be concealed from the monitors. To uniformly monitor earthquakes around the world the Worldwide Standardized Seismograph Network (WWSSN) was created in 1961. Studies of waves generated by earthquakes that travel through Earth's interior have provided important insights into the composition (particularly density) of the planet's interior. The use of computers to collate and analyze seismograph data has greatly

Charles F. Richter posing next to a device to measure the magnitude of earthquakes, 1966 (Bettmann/CORBIS)

reputable techniques for prevention exist, but Chinese scientists have devoted considerable effort to earthquake prediction since the 1960s and claimed partial success. Among their indicators are changes in groundwater levels and agitation among domestic animals and birds.

> **See also** Apollo Project; Cold War; Computers; Geology; Plate Tectonics; Satellites; Space Exploration and Space Science; Volcanoes
>
> **References**
> Bolt, Bruce A. *Earthquakes and Geological Discovery.* New York: Scientific American Library, 1993.
> ————. *Nuclear Explosions and Earthquakes: The Parted Veil.* San Francisco: W. H. Freeman, 1976.
> Hough, Susan Elizabeth. *Earthshaking Science :What We Know (and Don't Know) about Earthquakes.* Princeton: Princeton University Press, 2002.

Ecology

The study of biological organisms and their environment has been part of human inquiry from prehistoric times, but the effort gained an intellectual focus with the theory of natural selection developed by the English naturalist Charles Darwin (1809–1882). By the late nineteenth century, biologists, botanists, zoologists, and scientists from other disciplines were engaged in disciplined ecological studies. In the first half of the century, ecologists often viewed their subject matter in terms of an ecological community, though critics of this approach objected that ecological communities were difficult to define geographically because animal and plant species often participate in many different communities. Population ecology emerged as an alternative in the 1920s, attempting to create mathematical models of the population growth of individual species and predator-prey relationships. The idea of the carrying capacity of an environment to support certain levels of individuals of different species emerged. Experiments showed that microbes in solutions did not follow the mathematical models and that one species usually pushed out another species unless competition was

helped geologists. In the late 1990s, Global Positioning System (GPS) receivers placed around volcanoes revealed the existence of silent earthquakes. These earthquakes are significant movements of the Earth that take place over days rather than minutes, so they do not cause classic earthshaking and are not detected on seismographs.

The Apollo astronauts placed seismometers on the Moon's surface that recorded lunar earthquakes. Though the *Viking 2* lander on Mars carried a seismometer, noise from wind in the thin atmosphere interfered with its operation. Other planets and moons of the solar system obviously have earthquakes, since fault lines and rift valleys have been identified from photographs, but knowledge in this area is still limited.

The chief public concerns about earthquakes are prevention and prediction. No somehow reduced. This discovery led to the related ideas of ecological niches and competitive exclusion, the theory that natural selection tended to reduce the potential for competition by rewarding the subdividing of resources into niches.

The British ecologist Arthur Tansley (1871–1955) coined the term *ecosystem,* and after World War II the ecosystem approach became influential. This approach initially emphasized biogeochemistry, following the movement of energy, chemicals, water, and resources through the food chain in a given environment to support plant and animal species. In a classic study, the English-born limnologist G. Evelyn Hutchinson (1903–1991) showed that guano deposits in South America increased fish populations and when fishermen from Peru and Chile shot the birds eating the fish, in an effort to get rid of their competitors, the fish populations actually decreased because fewer birds meant less guano.

During World War II, scientists and engineers became used to working with automatic control systems, with their feedback loops and self-adjusting characteristics: aircraft control systems, for example, and mechanical computers and gun control systems. The American mathematician Norbert Wiener (1894–1964) coined the term *cybernetics* in 1948 to describe the study of such systems. Ecosystem ecologists recognized that ecosystems were a form of cybernetic system and often used similar metaphors. The two brothers Eugene Odum (1913–) and Howard Odum (1922–) promoted the ecosystem approach, and an influential textbook by Eugene, *Fundamentals of Ecology,* first published in 1953, promoted the ecosystem concept and later used cybernetic metaphors. The Odums also worked for the Atomic Energy Commission (AEC) studying the impact of nuclear testing on Pacific islands.

The classic example of evolutionary ecology came from studies in the 1950s by the British naturalist H. B. D. Kettlewell (1907–1979) of the peppered moth, *Biston betularia.* Kettlewell pointed out that the darker forms of the moth were more common downwind from industrial sites in England than in areas where pollution was not so common. He argued that pollution had stained lichens on trees with soot and that when moths rested on the trees, the light-colored moths were eaten by birds and the darker-colored moths had a greater chance of survival. This study became a common example of the principle of natural selection in action. In the 1980s and 1990s, scientists found that Kettlewell's technique contained flaws. Photographs from the study showed the moths on the trees, but Kettlewell used captured moths that he placed on trees, not realizing that that particular species of moth rarely lights on trees or that ornithologists doubted that the birds ate the moths when they were on the trees. The explanation that Kettlewell advanced is now abandoned, though the coloration frequency of moths in different areas has not been explained.

Spurred on by the writings of scientists like the biologist Rachel Carson (1907–1964), the environmental movement grew out of ecological concerns. As environmentalism gained strength, the demand for ecological studies increased and the field blossomed. In the 1970s, American universities began to grant degrees in ecology, whereas before, students interested in ecology graduated in biology, zoology, or botany. Ecology may be the most diverse and interdisciplinary of all scientific endeavors, with few scientists identifying themselves as just ecologists. The field is further subdivided by the type of organism primarily studied (plant ecology, animal ecology, or microbe ecology), the environment studied (marine ecology, tree ecology, or alpine ecology), or the methodology used (population ecology, community ecology, or ecosystem ecology).

The application of mathematical modeling, computer simulations, and thinking in terms of systems has grown since the 1940s to become a major part of the field of ecology. The web-of-life analogy so often used in teaching ecology has become a cliché, but it

Stetson Bank at Flower Garden Banks National Marine Sanctuary in Georgia, one of the northernmost coral reefs in the United States. Coral reefs are very intricate ecosystems. (National Oceanic & Atmospheric Administration / Frank and Joyce Burek)

is accurate in that it suggests that the many parts that make up the whole of an ecosystem are connected to each other; it is also useful in the sense that removing a thread or threads weakens the overall web.

Ecological studies are often small projects, but Big Science played a role in at least one project, the International Biological Programme (IBP), which began in 1961 in Amsterdam under the auspices of the International Union of Biological Sciences. The IBP began in earnest in 1968, and American ecologists obtained federal funding to participate in 1970. American ecologists explicitly used the ideas of cybernetics and systems ecology, and their usefulness to the environmental movement, to obtain the $40 million in funding. Much of the American funding went to creating computerized models of five biomes, such as the Eastern Deciduous Forest

Biome and the Grassland Biome. The National Science Foundation (NSF) continued to fund ecological research after the official end of the IBP in 1974.

Scientists came to recognize that biodiversity is the foundation of a resilient ecology. By 2000, only about 1.75 million species, most of them insect species, had been identified worldwide. Estimates of the total number of species range from 3 million to 100 million, most of them yet to be found in tropical rainforests. The island of Madagascar is particularly rich in species unique to the island.

At the 1992 Earth Summit in Rio de Janeiro, an international agreement, the Convention on Biological Diversity (CBD), was adopted. The pact promoted the sustainable use of natural resources, preservation of species, and the equitable use of genetic resources. An overwhelming majority of nations signed and ratified the CDB; the United States only signed, and did not ratify, the treaty. Some Third World nations and indigenous peoples have asserted their right under the CDB to prevent scientists from conducting studies on territory that they control. They fear that a plant or other genetic resource might be found and developed into a commercial product without the nation or people receiving any compensation. Under the guidelines of the CDB, scientists must get prior informed consent or sign a bioprospecting agreement before conducting surveys or studies.

Scientists thought that the fundamental source of energy in all ecosystems was solar energy, but the discovery of life around deep-sea hydrothermal vents in 1977 provided an alternate ecological model, as well as perhaps an explanation of how the first life on Earth developed. Where the food chain of most other life on Earth starts with the process of photosynthesis, which converts sunlight into energy, species of bacteria around the vents use chemosynthesis to create energy, converting sulfides into organic carbon.

The microbiologist Lynn Margulis (1938–), who reinvigorated the idea of symbiosis, in which two different organisms intimately associate with each other, has championed the idea that microbes are the dominant form of life on Earth. This perspective inverts the classic way of viewing all species as forming a kind of pyramid of life with human beings at the top and changes the way ecosystems are viewed. Perhaps the ultimate expression of the idea of an ecosystem as a cybernetic system came when the biophysicist and inventor James Lovelock (1919–) developed the Gaia hypothesis in the late 1960s, arguing that Earth was like a living organism, maintaining surface and atmospheric conditions so that chemical processes and the temperature would sustain life. Margulis found the idea compatible with her own promotion of symbiosis and has vigorously championed Gaia. Many other biologists remain suspicious of such a holistic notion and are uncomfortable with Gaia's mystical overtones.

See also Big Science; Biology and the Life Sciences; Carson, Rachel; Computers; Deep-Sea Hydrothermal Vents; Environmental Movement; Genetics; Global Warming; Lovelock, James; Margulis, Lynn; National Science Foundation; Oceanography; Symbiosis

References

Bocking, Stephen. *Ecologists and Environmental Politics: A History of Contemporary Ecology.* New Haven: Yale University Press, 1997.

Bowler, Peter J. *The Norton History of the Environmental Sciences.* New York: Norton, 1993.

Golley, Frank Benjamin. *A History of the Ecosystem Concept in Ecology: More Than the Sum of the Parts.* New Haven: Yale University Press, 1993.

Hagen, Joel B. *An Entangled Bank: The Origins of Ecosystem Ecology.* New Brunswick, NJ: Rutgers University Press, 1992.

Kingsland, Sharon E. *Modeling Nature: Episodes in the History of Population Ecology.* Chicago: University of Chicago Press, 1985.

Worster, Donald. *Nature's Economy: A History of Ecological Ideas.* Second edition. New York: Cambridge University Press, 1994.

Economics

The science of economics began with the work of the Scottish philosopher and political economist Adam Smith (1723–1790), who described the "invisible hand" of free markets promoting the most efficient use of economic resources and creating wealth. By encouraging the individual economic self-interest of people, a nation prospered. Later theorists, such as the British philosopher and economist John Stuart Mill (1806–1873), expanded on Smith's work, explaining how the forces of supply and demand led to efficiency, and these ideas were central in what became known as classical economics. Classical economics supported the system of capitalism that had become dominant in Western Europe with the advent of the Industrial Revolution. The German-born political theorist and failed revolutionary Karl Marx (1818–1883) lived in exile in London after 1849 and wrote his critique of capitalism, arguing for the ownership of the means of production by the working classes. The science of economics, called political economy at the time, influenced as it was by political ideology, was not just a scholarly pursuit, but formed the basis of government policy and inspired revolutionary movements.

The Great Depression of the 1930s challenged traditional economists to understand why the world economy had shrunk and how to recover from such a period of extended unemployment. The British economist John Maynard Keynes (1883–1946) published *The General Theory of Employment, Interest and Money* in 1936, contradicting classical economics and arguing that only massive government spending could end the Great Depression. Keynesian economics became influential as the world's economies emerged from the destructive chaos of World War II and began to rebuild.

Serving as government policy makers in the prosperous Western democracies, economists applied Keynesian principles by making credit easier to obtain during economic slowdowns and striving to maintain high employment. In the postwar period, communism expanded from its base in the Soviet Union, applying a modified form of Marxism to manage centralized economies with burdensome bureaucracies and constructing industry at heavy environmental costs. In the Western world, after the environmental movement gained prominence in the 1960s and 1970s, the ideal of growth at any ecological cost was questioned. Some economists responded by trying to develop ways of measuring environmental impacts and incorporate them into economic theory.

The United Kingdom fostered the best economic theorists before World War II, and in the postwar era, economists in the United States often took the lead. Beginning in the 1930s, economists had increasingly applied mathematics to their theories, creating equation-driven models of macroeconomics and microeconomics. In 1951, a famous article by the American economist Kenneth Arrow (1921–) and French economist Gerard Debreu (1921–) showed mathematically that Smith's "invisible hand" really existed, a method of arriving at a "general equilibrium," a complex theory explaining how individuals make economic decisions. This mathematical underpinning has given economics a firmer claim to status as a science, even though it remained hamstrung by its study of human-created behaviors, rather than the natural phenomena that the harder sciences, such as physics and chemistry, study. The Bank of Sweden funded an annual economics prize to be awarded by the Nobel Foundation, with the first prize being awarded in 1969 to the Norwegian Ragnar Frisch (1895–1973).

The Hungarian-born mathematician John von Neumann (1903–1957), besides laying the theoretical foundations of the modern electronic computer, developed the concept of "game theory" with the economist Oskar Morgenstern (1902–1976) in 1944. Game theory describes how to apply rational decisions to games and, by extension, to economic or political behavior. The brilliant mathematician John Nash (1928–) expanded on the work of von Neumann and Morgenstern to better understand noncooperative games, cooperative games, and two-person bargaining games, by focusing on mutually optimal threat strategies in these types of games, and created what became known as Nash equilibria. After other theorists expanded on Nash's work, game theory became an integral part of economic theory in the 1960s. Game theory becomes particularly interesting when actors in the game do not have access to the same information and are thus prevented from making rational decisions.

The Keynesian model began to break down in the 1960s as economists disproved some of its assumptions. Milton Friedman (1912–), one of the new neoclassical economists who taught at the University of Chicago, emphasized the importance of monetary policy and thought that government intervention in the economy had harmful effects. Friedman has since become a patron saint of the resurgent conservative parties in the United States and Europe.

In the 1970s, some economists began to work with psychologists and to use insights from the behavioral sciences in general, recognizing that not all economic behavior is rational. For instance, stock trading is often driven by ego and exuberance rather than rational calculation. By the 1990s, behavioral economics had become an accepted approach in the field.

If it is seen only as the economic theory behind communism, Marxism failed in the last two decades of the twentieth century. On the other hand, many nations have been influenced by Marxist thought in the shaping of their socialist or semisocialist economies, and some of these economies have been successful. Recently, critics have argued that economists are good at explaining what is happening when the economy is functioning normally, but are not so good at explaining how to handle a serious economic crisis or how new technologies and changing work patterns might affect the economy. Despite such criticisms, economists remain influential. Noted economists often advise governments around the world on economic matters, drawing on economic models, increasingly sophisticated reporting of economic statistics, their personal ideological inclinations, and personal experience born of long study.

See also Environmental Movement; Mathematics; Nash, John F.; Nobel Prizes; Soviet Union; Neumann, John von

References

Blaug, Mark. *The Methodology of Economics: Or How Economists Explain*. Second edition. New York: Cambridge University Press, 1992.

Fox, Justin. "What in the World Happened to Economics?" *Fortune,* March 15, 1999, 91–102.

Heilbroner, Robert L. *The Worldly Philosophers: The Lives, Times and Ideas of the Great Economic Thinkers.* Seventh edition. New York: Touchstone, 1999.

Mirowski, Philip. *Machine Dreams: Economics Becomes a Cyborg Science.* New York: Cambridge University Press, 2001.

Morgan, Mary S. *The History of Econometric Ideas.* Cambridge: Cambridge University Press, 1990.

Weintraub, E. Roy. *How Economics Became a Mathematical Science.* Durham, NC: Duke University Press, 2002.

Edelman, Gerald M. (1929–)

The biochemist Gerald M. Edelman made significant contributions to our understanding of antibodies and morphogenesis. Gerald Maurice Edelman was born in New York City, where his father was a physician. Edelman entertained early ambitions to be a violinist, but later he turned to science and graduated from Ursinus College in 1950 with a B.S. in chemistry. Following his father into medicine, he graduated from the University of Pennsylvania medical school in 1954. After serving his obligation to the U.S. Army for two years in Paris, he became convinced that he wanted to dedicate his life to research. The problems of immunology particularly fascinated him. Edelman entered the Rockefeller Institute in New York (later Rockefeller University) to study biochemistry. After receiving a doctorate in 1960, he remains at Rockefeller as a faculty member. Edelman married in 1950 and fathered two sons and a daughter.

Edelman's work in immunology focused on antibodies, large molecules formed by the immune system to destroy viruses and bacteria by binding to them. Edelman's first insight was that immunoglobulins (antibodies) were really peptides made up of heavy and light chains. The 1960s were an exciting time, as scientists closed in on the discovery of how antibodies formed and operated. A team led by Rodney R. Porter (1917–1985) of Oxford University competed with Edelman's team in the contest to decipher the structure of a myeloma globulin antibody molecule. They chose to work on the myeloma antibody because populations of these antibodies were more homogeneous than those of other antibodies. Edelman's team achieved their goal first, but the 1972 Nobel Prize in Physiology or Medicine was awarded to both biochemists. Edelman and his team also promoted a theory that showed that each immune system contained many different types of antibodies and when an antigen was detected, those antibodies most likely to counteract the antigen were selected for in a

Gerald M. Edelman, biochemist, 1969 (Bettmann / Corbis)

process analogous to Charles Darwin's principle of natural selection.

After receiving the Nobel prize, Edelman turned his attention to morphogenesis, the process in which individual cells come together to form tissues. Edelman and his research team discovered three different types of cell adhesion molecules (CAM) that provided the cement to keep cells together in a tissue. They also developed a theory that showed that cells develop into one of many possible forms based on interaction with neighboring cells. The discovery of CAM revolutionized the field of developmental biology.

In the late 1970s, Edelman again shifted his focus and turned to the study of the brain. He formed the Neurosciences Institute at the Rockefeller University in New York City in 1981 and in 1993 moved to the Scripps Research Institute in La Jolla, California, to further continue his research in neurology. Edelman developed a theory of the brain that again drew on the principle of natural selection. In his theory, the connections within some groups of neurons are selected from those within other groups because they are more effective at producing consciousness, memory, and other higher functions of the mind. He has published four books on his theory, including *The Remembered Present: A Biological Theory of Consciousness* (1990). His theory is not widely accepted by other neurologists.

See also Cognitive Science; Evolution; Medicine; Neuroscience; Nobel Prizes

References

Edelman, Gerald M. *Bright Air, Brilliant Fire: On the Matter of the Mind.* New York: Basic Books, 1992.

Nobel *e*-Museum. "Gerald M. Edelman— Biography." http://nobelprize.org/medicine/ (accessed February 13, 2004).

Edwards, Robert (1925–)

Robert Edwards and Patrick Steptoe (1913–1988) perfected the process of in vitro fertilization (IVF) and created the first human test-tube baby. Robert Geoffrey Edwards was born in a small town in Yorkshire and grew up in the city of Manchester, England. Edwards served in the British Army from 1944 to 1948, then received his advanced education at the University of Wales and his doctorate in physiology in 1957 at the Institute of Animal Genetics at Edinburgh University. His doctoral research concentrated on mouse embryos, and in collaboration with a fellow student, Ruth Fowler, he developed the Fowler-Edwards method, used to treat female mice with hormones and regulate the number of eggs produced and their time of ovulation. He also married Fowler. A fellowship at the California Institute of Technology brought Edwards to America for a year. He returned to the United Kingdom, spending 1958 to 1962 at the National Institute of Medical Research at Mill Hill in London, then a year at the University of Glasgow, before joining the faculty at Cambridge University in 1963.

Other scientists had already perfected in vitro fertilization of mouse embryos, and Edwards conducted research in this area. He realized that the basic problems of the procedure were the same from mammal to mammal and turned to working with human embryos. He found it difficult to obtain samples from ovaries until the gynecologist who had delivered two of his daughters offered to provide occasional samples and the rare human egg. Even with this material, his research suffered from the lack of material to work with. He read about the pioneering laparoscopy techniques of Patrick Steptoe, an obstetrician and gynecologist in practice in Oldham, England. Edwards contacted Steptoe in 1968, and the two men began a long collaboration.

Steptoe used laparoscopy to obtain human eggs, and Edwards fertilized them in the laboratory. In 1972, Steptoe implanted a fertilized egg into a uterus for the first time. None of the approximately thirty women in their test group who became pregnant carried the babies for more than a trimester. Funding for their experiments was cut amidst a general outcry from Members of Parliament, religious leaders, and other scientists. Among the objections was fear that a successful technique would not stop at helping infertile couples, but would lead to genetically engineered babies and surrogate mothers. Edwards was willing to engage in discussions on the ethics of his work and published with the lawyer David Sharpe the first ethics paper on IVF in 1971.

Continued criticism from their colleagues caused Edwards and Steptoe to suspend reporting on their research. Steptoe financed the continuation of their research from his practice, including legal abortions, and the two scientists moved their work to a smaller hospital in Oldham, Dr. Kershaw's Cottage Hospital. They persisted amid many failures until late 1977, when Lesley Brown was impregnated with her own egg fertilized by her husband's sperm. Steptoe and Edwards decided to implant the egg after it had divided only three times and was only eight cells large, instead of waiting for it to get larger as they had done up till then. On July 25, 1978, a daughter name Louise Joy was delivered by caesarian section. The world press proclaimed the world's first test-tube baby. Steptoe and Edwards continued their collaboration, and their clinic succeeded in producing over a thousand babies for infertile couples. Their technique of IVF and successful implantation of the egg into a uterus is now used throughout the industrialized world. Edwards retired from Cambridge University in 1989.

See also Biotechnology; In Vitro Fertilization; Medicine; Steptoe, Patrick
References
Edwards, Robert G. "The Bumpy Road to Human In Vitro Fertilization." Nature Medicine 7, no. 10 (October 2001): 1091–1094.

Ehrlich, Paul R. (1932–)

In the 1960s, the entomologist Paul R. Ehrlich became the best known of those scientists and writers sounding a warning call over human population growth. Paul Ralph Ehrlich was born in Philadelphia, Pennsylvania, where his father was a salesman and his mother taught high school Latin. The family moved to Maplewood, New Jersey, where Ehrlich indulged his passion for the outdoors. In high school he read *Road to Survival,* the prophetic 1948 book by William Vogt on the dangers of overpopulation. Ehrlich received his B.A. in zoology from the University of Pennsylvania in 1953, then moved to the University of Kansas, where he received his M.A. in 1955 and Ph.D. in 1957, both in zoology. He concentrated in entomology and did fieldwork on biting flies in the Arctic and parasitic mites. He became an assistant professor of biology at Stanford University in 1959, and rapidly moved up the academic ranks, becoming an associate professor in 1962 and a full professor of biology in 1966. His research at Stanford concentrated on butterflies. He married Anne Fitzhugh Howland in 1954, and his wife worked closely with him in their research. They had one daughter.

Paul R. Ehrlich, entomologist (Bettmann / Corbis)

Always conscious on an intellectual level of the problem of burgeoning human population growth, Ehrlich was with his family in India in 1966, riding in a taxi through a crowded city, when he had an emotional epiphany that provoked a lifelong crusade against population growth. In 1968 he published *The Population Bomb,* drawing attention to the coming crisis when there would not be enough food to feed the world's growing population. Increasing food production through dramatic efforts like the agricultural Green Revolution would serve only as a "stay of execution," if humanity did not learn to curb its population growth. The world's population had grown from 1.6 billion in 1900 to 3.6 billion in 1968. Ehrlich felt that 1.5 billion was the reasonable human population for the planet. Ehrlich's book became a bestseller, and he received considerable media exposure, including a full hour on the popular Johnny Carson late-night television show.

Ehrlich's concern with population, combined with concern over the environment, led him to found a political advocacy organization called Zero Population Growth, which was later renamed Population Connection. The organization sought increased birth control, legalized abortion (before it was legalized nationwide in the United States in 1973), government support for smaller families, and increased protection of the environment. His wife and he coauthored other books on the subject, including *The Population Explosion* (1990) and *Betrayal of Science and Reason: How Anti-Environmental Rhetoric Threatens Our Future* (1996). Ehrlich became the Bing Professor of Population Studies at Stanford in 1976. Among his many honors, Ehrlich shared the 1990 Crafoord Prize with the zoologist Edward O. Wilson (1929–). The Crafoord Prize is awarded by the Royal Swedish Academy of Sciences in those areas not covered by the Nobel prizes.

See also Agriculture; Biology and the Life
  Sciences; Ecology; Environmental Movement;
  Population Studies; Wilson, Edward O.

References

"Paul R. Ehrlich." http://www.stanford.edu/
  group/CCB/Staff/paul.htm (accessed
  February 13, 2004).

Population Connection. http://www.
  populationconnection.org/ (accessed February
  13, 2004).

El Niño

For centuries fishermen off the equatorial coast of Peru noticed that the water warmed around the time of Christmas, and they called the phenomenon El Niño, which means "The Little One" in Spanish, a reference to the infant Jesus. This phenomenon came to the attention of scientists in the late nineteenth century, who realized that there was a relationship between the warming water along the equator in the Pacific Ocean and heavier seasonal rains in South America. In 1923, Sir Gilbert Walker (1868–1958), who headed the Indian Meteorological Service, observed that when the air pressure over the Pacific Ocean was high, the air pressure over the Indian Ocean was low,

and vice versa. He named this phenomenon the Southern Oscillation. In 1969, the Norwegian meteorologist Jacob Bjerknes (1897–1975), who taught at the University of California at Los Angeles, tied the two phenomena together in his description of the El Niño Southern Oscillation (ENSO). La Niña (The Little Girl), also sometimes called El Viejo (The Old One), is the opposite end of the oscillation, when the Pacific water along the equator turns cooler than normal.

The strong El Niño of 1982–1983 attracted the attention of scientists, and they began to study the ENSO more intensely. The use of satellites, buoys, and other sensors led to more accurate temperature and pressure readings. As they followed the cycle, which lasted about four to five years, meteorologists came to partially understand the effects of El Niño. Although scientists still cannot predict when an El Niño episode will occur, after they detect its beginning they can make accurate six-month weather forecasts of its effects on South America and North America. The strength of El Niño episodes varies, with mild episodes raising the temperature of the water

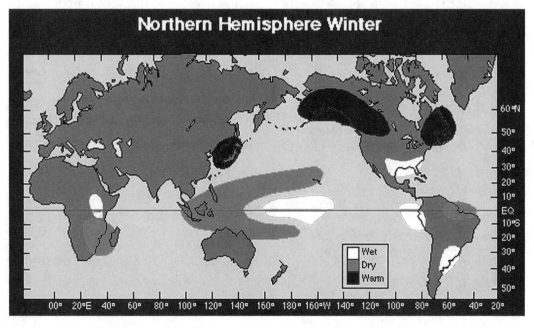

A graph at the Max Planck Institute in Hamburg shows the abnormal temperatures caused by El Niño in 1997. (Pandis Media/Corbis Sygma)

only 2 to 3 degrees Celsius (4–5 degrees Fahrenheit) and the most intense episodes raising the temperature of the water 10 degrees Celsius (18 degrees Fahrenheit).

Though the eruption of El Chichón in Mexico preceded the El Niño of 1982–1983 and the eruption of Mount Pinatubo in the Philippines preceded the El Niño of 1991–1992, scientists believe that these are coincidences and that volcanoes are not significant actors in the ENSO phenomenon. Tree-ring data have shown that the ENSO has existed for at least the past 750 years. El Niño episodes cause fish populations to decline off the coast of South America because the thick layer of warmer water prevents the upwelling of nutrients. ENSO also has a major effect on rain patterns across the Pacific, on the monsoons of the Indian Ocean, and on the rainfall in South America and North America. Drought patterns often follow the ENSO cycle.

The 1997–1998 El Niño became the largest and warmest in the twentieth century, generating extensive news media coverage. Some scientists and politicians speculated that global warming caused the magnitude of that particular El Niño. This supposition remains unproven, and computer models that combine global warming and the ENSO are so far unsatisfactory. Scientists do recognize that the equatorial Pacific provides carbon dioxide to the atmosphere and that during La Niña episodes more carbon dioxide is released than during El Niño episodes.

See also Computers; Environmental Movement;
   Global Warming; Meteorology; Oceanography;
   Volcanoes

References

Changnon, Stanley A., editor. *El Niño 1997–1998: The Climate Event of the Century.* New York: Oxford University Press, 2000.

Fagen, Brian. *Floods, Famines, and Emperors: El Niño and the Fate of Civilizations.* New York: HarperCollins, 2000.

Glantz, Michael H. *Currents of Change: El Niño's Impact on Climate and Society.* Cambridge: Cambridge University Press, 1996.

Nash, J. Madeleine. *El Niño: Unlocking the Secrets of the Master Weather-maker.* New York: Warner Books, 2002.

Elion, Gertrude B. (1918–1999)

The pharmacologist Gertrude B. Elion helped pioneer a new way to search for drugs and discovered several important drugs to treat leukemia, suppress the immune system, and treat viral infections. Gertrude Belle Elion was born in New York City to Jewish immigrants. Her father, a dentist, came from Lithuania, and her mother came from Russia. A bright student, she skipped four grades before graduating from high school at the age of fifteen. Because of the stock market crash of 1929, her parents could not send her to a private college, so Elion turned to the tuition-free, women-only Hunter College in New York City. A voracious reader, skilled at many different academic disciplines, she could not decide on a career choice until her grandfather died painfully of stomach cancer the summer before she entered college. She resolved to pursue a cure for cancer, choosing to major in chemistry because she did not like dissecting animals in biology.

Obtaining her A.B. degree summa cum laude in 1937, she wanted to go to graduate school. Even though she had excellent grades, the financial hardship of the Depression and discrimination against women in science proved to be difficult hurdles, and she could not find a graduate assistantship. She went to work as an unpaid assistant to a chemist at a small pharmaceutical company. The job eventually started to pay a little money, and she found additional employment as a doctor's receptionist and a substitute teacher in secondary schools. This allowed her to attend New York University and graduate with an M.Sc. in chemistry in 1941. World War II opened up employment opportunities, and after several jobs she went to work for the Burroughs Wellcome Company, a pharmaceutical company. The research chemist George H. Hitchings (1905–1988) hired her as his assistant for fifty dollars a week.

Hitchings believed in a new way of searching for drugs. The old method was based on a hit-or-miss process of trying many different compounds on many different diseases and seeing what happened. The new way, based on what was called antimetabolite theory, consisted of learning how the disease worked, then trying to find a compound that interfered with the disease process. Elion proved to be an excellent laboratory worker, and Hitchings gave her considerable freedom. In 1948 a compound that she isolated, 2,6-diaminopurine, was tested on three leukemia patients with some success. Elion pursued work with other purine compounds and isolated 6-mercaptopurine (6-MP), which proved useful in extending the life of leukemia patients, though not curing them. Elion continued to work on 6-MP, and her work was noticed by others, who realized that 6-MP could suppress the immune system. In 1962, the surgeon Joseph E. Murray (1919–) used the 6-MP derivative azathioprine to suppress the immune system of the recipient of the first successful kidney transplant from a nonrelated donor. Murray later shared the 1990 Nobel Prize in Physiology or Medicine for this accomplishment. Elion's research on 6-MP also led to an enzyme that became the drug allopurinol, which successfully treated gout and a number of other diseases.

As Hitchings rose up through the ranks of Burroughs Wellcome, his protégé followed, and in 1967, Elion became head of the Department of Experimental Therapy. She remained heavily involved in research, helping in the development of acyclovir, a treatment for shingles, chicken pox, and other herpesvirus conditions. Elion retired in 1983, but remained active in the field and continued to partially oversee projects for Burroughs Wellcome, including the development of azidothymidine (AZT), used to treat acquired immune deficiency syndrome (AIDS).

After her fiancé died in the late 1930s, Elion chose not to marry; she felt that her career in research consumed too much of her time to entertain the possibility of children. She also never completed a doctorate because of time constraints. The 1988 Nobel Prize in Physiology or Medicine was awarded to Sir James W. Black, Hitchings, and Elion. Black was a British pharmacologist who had developed beta-blocking agents. Elion took out forty-five patents; she was awarded a National Medal of Science, and she became the first woman inducted into the National Inventors Hall of Fame. She died in Chapel Hill, North Carolina.

> *See also* Medicine; Nobel Prizes; Organ
>   Transplants; Pharmacology
> *References*
> Elion, Gertrude B. "The Quest for a Cure." *Annual
>   Review of Pharmacology and Toxicology* 33 (1993):
>   1–23.
> Holloway, Marguerite. "Profile: Gertrude Belle
>   Elion: The Satisfaction of Delaying
>   Gratification." *Scientific American* 265, no. 4
>   (October 1991): 40, 43–44.

Enders, John (1897–1985)

The American microbiologist John Enders developed with his coworkers a better way to culture mammalian viruses, leading to effective antiviral vaccines. John Franklin Enders was born in West Hartford, Connecticut, where his father was a banker. He attended prestigious private schools before entering Yale University in 1915. During World War I, he joined the navy and learned to fly. After the war Enders returned to school and graduated in 1920 with a B.A. in English. He worked as a real estate agent, but decided that the world of commerce was not for him and entered Harvard University. In 1922, he earned an M.A. in English literature. He remained at Harvard, studying literature and ancient languages, intending to receive a doctorate in those areas. A fellow student introduced him to the microbiologist Hans Zinsser (1878–1940), and Enders changed his direction, working in Zinsser's laboratory and graduating in 1930 with a Ph.D. in microbiology. Harvard retained his services as an instructor, and Enders conducted research into bacterial immunity.

In 1937, Enders turned to studying viruses. At that time, before the discovery of DNA, viruses were known to be smaller than bacteria, and they could not be grown in normal cultures. Viruses required animal cells to reproduce. During World War II, Enders served as a consultant advising the military on epidemic diseases. After the war, he created an Infectious Diseases Research Laboratory at Boston Children's Hospital, where he remained for the rest of his professional life, while also remaining a member of the Harvard faculty. The medical doctors Thomas H. Weller (1915–) and Frederick C. Robbins (1916–2003) joined Enders in his research. In 1947, the three scientists developed a process called continuous culture, which allowed them to grow mumps virus in a medium of cultured chicken cells. They used the new antibiotics of streptomycin and penicillin to kill the bacteria that had contaminated previous efforts.

In 1948, the three scientists succeeded in culturing the polio virus in normal human cells. Up to that time, the polio virus had only been cultured in human nerve cells and in live monkeys. Enders, Weller, and Robbins received the 1954 Nobel Prize in Physiology or Medicine for their discovery. Albert Sabin (1906–1993) and Jonas Salk (1914–1995) used the new technique to create their different polio vaccines.

Enders continued his research and found, with his coworkers, a strain of measles that conferred immunity without the ravages of the disease. This strain became the basis of the successful measles vaccine in 1963. Enders married in 1927 and had a son and a daughter. After his wife died in 1943, he remarried in 1951. He retired in 1967 and died at his home in Waterford, Connecticut, in 1985.

See also Medicine; Nobel Prizes; Pharmacology; Sabin, Albert; Salk, Jonas

References

Tyrrell, D. A. J. "John Franklin Enders: February 10, 1897–September 8, 1985." *Biographical Memoirs of the Fellows of the Royal Society* 33 (1987): 213–233.

Weller, Thomas H., and Frederick C. Robbins. "John Franklin Enders: February 10, 1897–September 8, 1985." *Biographical Memoirs of the National Academy of Sciences* 60 (1991): 47–65.

Engelbart, Douglas (1925–)

The computer scientist Douglas Engelbart developed a system in the 1960s that pioneered many of the technological innovations that led to the personal computer and World Wide Web. Douglas C. Engelbart was born in Portland, Oregon; his father died when he was nine years old, and he was raised on a farm. He entered Oregon State College in 1942, majoring in electrical engineering under a military deferment program during World War II. After two years, the military ended the deferment program because the immediate need for military personnel took precedence over the long-term need for engineers. Engelbart elected to join the navy and became a technician, learning about radios, radar, sonar, teletypes, and other electronic equipment. He missed the fighting and returned to college in 1946. Two years later he graduated and went to work for the National Advisory Committee for Aeronautics (NACA), a precursor to the National Aeronautics and Space Administration (NASA). He married in 1951, and feeling dissatisfied with his work at NACA, he sought a new direction in his life.

After considerable study he realized that the amount of information was growing so fast that people needed a way to organize and cope with the flood, and computers were the answer. Engelbart was also inspired by the seminal 1945 article, "As We May Think," by the electrical engineer Vannevar Bush (1890–1974), who directed the American Office of Scientific Research and Development during World War II. Bush envisioned the use of computers to organize information in a linked manner that we now recognize as an early vision of hypertext. Electronic computers in 1951 were in their infancy, with only a few dozen in existence. Engelbart entered the University of California at Berkeley, and earned a master's degree in 1952 and a Ph.D. in electrical engineering in 1955, with a specialization in computers.

Engelbart became an employee at the Stanford Research Institute (SRI) in 1957 and after two years obtained his own laboratory, becoming director of the Augmentation Research Center. After developing a theoretical framework for the NLS (oNLine System) that he wanted to create, Engelbart and his team of engineers and psychologists worked through the 1960s on realizing the dream. Engelbart wanted to do more than automate previous tasks like typing or clerical work; he wanted to use the computer to fundamentally alter the way people think. In a demonstration of NLS at the Fall Joint Computer Conference in December 1968, Engelbart showed the audience on-screen videoconferencing with another person back at SRI, thirty miles away, as well as an early form of hypertext, the use of windows on the screen, mixed graphics and text files, structured document files, and the first mouse. Although he is often credited only with inventing the mouse, Engelbart had developed the basic concepts of groupware and networked computing. Engelbart's innovations were ahead of their time, requiring expensive equipment that retarded his ability to innovate. A computer of Engelbart's at SRI became the second computer to join the ARPAnet in 1969, an obvious expansion of his emphasis on networking. ARPAnet later evolved into the Internet.

In the early 1970s, several members of Engelbart's team left to join the Palo Alto Research Center (PARC), which Xerox Corporation had set up, where ample funding led to rapid development of the personal computer, local area networking, laser printers, graphical user interfaces, and object-oriented programming. Ironically, Xerox failed to capitalize on these innovations, and other companies, including 3Com and Apple, took the innovations to market, making billions of dollars and transforming the computer industry. SRI eventually sold NLS to a small company called Tymeshare, and Engelbart followed his creation and tried to turn it into a commercial product. In 1984, McDonnell Douglas bought Tymeshare, and Engelbart became a senior scientist for the aerospace company, but failed to convince them to take

advantage of his ideas before his retirement in 1989. In 1990, Engelbart founded the Bootstrap Institute in Palo Alto, California, to continue to promote his ideas of how organizations could cope with too much data. In 1991, the English computer programmer Tim Berners-Lee (1955–) used the ideas of hypertext, by then much further developed by engineers other than Engelbart, to create the World Wide Web.

In the mid-1980s, people in the computer industry began to take notice of Engelbart's contributions, and the awards began to flow. Among his numerous awards are a lifetime achievement award in 1986 from *PC Magazine,* a 1990 ACM Software System Award, the 1993 IEEE Pioneer award, the 1997 Lemelson-MIT Prize, with its $500,000 stipend, and the National Medal of Technology in 2000.

See also Berners-Lee, Timothy; Computers; Internet

References

Bardini, Thierry. *Bootstrapping: Douglas Engelbart, Coevolution, and the Origins of Personal Computing.* Stanford, CA: Stanford University Press, 2000.

Bootstrap Institute. http://www.bootstrap.org/ (accessed February 13, 2004).

Environmental Movement

The environmental movement grew out of the older conservation and preservation movements of the late nineteenth and early twentieth centuries. The conservationists sought to manage public resources by applying scientific principles to rational planning of the development of natural resources. The preservationists tried to preserve unique natural treasures by establishing national parks, trusts, and reserves. As the twentieth century progressed, population grew quickly and industrialization spread, increasing resource consumption. The Green Revolution in agriculture, which increased crop yields through the increased use of artificial fertilizers and synthetic pesticides and new crop strains, staved off starvation in Third World countries.

In 1952, a London fog trapped enough pollution from coal furnaces to kill 4,000 people, leading to a 1956 Clean Air Act and conversion of home heating in London from coal to gas and electricity. In the 1940s and 1950s, automobiles in the city of Donora, near Pittsburgh, sometimes stalled on the street for lack of sufficient oxygen. The passage of the Clean Air Act (1970) and Clean Water Act (1972) in the United States mirrored efforts in other nations. Although not perfect in their efforts by any measure, Westernized nations cleaned up the more egregious examples of air and water pollution in their nations by the 1980s. Growing Third World cities have continued to suffer from air pollution, with the megalopolis of Mexico City leading the indicators, and seven of the top ten most polluted cities in the world being found in the newly industrialized nation of China.

The biologist Rachel Carson (1907–1964) published her prophetic classic, *Silent Spring*, in 1962, a well-documented polemic on the dangers of synthetic pesticides. Among the pesticides she described was DDT (dichloro-diphenyl-trichloro-ethane), which is toxic in minute amounts, and Carson described the research that showed that exposure to DDT poisoned agricultural workers, damaged bird populations, and remained persistent enough to pass through the food chain up to animals and humans. Carson argued that modern agriculture should turn to alternatives, such as biological controls and the release of sterile insect males.

The counterculture ferment of the 1960s in Western nations included environmental concerns among its signature issues. Scientists like the American entomologist Paul R. Ehrlich (1932–) raised the alarm about population increases. Ehrlich's 1968 book, *The Population Bomb*, was published when world population stood at 3.6 billion. In 1999, world population passed 6 billion. The pictures of Earth taken by the Apollo astronauts also contributed to a sense that our world was a limited, fragile, and unique habitat. The

French-born microbiologist René Dubos (1901–1982) co-authored the Pulitzer Prize–winning *Only One Earth: The Care and Maintenance of a Small Planet* in 1972, exemplifying a new attitude toward nature and Earth. Twenty million Americans participated in the first Earth Day, held in 1970. The Earth Day held in 1990 attracted 200 million participants worldwide. The federal government established the Environmental Protection Agency in 1970. The United Nations established the UN Environmental Programme in 1972. Green political parties, emphasizing environmentalism, human rights, and other causes, have arisen in many countries and gained a measure of influence in some European countries. Environmental movements also grew in poorer Third World nations after 1980, with government ministries being established and private organizations being founded.

Concerns over declining biodiversity and species extinction have led to many private and government attempts to protect individual species and species habitat, since some species now only exist in zoos or other collections. How many species have been lost is difficult to measure, since a complete catalogue of the world's species, especially insects, has not been compiled. Certainly the rate is at least one species a day. In 1900, at least 150,000 blue whales lived in the southern ocean, and in 1989, only 500 remained. Most other whale species faced similar decimation. A vocal and effective campaign to "Save the Whales" led to most whaling being banned worldwide in 1986. A few countries still whale under the guise of taking scientific specimens, though the harvest of hundreds of whales a year by whaling factory ships cannot be justified under any scientific rationale. Some aboriginal peoples also have the right to harvest a limited number of whales.

Concerns over nuclear weapons testing and pollution from commercial nuclear reactors became powerful motivators in the global environmental movement. Open-air nuclear testing concerned many scientists be-

Sewage discharges into the ocean near Rabat in Morocco. (Chinch Gryniewicz; Ecoscene/Corbis)

cause of the documented increase of radioactive fallout in the upper atmosphere and downwind from nuclear testing sites.

The Nobel laureate Linus Pauling (1901–1994) and other scientists campaigned for the United States and Soviet Union to at least stop open-air testing in the interest of public health. Their efforts helped lead to the 1963 Nuclear Test Ban Treaty, and Pauling received the 1962 Nobel Peace Prize. Nuclear testing and manufacture during the cold war (1945–1991) also led to large areas within the United States and the Soviet Union, as well as some islands in the South Pacific, being poisoned by radioactive debris. Cold war soldiers, sailors, and airmen, in both the United States and the Soviet Union, unrestrained by environmental regulations, transformed their military bases into toxic waste dumps.

A 1979 nuclear accident at Three Mile Island in Pennsylvania led to a minor release of radiation and accentuated concerns about the environmental wisdom of nuclear reactors.

Although commercial nuclear reactors provide clean electricity with little pollution, they are expensive to build and generate radioactive waste that is difficult to transport and store for the thousands of years necessary for the waste to become safer. The long-feared accident occurred in 1986 at a Soviet plant near Chernobyl in the Ukraine with explosions, fire, and a partial meltdown of the core of one of the four reactors at the site. No expensive containment buildings, such as are usually found in Western countries, surrounded any of the reactors. A Soviet cover-up failed because sensors in Western Europe detected the radioactive fallout in the wind. Dozens of people died in the immediate aftermath of the accident, with towns and farmland around the reactor permanently abandoned, and the local population experienced long-term radiation sickness and increases in various forms of cancer.

After the collapse of the Soviet Union and communist regimes of Eastern Europe, it became apparent that the large state-directed industrial and agriculture enterprises in those countries had created environmental disasters. The lack of a democratic process had prevented environmental activists from being effective. Polluted rivers, poisoned farmland, and toxic industrial areas led to documented health effects and reduced life expectancy. East Germany created more sulfur dioxide emissions per capita than any other nation on the planet. The Aral Sea in central Asia shrank to half of its previous size by the mid-1990s and continues to shrink because of an ill-conceived mass irrigation project that drew away water that normally drained into the inland sea. The Soviets disposed of their nuclear waste by draining it into the Arctic Ocean and sunk old submarine and ship nuclear reactors under the ocean as a way to dispose of them.

Before 1980, most industrial enterprises in the Western world also disposed of toxic wastes, a common by-product of industrial processes, in a haphazard manner. In a famous incident, the residential area of Love Canal in upstate New York was evacuated from 1976 to 1980, as toxic waste buried over two decades earlier percolated to the surface, leading to a marked increase in cancer and birth defects among the local population. Laws on toxic waste disposal in Western countries encouraged corporations to engage in the often illegal practice of exporting toxic waste to poorer countries. An irony of the reunification of East Germany with West Germany in 1990 was that for years West Germany had exported its toxic waste to its eastern neighbor and after reunification had to deal with the resulting toxic waste deposits.

Concern over the environment also led to increased emphasis on the science of ecology, so that scientists and laypersons might better understand the intricate connections between different elements of the natural world. The Gaia hypothesis of the biophysicist James Lovelock (1919–), which proposes that Earth is like a living organism, maintaining surface and atmospheric conditions so that the chemistry and temperature will sustain life, resonated with activists in the environmental movement. Many biologists are uncomfortable with the idea's mystical overtones and remain suspicious of such a holistic notion.

In the mid-1970s, research by the Dutch chemist Paul Crutzen (1933–), the Mexican chemist Mario Molina (1943–), and the American chemist F. Sherwood Rowland (1927–) showed that artificially manufactured chlorofluorocarbons were eroding the ozone layer in the upper atmosphere. A major success story for the environmental movement came as the Montreal Protocol international agreements of 1987, 1990, and 1992 agreed to phase out the use of chlorofluorocarbons. Chlorofluorocarbons are also thought to contribute to global warming.

In the 1980s, scientists determined that the average temperature on Earth was rising, on the order of approximately 0.6 degree Celsius (1 degree Fahrenheit) during the twentieth century, with about a third of that change coming in the last twenty-five years. Whether this global warming is part of a long-term natural pattern or an effect of human industrial activity became a heated issue in the 1990s and has remained so. By the end of the century, global warming had become an important rallying point for environmental groups. The 1992 United Nations Conference on Environment and Development in Rio de Janeiro, also known as the Earth Summit, included global warming among many other environmental and economic development issues. The United Nations Conference on Climate Change in 1997 in Kyoto, Japan, led to the Kyoto Protocol, an international treaty that required nations to reduce their carbon dioxide emissions. The United States has refused to ratify the treaty, leaving the status of the Kyoto effort partially crippled. A study by the National Science Foundation in 1997 concluded that environmental monitoring and environmental research had become the largest research program of all time, spread out across the world in more than 2,000 organizations. Despite considerable progress up to the end of the century, population continues to grow, natural resource consumption increases, the world's oceans grow more polluted, coral reefs are dying, and the rainforests of Central and South America, particularly in the Amazon River basin, are being quickly destroyed.

See also Apollo Project; Carson, Rachel; Cold War; Crutzen, Paul; Ecology; Ehrlich, Paul R.; Global Warming; Green Revolution; Lovelock, James; Molina, Mario; National Science Foundation; Oceanography; Ozone Layer and Chlorofluorocarbons; Pauling, Linus; Population Studies; Rowland, F. Sherwood; Wilson, Edward O.

References

Cronon, William, editor. Uncommon Ground: Toward Reinventing Nature. New York: Norton, 1995.

Dowie, Mark. Losing Ground: American Environmentalism at the Close of the Twentieth Century. Cambridge: MIT Press, 1995.

Hutton, Drew, and Libby Connors. A History of the Australian Environment Movement. Melbourne, Australia: Cambridge University Press, 1999.

McNeill, John Robert. Something New under the Sun: An Environmental History of the Twentieth-Century World. New York: Norton, 2000.

Mould, Richard F. Chernobyl Record: The Definitive History of the Chernobyl Catastrophe. Philadelphia: Institute of Physics, 2000.

Wilder, Robert Jay. Listening to the Sea: The Politics of Improving Environmental Protection. Pittsburgh: University of Pittsburgh Press, 1998.

Ethics

The Piltdown Man, discovered in 1912, led scientists to believe that England was the birthplace of humanity. In 1953, a researcher showed conclusively that the human cranium combined with the jaw of an orangutan was a deliberate hoax. The practice of science is based on the expectation that other scientists are truthful. Because science builds on previous experiments that have to test a real world, mistakes and fraud will eventually be discovered. Yet fraud is often difficult to prove, especially when the researcher claims to have just been sloppy.

In the context of science, the discipline of ethics addresses how to behave as an individual scientist and how scientists should use their knowledge to benefit humanity. A code of ethics, sometimes formal, often informal, has developed. Scientists are expected to behave in their students' best interest when acting as mentors; scientists are expected to keep good records of their work and not falsify any data or conceal data that do not support their assertions; and scientists are expected to participate honorably in the peer review system. As the financial and professional stakes get ever larger, scientists face greater temptation to commit fraud. Conflict-of-interest problems can also arise when scientists have a substantial financial stake in the results of their own research or in the results of research done by others that they can influence. Private funding sources and private industry are now working more closely with academic institutions, raising issues of who owns the intellectual property rights that might arise from the research.

Another aspect of scientific ethics arises from the simple fact that there are consequences to every scientific and technological development. Some scientists and engineers who have developed powerful new weapons, such as machine guns, chemical weapons, and atomic weapons, have thought that their inventions would actually end war because the weapons would be too horrible to use. Only in the case of atomic bombs has this kind of effect possibly been proven true; the nuclear standoff between the Soviet Union and United States during the cold war restrained that ideological conflict from igniting into a third world war.

Nevertheless, nuclear weapons did place many scientists in serious ethical quandaries. The physicist J. Robert Oppenheimer (1904–1967), the "father of the atom bomb," opposed building the hydrogen bomb on both moral and technical grounds. Oppenheimer and others saw the hydrogen bomb as being so powerful that its only possible targets were cities, whereas atomic bombs might be used to destroy many types of military targets. Linus Pauling (1901–1994), a Nobel laureate in chemistry and a pacifist, refused to participate in the Manhattan Project, successfully campaigned against open-air nuclear weapons testing, and vigorously pursued antinuclear activities. Edward Teller (1908–2003), on the other hand, driven by anticommunism, encouraged the development of nuclear weapons to protect the United States during the cold war.

Since the late 1960s, bioethics has grown into a major branch of philosophy and medicine. The issues of abortion, in vitro fertilization, organ transplants, cloning, stem cell research, suicide, and the use of human subjects in research have all become subjects of bioethical discussions. Developments in genetic engineering and within the biotechnology industry will force scientists, ethicists, and laypeople to confront serious ethical questions, such as what is life, when does life start, when is a person dead, and should health care be rationed because of cost? Moral questions about the use of deception in social psychology and medical clinical trials, in which patients in the control group do not receive what may be a new lifesaving treatment, are often difficult to answer.

The Australian-born, Oxford-educated moral philosopher Peter Singer (1943–) wrote *Animal Liberation* in 1975, which is often credited with reviving the modern animal rights movement. This movement strongly objects to the use of animals in scientific research, and Singer himself also claims that "speciesism" is similar to racism, in that we value human life over animal life. Most medical scientists are adamant that animal experimentation is the foundation of their work and is the only way that they can make medical progress.

Many scientists, imbued with scientific materialism, argue that the basis of all knowledge must be science and scientific materialism. Even moral reasoning must use the scientific process to arrive at its conclusions. The biologist Edward O. Wilson (1929–) has advanced this argument forthrightly. Other scientists argue that science is only a method and that values should be derived from other sources, such as religious beliefs or secular humanism.

See also Bioethics; Cold War; Evolution; Hydrogen Bomb; Oppenheimer, J. Robert; Pauling, Linus; Philosophy of Science; Teller, Edward; Wilson, Edward O.; War

References
Bronowksi, Jacob. *Science and Human Values.* Revised edition. New York: Julian Messner, 1965.
Erwin, Edward, Sidney Gendin, and Lowell Kleiman. *Ethical Issues in Scientific Research: An Anthology.* New York: Garland, 1994.
Macrina, Francis L. *Scientific Integrity: An Introductory Text with Cases.* Washington, DC: American Society for Microbiology, 1995.
Spier, Raymond E., editor. *Science and Technology Ethics.* New York: Routledge, 2002.

European Space Agency

The European Space Agency (ESA) is an example of Big Science; member nations have pooled their financial resources to create a larger space program than would be possible for the individual nations. The United States and the Soviet Union (in its space age heyday) had had the largest programs. In the 1990s, the Soviet program became the Russian program and declined due to lack of funding, leaving the ESA the second biggest space program in the world.

The ESA was created in 1975 with the merger of two earlier organizations, the European Space Research Organization (ESRO) and the European Launcher Development Organization (ELDO). ESRO and ELDO were both created in the early 1960s. The ten founding nations of the ESA were France, West Germany, Italy, Spain, the United Kingdom, Belgium, Denmark, the Netherlands, Sweden, and Switzerland. Five other nations joined later: Ireland, Austria, Norway, Finland, and Portugal. Canada is also associated with the current fifteen nations as a cooperating state. The ESA is headquartered in Paris, France.

Scientific inquiry has always been a major mission of the ESA. They contributed funding, materials, and expertise to the Hubble Space Telescope. They sponsor the Spaceguard organization, headquartered in Rome,

The French Ariane-5 *rocket, used by the European Space Agency (Reuters/Corbis)*

Italy, which coordinates the search for asteroids and comets. The joint ESA-USA spacecraft *Ulysses,* launched in 1990, used a gravitational assist past Jupiter to enter a solar orbit that flew over the poles of the Sun, exploring sections of space outside the ecliptic plane. The ERS-1 and ERS-2 (European Remote Sensing) satellites have surveyed Earth's resources with radar, infrared cameras, and microwaves. ERS-2 has paid special attention to monitoring the ozone hole in the atmosphere over Antarctica. The Solar and Heliospheric Observatory spacecraft, launched in 1995, continues to monitor the Sun. The XMM-Newton orbiting telescope has provided x-ray observations since 1999. Numerous other scientific missions are planned, rivaling those of the American National Aeronautics and Space Administration (NASA) in their breadth.

The Spacelab module, which flew on several space shuttle missions, provided a platform for many short-term scientific experiments that took advantage of the zero gravity of outer space. ESA astronauts have flown on the American space shuttle and in Russian capsules. The ESA is also participating in the current International Space Station as a full partner. In order to promote international cooperation and share the costs, the ESA and NASA are cooperating on many planned future scientific spacecraft.

Desiring their own launchers, the ESA created the successful Ariane series of rockets. The Ariane rockets are launched from Kourou, French Guiana, a launch center founded by the French Centre National d'Études Spatiales in 1964. Its location on the equator allows rockets launched there to carry more payload than rockets launched from the United States or the former Soviet Union. Ariane rockets carry most ESA and European commercial satellites into orbit.

See also Big Science; National Aeronautics and Space Administration; Space Exploration and Space Science; Telescopes

References

European Space Agency. http://www.esa.int/ (accessed February 13, 2004).

Krige, John, and Arturo Russo. *SP-1235—A History of the European Space Agency, 1958–1987.* Volume 1, *ESRO and ELDO, 1958–1973.* Noordwijk, The Netherlands: European Space Agency Publications Division, 2000.

Krige, John, Arturuo Russo, and Lorenza Sebesta. *SP-1235—A History of the European Space Agency 1958–1987.* Volume 2, *The Story of ESA, 1973–1987.* Noordwijk, The Netherlands: European Space Agency Publications Division, 2000.

Evolution

The British naturalist Charles Darwin (1809–1882) published the theory of natural selection in his famous book *On the Origin of Species by Means of Natural Selection* (1859). Darwin argued that the variety of species in the world arose from small random mutations among earlier species that led to the mutated animals surviving to pass on those mutations to their children. He used the term *natural selection* in contrast to the term *artificial selection,* which referred to the practice of dog breeders selecting for traits when they bred dogs and developing new varieties of dogs, or the selections by herders and farmers that led to more desirable domesticated animals and plants. Darwin saw natural selection at work in nature, but he could not explain the underlying mechanism. The science of genetics had been simultaneously developed, by the Austrian botanist Gregor Mendel (1822–1884), but no one recognized its importance until the turn of the century. The science of genetics grew up in the early twentieth century, while Darwin's theory of natural selection fell out of favor with scientists, replaced by a vague theory of the inheritance of acquired characteristics. Still, the idea of evolution had become firmly established among scientists, despite their doubts about Darwin, and Social Darwinists used natural selection to justify racist and imperialist assumptions about human nature.

The Russian-born zoologist Theodosius Dobzhansky (1900–1975), the ornithologist Ernst Mayr (1904–), and others in the 1930s and 1940s returned to the theory of natural selection and Darwin's emphasis on random mutations. Dobzhansky also successfully promoted the use of mathematical models of population within species populations. Mayr revived an old theory of geographic speciation and argued that small mutations in geographically isolated populations led to rapid evolution. Mayr also proposed the founder effect, which occurs when some geographic quirk or partial die-off has caused a population to be founded by a few individuals. Dobzhansky, Mayr, and like-minded scientists created what became known as the modern synthesis by combining the theory of natural selection with Mendelian genetics. The discovery of the structure of DNA in 1953 by the zoologist James D. Watson (1928–) and the biologist Francis Crick (1916–2004) led to a revolution in genetics that finally allowed a solid understanding of the underlying mechanism for natural selection. As genetics developed, evolutionists argued that species should be classified by their evolutionary relationship rather than the traditional system developed centuries ago by the Swedish botanist Carolus Linnaeus (1707–1778).

The Harvard-based paleontologist Stephen Jay Gould (1941–2002) became an important theorist and popularizer of the theory of evolution. In 1972, Gould and Niles Eldredge (1943–) published "Punctuated Equilibria: An Alternative to Phyletic Gradualism," which argued that evolution did not always occur in gradual steps over long periods of time. The lack of intermediate fossils for many species in the geological record indicated that evolution often occurred in spurts, with new species evolving in as little as a couple of thousand years, rather than in the longer time periods favored by the more strictly gradualist Darwinists. The theory of punctuated equilibria was not widely accepted but did provoke considerable intellectual ferment.

The biologist Edward O. Wilson (1929–), based at Harvard University, became the world's recognized authority on ants and consistently applied the principles of natural selection in his work. His landmark book, *Sociobiology: The New Synthesis* (1975), asserted that what biologists had learned about insect and animal societies, along with the insights of the theory of natural selection, should also be applied to understanding human behavior and human society. This manner of applying the theory of natural selection to human and animal behavior led to the growth of a subdiscipline called evolutionary psychology. The English zoologist Richard Dawkins (1941–), based at Oxford University, became a significant popularizer of evolutionary psychology; he argued that evolution occurs on the level of genes, rather than the level of individuals or species. This approach offered a solution to the vexing problem of altruism, which is the philosophical puzzle of why people engage in generosity without any desire for a reward. The work of animal behavioral scientists, such as Konrad Lorenz (1903–1989), Jane Goodall (1934–), and Dian Fossey (1932–1985), helped scientists understand animal behaviors that might have a genetic component.

Scientific advocates of the theory of evolution through natural selection have regularly spoken out against any revival of Social Darwinism. Those same advocates, with Wilson, Gould, and Dawkins acting as prominent voices, have condemned creationism. Although creationism is simply the belief that a deity created the universe and life within the universe, most creationist science is focused primarily on a literal interpretation of the Judeo-Christian Book of Genesis. Some scholars have pointed out that the attacks of creationists on evolution have spurred contemporary evolutionists in their efforts to refine and communicate their theories.

As cosmological theories developed, they described the development of the solar system and a young Earth. How did life start on Earth? In the 1920s, the Russian biochemist Aleksandr Oparin (1894–1980) and the British scientist J. B. S. Haldane (1892–1964) argued that an early Earth atmosphere composed of water vapor, methane, hydrogen, and ammonia formed a primeval soup that could lead to organic compounds. Stanley L. Miller (1930–), a chemistry graduate student under the guidance of Nobel laureate Harold C. Urey (1893–1981), supported the Oparin-Haldane theory with a successful 1953 experiment that simulated the early Earth's atmosphere and spontaneously created amino acids, an essential building block of life. Fred Hoyle (1915–2001) and a few other scientists have argued for panspermia, which argues that life did not spontaneously originate on Earth, but came from biological material contained in interstellar dust clouds. This theory is not widely accepted. The discovery of early microfossils by the paleobotanist Elso Barghoorn (1915–1984) and others showed that early Earth, up to 3.5 billion years ago, contained microbial life.

With her serial endosymbiosis theory

Charles Darwin, biologist and advocate of evolution (Library of Congress)

(SET), first published in 1967, the microbiologist Lynn Margulis (1938–) proposed an additional twist to evolution. While still using the principle of natural selection, Margulis argued that eukaryotic cells, which contain nuclei, evolved when non-nucleated bacteria fused together billions of years ago. She also demonstrated that microbes could transmit induced characteristics, and that organelles inside cells were really former bacteria that had fused with other cells. These ideas have come to be accepted, though her theory that most evolutionary change is the result of symbiogenesis and her support of the Gaia hypothesis are not widely accepted. The Gaia hypothesis, developed by the scientific polymath James Lovelock (1919–) in the late 1960s, proposed that Earth was like a living organism, maintaining surface and atmospheric conditions so that the chemistry and temperature would sustain life.

As Darwin pointed out in his *Descent of Man* (1873), if natural selection is a valid scientific theory, then humans are also descended from previous species. Scientists sought fossils of precursor species and found

Java Man, a partial skeleton of what is now known as *Homo erectus,* in 1891 at Trinil, Java. Peking Man, also *Homo erectus,* was found in China in 1927. The Piltdown Man, a fossilized skull found near Lewes, England, at Piltdown Common in 1912, played on racist and nationalist assumptions that modern humans had evolved in England. With the Piltdown Man finally exposed as a hoax in 1954, scientists looked elsewhere for the origins of humanity.

The paleoanthropologist Louis S. B. Leakey (1903–1972) argued that Africa was the birthplace of humanity and proved it with the discoveries made by his wife, Mary Leakey (1913–1996), at Olduvai Gorge in northern Tanzania. She found hundreds of fragments of a hominid skull in 1959, later determined to be of a separate line of hominids named *Australopithecus boisei.* Mary also found, at Laetoli in 1978, the fossilized footprints of three hominids dated 3.6 million years old. The three are thought to be the footprints of a man, a woman or a smaller man, and a child. The gait of the footprints clearly showed that these hominids walked upright much sooner than anyone had suspected. In 1974, Donald Johanson (1943–) found the first nearly complete skeleton of *Australopithecus afarensis,* which he named "Lucy."

Discoveries at Lake Turkana by a team led by Richard Leakey (1944–), the son of Mary and Louis, showed that at least three different species of hominids lived together in close proximity millions of years ago, though controversy remains over which line of hominids led to modern humans. The meaning of these discoveries remains elusive. The so-called missing link has not been found, and the continuing profusion of extinct hominid species sows confusion in the continuing efforts to construct a solid line of evolution to modern humans. Though numerous skeletons of Neanderthals have been discovered, scientists argued for much of the last century over whether this recent species was a precursor to modern humans or a separate species. By the end of the century, a consensus had gradually grown that Neanderthals were a separate sapient species with their own ability to make tools, art, and graves for their dead. Neanderthals also coexisted with modern humans, *Homo sapiens sapiens,* during the last ice age. Recent finds of Neanderthal skeletons in the late 1990s allowed scientists to extract mitochondrial DNA, showing that Neanderthals were a separate species, but crossbreeding with modern humans was possible and may have occurred.

See also Anthropology; Barghoorn, Elso; Biology and the Life Sciences; Creationism; Crick, Francis; Dawkins, Richard; Dobzhansky, Theodosius; Fossey, Dian; Genetics; Geology; Goodall, Jane; Gould, Stephen Jay; Hoyle, Fred; Leakey Family; Lorenz, Konrad; Lovelock, James; Margulis, Lynn; Mayr, Ernst; Miller, Stanley Lloyd; Sociobiology and Evolutionary Psychology; Symbiosis; Watson, James D.; Wilson, Edward O.

References

Eldredge, Niles, and Stephen Jay Gould. "Punctuated Equilibria: An Alternative to Phyletic Gradualism." Pp. 82–115 in *Models in Paleobiology.* Edited by Thomas Schopf. San Francisco: Freeman, Cooper, 1972.

Hull, David L. *Science as a Process: An Evolutionary Account of the Social and Conceptual Development of Science.* Chicago: University of Chicago Press, 1988.

Ruse, Michael. *The Evolution Wars: A Guide to the Debates.* Santa Barbara, CA: ABC-CLIO, 2000.

Tattersall, Ian. *The Fossil Trail: How We Know What We Think We Know about Human Evolution.* New York: Oxford University Press, 1995.

Weiner, Jonathan. *The Beak of the Finch: A Story of Evolution in Our Time.* New York: Knopf, 1994.

Extrasolar Planets

Once astronomers came to understand the nature of our solar system and that stars were really faraway suns, the question naturally arose whether planets around other stars existed. Was our solar system created by an interstellar calamity, the result of matter being pulled from the Sun by another star passing nearby? Or was planet formation a common by-product of star formation? Both theories were accepted at one time or another, with the latter becoming ascendant. Stars are too far away for planets around them to be seen using conventional techniques, even with the power of the orbiting Hubble Space Telescope.

The astronomer Peter Van de Camp began observing Barnard's Star in 1938. Barnard's Star is only six light-years away, and only the three stars making up the Alpha Centauri system are closer to Earth. He gradually accumulated over 2,000 photographic plates and in 1963 announced that a wobble in the images of Barnard's Star indicated the presence of a large unseen planet. This proposed planet had an extreme elliptical orbit, and after further observation, in 1969 Van de Kamp decided that the wobble was really caused by two planets similar in size to our gas giants, Jupiter and Saturn. Two other astronomers measured the wobble much more accurately and published their conclusion in 1973 that there were no planets around Barnard's Star.

The entire debate was revitalized in October 1995 by Michel Mayor (1942–) and Didier Queloz (1966–), two Swiss astronomers at the Geneva Observatory. Mayor had spent years searching for Doppler shifts that would indicate hidden planets. They found strong evidence that a planet at least half the mass of Jupiter orbited 51 Pegasi, a star similar to the Sun about forty-five light-years away. The news received wide publicity, and commentators immediately realized that the existence of extrasolar planets boded well for those who argued for extraterrestrial life on planets around other stars. It was then only a small step to arguing that the search for extraterrestrial intelligence (SETI) should be expanded.

Since 1995, numerous other extrasolar planets have been discovered by observations of subtle Doppler shifts. The discoveries happened so quickly that one astronomer joked about a Planet-a-Month club. At times, the frequency of discoveries made a joke about a Planet-a-Week club more appropriate. The planets discovered so far are all the size of gas giants, but smaller than brown dwarfs. The National Aeronautics and Space Administration has proposed launching an observatory that will be able to see extrasolar planets around nearby stars as dots of light.

See also Astronomy; National Aeronautics and Space Administration; Search for Extraterrestrial Intelligence; Space Exploration and Space Science; Telescopes

References

Boss, Alan. *Looking for Earths: The Race to Find New Solar Systems.* New York: Wiley, 1998.

Clark, Stuart. *Extrasolar Planets: The Search for New Worlds.* New York: Wiley, 1998.

Croswell, Ken. *Planet Quest: The Epic Discovery of Alien Solar Systems.* New York: Free Press, 1997.

Goldsmith, Donald. *Worlds Unnumbered: The Search for Extrasolar Planets.* Sausalito, CA: University Science Books, 1997.

Lewis, John S. *Worlds without End: The Exploration of Planets Known and Unknown.* Cambridge, MA: Perseus, 1998.

F

Feminism

Modern feminism began during the counterculture movement of the 1960s, building on the foundation laid by the earlier women's rights movement of the nineteenth and early twentieth centuries. The earlier movement in Westernized countries often gained women the right to vote, legal rights, access to birth control, and some measure of social mobility. Feminism as applied to science has engaged principally in two areas. First, feminists have sought ways to break down gender barriers, in order to give women completely equal opportunity to engage in scientific work. Second, feminists have studied the ways gender bias in male-dominated fields has affected the practice and actual content of the sciences.

Strong barriers existed in the mid-twentieth century to the entry of women into science. Cultural attitudes often placed women in the home rather than the workforce; educational opportunities were limited, and academic appointments were often restricted from women. A study of Nobel prize–winning women in the last century found that they had often had to overcome considerable obstacles, including hostility, to leaving gender roles. Only a determined commitment to science, support by relatives, and often a supportive husband enabled them to be pioneers

not only in science, but in the cause of eroding gender barriers. Beginning in the 1960s, feminist activists helped break down many of the educational and cultural barriers.

Although gender bias still exists in the practice and profession of science, decades of progress have led to a situation in which many women now hold prominent positions and are in a position to make important contributions. Even so, female scientists are generally less productive on average than male scientists. Scholars have explained this discrepancy as a result of the distractions of children and family, the tendency of men to be more competitive, and a sense that women have to achieve proportionally more in order to be treated equally by their male counterparts. Studies also show that women scientists in Westernized countries still tend to be paid less than their male counterparts, even when other factors, such as age, rank, and field of specialization, are compensated for statistically.

On the other hand, women have sometimes had an advantage because they were women, in that they have been able to make distinctive contributions. For example, early primatologists tended to see primates in masculine terms, making aggressive and territorial males the primary focus of study. Female primatologists, such as Jane Goodall (1934–) and Dian Fossey (1932–1985), helped change that perspective through patient fieldwork in the 1960s and 1970s, and now the scientific understanding of primates is much more nuanced. As more women archaeologists entered the field of archaeology, archaeologists became more sensitive to finding tools, such as digging sticks, baskets, textiles, and slings for carrying babies, used heavily by women, rather than only the ax heads, arrowheads, spear points, and other hunting tools used by men. Women researchers have also challenged the theory that hunting activities drove human evolution, rather than the gathering activities more common among women.

Medical researchers in pharmacology are now aware that drugs react differently within women's bodies, and greater numbers of women must be included in clinical trials, especially since women on average take more medicinal drugs than men do. Women also interact differently with physicians and require different approaches in medical care. The fields of psychology and psychiatry used to use masculine terminology and male-centered concepts to explain human behavior and to define mental health. Pressure from feminist activists and a greater influx of trained women into the field have changed this approach considerably.

Studies have shown that women are often drawn in greater numbers to some sciences than others: biology rather than physics, for example. Some feminist scholars have argued that the idea of symbiosis and various subdisciplines involved in ecology profit from women's inclination to view nature in terms of cooperation rather than competition. Feminist-inspired studies have also explained how to change the pedagogy of the sciences and various technical fields to draw more women into those fields. More radical feminist critiques have drawn support from social constructionism and postmodernism to question much of the entire enterprise of science, though this questioning is not usually shared by practicing female scientists.

See also Archaeology; Birth Control Pill; Fossey, Dian; Goodall, Jane; Medicine; Psychology; Social Constructionism; Sociology of Science; Symbiosis

References

Creager, Angela N. H., Elizabeth Lunbeck, and Londa Schiebinger, editors. *Feminism in Twentieth-Century Science, Technology, and Medicine.* Chicago: University of Chicago Press, 2001.

Kass-Simon, Gabriele, and Patricia Farnes, editors. *Women of Science: Righting the Record.* Bloomington: Indiana University Press, 1990.

McGrayne, Sharon Bertsch. *Nobel Prize Women in Science.* Second edition. Secaucus, NJ: Carol, 1998.

Rossiter, Margaret W. *Women Scientists in America: Before Affirmative Action 1940–1972.* Baltimore: Johns Hopkins University Press, 1995.

Schiebinger, Londa. *Has Feminism Changed Science?* Cambridge: Harvard University Press, 1999.

Wyer, Mary, and others, editors. *Women, Science, and Technology: A Reader in Feminist Science Studies.* New York: Routledge, 2001.

Feynman, Richard (1918–1988)

Richard P. Feynman, perhaps the most brilliant American physicist of the twentieth century, refined quantum electrodynamics (QED), giving it its modern form. Richard Phillips Feynman was born in New York City, where his father, a sales manager, encouraged his son to become a scientist. Feynman taught himself calculus while still a child and repaired radios during high school for extra money. He earned a B.S. degree with honors in physics from the Massachusetts Institute of Technology in 1939. His doctorate in physics followed three years later from Princeton University, where John A. Wheeler (1911–) served as his advisor and mentor. The Manhattan Project recruited him, and his work with Hans Bethe (1906–) led to a formula for calculating the yield of an atomic explosion. Bethe later received the 1967 Nobel Prize in Physics. Feynman witnessed the first atomic explosion at Trinity, New Mexico, in 1945. His first marriage in 1941 ended with the death of his wife from cancer four years later. His second marriage ended in divorce, and his third marriage resulted in a son and daughter.

After the war, Feynman taught at Cornell University, before accepting an appointment at the California Institute of Technology (Caltech), where he remained for the rest of his life. At Cornell, Feynman returned to his interest in QED, where he developed a new theory to renormalize quantum electrodynamics, fixing flawed equations that led to infinite results and problems with how to handle the self-energy of particles. Feynman was part of the second generation that refined the theory. Julian Schwinger (1918–1994), a fellow New Yorker, independently developed a similar theory, as did Shinichiro Tomonaga (1906–1979), working in war-imposed isolation in Japan. Freeman Dyson (1923–) showed that the three theories were mathematically equivalent, and the 1965 Nobel Prize in Physics was shared among Feynman, Schwinger, and Tomonaga. Feynman's mathematical notation system for QED was accepted by the physics community over the systems developed by Schwinger and Tomonaga, and his method of graphically drawing particle interactions, in what came to be dubbed Feynman diagrams, was widely adopted.

At Caltech, Feynman worked with Murray Gell-Mann (1929–) to create a theory that explained the weak nuclear force, one of the four fundamental physical forces. Later Feynman worked on the strong nuclear force, another of the four fundamental physical forces, proposing hypothetical particles called partons that make up protons and neutrons. He also explained on a quantum level the superfluid behavior of liquid helium, based on the work of the Soviet physicist Lev Landau (1908–1968).

A gregarious man known for his infectious enthusiasm for mathematics and physics, Feynman was widely admired by his peers. He was always ready to seek out his own answers and always spoke his mind. When he was appointed to a panel to investigate the causes of the 1986 explosion of the space shuttle *Challenger,* he conducted his own separate investigation and produced a separate

Richard Feynman, physicist, 1986 (Bettmann / Corbis)

report castigating NASA management. Unlike many academics of his stature, he excelled at undergraduate teaching.

> ***See also*** Dyson, Freeman; Gell-Mann, Murray; Grand Unified Theory; National Aeronautics and Space Administration; Nobel Prizes; Particle Physics; Physics; Schwinger, Julian; Wheeler, John A.
> ***References***
> Feynman, Richard P. *The Pleasure of Finding Things Out: The Best Short Works of Richard P. Feynman.* Cambridge, MA: Perseus Books, 1999.
> Gleick, James. *Genius: The Life and Science of Richard Feynman.* New York: Pantheon, 1992.

Fossey, Dian (1932–1985)

Dian Fossey became the leading expert on the mountain gorilla through years of difficult field research and campaigned for the preservation of their African habitat. Fossey was born in San Francisco, where her love of animals sustained her during an unhappy childhood. When Fossey entered the University of California at Davis, she intended to become a

Dian Fossey, zoologist, 1985 (Yann Arthus-Bertrand/Corbis)

veterinarian, but the required courses in chemistry and physics convinced her that she should choose another path. Fossey graduated from San Jose State College in 1954 with a B.A. in occupational therapy. She moved to Kentucky and worked in Louisville as director of the occupational therapy department at the Kosair Crippled Children's Hospital. Taking out a loan, she traveled to Africa in 1963 with the desire to meet the famous Louis S. B. Leakey (1903–1972) and see the mountain gorillas. That species of gorilla lives only in the Virungas Mountains, located at the junction of the nations of Rwanda, Uganda, and Congo (renamed Zaire in 1971, then renamed the Democratic Republic of the Congo in 1997). She was fascinated by the large gentle animals.

In 1966, Leakey visited Fossey in Louisville and convinced her to study the animals, even though she had no field experience or relevant training. Leakey persuaded his friend Leighton A. Wilkie to fund Fossey's

study, just as Wilkie had funded Jane Goodall's study of chimpanzees in the wild. After a brief visit with Goodall (1934–) to learn the rudiments of fieldwork, Fossey set up camp in Zaire. The gorillas normally fled from strangers, forcing her to slowly work her way into their confidence until she was allowed to approach close enough to observe the different family groups. Political turmoil in Zaire led to her capture by government troops. After escaping from her captors, she fled the country and restarted her research in Rwanda.

Encouraged by Leakey, Fossey left her research long enough to attend Cambridge University from 1970 to 1974 and earn a Ph.D. in zoology. The National Geographic Society began funding Fossey in 1968 and provided other forms of assistance and publicity for Fossey that made her into a celebrity. Following the death of a favorite gorilla at the hands of poachers in 1978, Fossey began a campaign to publicize the diminishing numbers of mountain gorillas and the depredations of poachers. In 1980 she also interrupted her research to teach at Cornell University. This return to America gave her time to complete *Gorillas in the Mist* (1983), which reported on mountain gorillas for a popular audience. Fossey was more comfortable in the company of her beloved gorillas than with other people. Her anthropomorphic accounts of the gorillas treated the primates as individuals, not just research subjects.

Fossey returned to Africa to continue her research, at the same time organizing antipoaching patrols and aggressively seeking to protect the mountain gorilla habitat from the burgeoning local human population. Fossey was found dead in her Rwanda camp the day after Christmas in 1985, presumably slain by poachers.

See also Anthropology; Environmental
Movement; Goodall, Jane; Leakey Family;
National Geographic Society; Zoology
References
Mowat, Farley. *Woman in the Mists: The Story of Dian Fossey and the Mountain Gorillas of Africa.* New York: Warner Books, 1987.
Torgovnick, Marianna. *Primitive Passions: Men, Women, and the Quest for Ecstasy.* New York: Knopf, 1997.

Foundations

Private foundations have made major contributions to scientific research, especially in the area of public health. In the late nineteenth and early twentieth centuries, corporations based on large amounts of capital and new methods of management came to dominate the economies of the United States and Great Britain. Some of the wealthiest of the "robber barons" who founded these corporations devoted considerable parts of their fortunes to creating private philanthropic foundations. These foundations functioned on a different level from other charities, both because of their size and because of their emphasis on larger projects. Among the most notable of the American foundations are the Carnegie Corporation (1911), the Rockefeller Founda-

tion (1913), and the Ford Foundation (1936), created from fortunes built respectively in the steel, oil, and automobile industries.

The sciences have benefited from these private foundations. In the period between World War I and World War II, a patron-client relationship developed between scientists and the foundations. Scientists became used to applying for grants and directing their research into those areas that foundations wanted to emphasize. In the United States, private foundations provided much of the funding for basic research during the interwar period. Foundation funding created the practice of postdoctoral fellowships to further educational and research opportunities for scientists. The foundations were just as interested in the social sciences during this period as in the harder sciences. Rockefeller funding was so generous that in the 1930s the majority of cultural anthropologists working in the field received funding from the Rockefeller Foundation. After World War II, the federal government, through organizations like the National Science Foundation, National Institutes of Health, and defense-related initiatives, took over funding basic research. The foundations moved to more applied studies, where trustees and managers saw opportunities to be effective.

From the beginning, health issues have received considerable emphasis from the foundations. The Rockefeller Foundation has been instrumental in numerous health initiatives around the world, promoting medical education, seeking ways to alleviate chronic diseases like yellow fever and malaria, and funding medical research. The heiress Katherine Dexter McCormick (1875–1967) provided the funds in the 1950s that led to the first effective birth control pill. The National Foundation for Infantile Paralysis successfully funded the separate efforts of virologists Jonas Salk (1914–1995) and Albert Sabin (1906–1993) to develop effective polio vaccines. The philanthropist John D. Rockefeller III (1906–1978) founded The Population Council in 1952 to promote research into population control, and he aggressively promoted family-planning programs and contraceptive use throughout the world. This was an early expression of the concern with population growth that became a dominant focus of thought and action among scientists in the 1960s.

The Green Revolution that transformed twentieth-century agriculture occurred as a result of deliberate foundation policies. The Rockefeller Foundation hired the American plant specialist Norman E. Borlaug (1914–) to go to Mexico in 1944 to work on a project to create new strains of wheat. The resulting strains produced higher yields and resisted diseases better, though they also required more water and depleted the soil more quickly, requiring fertilizers to help them grow. In 1962, the Ford Foundation joined the Rockefeller Foundation to found the International Rice Research Institute in the Philippines at Los Banos, where Robert Chandler and his team quickly created new strains of rice that proved as successful as Borlaug's wheat strains. In 2000, the value of the Ford Foundation stood at $13 billion, making for about $500 million a year being spent. Forty percent of the spending went to developing countries (Ford Foundation Web site).

In Great Britain, the American expatriate Henry Wellcome (1853–1936) cofounded a pharmaceutical company in 1880. As his company became successful, Wellcome indulged in his passion for medical history, collecting instruments and archival material. His will converted his holdings in the company into a private foundation, the Wellcome Trust. The trust promoted research in the history of medicine and medical research. By the end of the century, the trust was worth some 9 billion pounds, and some 400 million pounds a year supported medical research, chiefly in Great Britain. The Wellcome Library became one of the premier medical history facilities in the world.

Although other foundations exist around the world, working in areas applicable to science, the foundations in the United States and the Wellcome Trust are unique in their size. Critics have often accused private foundations of promoting the social and cultural agendas of their founders. Although that is true to a degree, foundations have usually followed general cultural trends already emerging.

The emphasis on health and developing nations continued in 2000 when the cofounder of Microsoft Corporation and his wife created the Bill and Melinda Gates Foundation, with an initial endowment of approximately $25 billion. The Gates Foundation works to promote global health equity, bringing health education and medical help to Third World countries. The Gates Foundation also promotes technology education, in order to minimize the "digital divide" that separates the computer literate, who have ready access to education and computers, and the computer illiterate, who do not have ready access to either the knowledge or actual machines.

See also Birth Control Pill; Green Revolution; National Institutes of Health; National Science Foundation; Pharmacology; Population Studies; Sabin, Albert; Salk, Jonas

References

Bill and Melinda Gates Foundation. http://www.gatesfoundation.org/ (accessed February 13, 2004).

Ford Foundation. http://www.fordfound.org/ (accessed February 13, 2004).

Jonas, Gerald. *The Circuit Riders: Rockefeller Money and the Rise of Modern Science.* New York: Norton, 1989.

Rockefeller Foundation. http://www.rockfound.org/ (accessed February 13, 2004).

Schneider, William H., editor. *Rockefeller Philanthropy and Modern Biomedicine: International Initiatives from World War I to the Cold War.* Bloomington: Indiana University Press, 2002.

The Wellcome Trust. http://www.wellcome.ac.uk/ (accessed February 13, 2004).

Franklin, Rosiland (1920–1958)

Rosiland Franklin specialized in x-ray diffraction methods, and her photographs of deoxyribonucleic acid (DNA) helped Francis Crick (1916–2004) and James D. Watson (1928–) determine the double-helix nature of the molecule. Rosiland Elsie Franklin was born in London, England, where her father was a wealthy Jewish banker. Even as a child, she proved to be headstrong and impatient with the limitations that cultural attitudes imposed on women. Determined to pursue a career as a scientist, she earned admission to Cambridge University, though her father only grudgingly agreed to fund her education, since he did not believe that women should be educated at universities. She attended Newnham College, a women's college at Cambridge. She graduated with a bachelor's degree in chemistry in 1941 and went to work at the British Coal Utilization Research Association. Her research on how burning coal turns into graphite resulted in five published papers and earned her a Ph.D. in physical chemistry from Cambridge in 1945. From 1947 to 1950, she worked at the Laboratoire Central des Services Chimiques de l'État in Paris.

In 1951, Franklin returned to England to work with the Medical Research Council at King's College at the University of London. Franklin mastered the new field of x-ray crystallography, in which reflected x-rays through crystals resulted in photographs. She worked with Maurice H. F. Wilkins (1916–), applying her skills by taking pictures of the DNA molecule. Wilkins showed his friend James D. Watson one of Franklin's best photographs, which helped Francis Crick and Watson to determine that the DNA molecule was a double helix. Crick and Watson published their discovery in a letter to *Nature* in April 1953. Franklin's photographs and her own article were published in the same issue. Franklin later became a close friend of Crick's, but Watson's account of the discovery, *The Double Helix* (1968), portrayed her in a negative light, and did not give her the credit that was her due.

In 1953, Franklin moved to the Crystallography Laboratory at Birkbeck College in London. She turned her attention to divining the molecular structure of the tobacco mosaic virus and studies of ribonucleic acid

(RNA). Franklin died of cancer at the age of thirty-seven. Four years later, Crick, Watson, and Wilkins received the 1962 Nobel Prize for Physiology or Medicine. The Nobel is not awarded posthumously, nor is a single prize awarded to more than three persons at a time. One of the more interesting what-ifs of history is the question of whether Franklin would have also received the prize if she had still been alive.

See also Crick, Francis; Genetics; Nobel Prizes; Watson, James D.

References

Maddox, Brenda. *Rosiland Franklin: The Dark Lady of DNA*. New York: HarperCollins, 2002.

McGrayne, Sharon Bertsch. *Nobel Prize Women in Science*. Second edition. Secaucus, NJ: Carol, 1998.

Fuller, Buckminster (1895–1983)

Buckminster Fuller was one of the most innovative thinkers of the twentieth century. Richard Buckminster Fuller ("Bucky") was born in Milton, Massachusetts, to a distinguished family. His father's male ancestors of the Fuller line had all attended Harvard University since 1760. Fuller entered Harvard in 1913, was expelled a year later for spending his tuition money on a lavish party in New York City, worked as an apprentice mechanic at a textile mill in Canada for a time, then returned to Harvard, to be expelled again for not concentrating on his studies. Fuller found work as a laborer, and with the coming of World War I in 1917, served in the U.S. Navy as an officer. He also married the daughter of a prominent architect in 1917.

Fuller continued to work odd jobs, until he joined his father-in-law in a business manufacturing a fibrous building block that his father-in-law had invented. The death of his four-year-old daughter sent him into a deep depression, and when the business suffered, the stockholders fired him. In 1927, he found himself in Chicago, living with his wife and a new daughter, but on the edge of a precipice, drinking heavily and even considering sui-

Buckminster Fuller, stands before a geodesic dome at the 1967 World's Fair in Montreal, Quebec. (Bettmann/Corbis)

cide. Fuller survived by finding a personal mission, the improvement of the human condition through new design principles. These principles applied to machines, architecture, technology, and a philosophy of life. His designs aimed to develop ways to be more efficient in all things.

Fuller founded Dymaxion Corporation (combining the words *dynamic, maximum,* and *ion*) and proceeded to spew forth inventions: a prefabricated house, a revolutionary car, a type of map that projected the globe onto a flat surface without visible flaws, and a portable shelter for use by the military during World War II. Many of these inventions derived from innovative ways of using triangles, circles, and tetrahedron-shaped objects. His inventions aroused interest, but did not

result in many commercial successes. During World War II, Fuller served as chief of the mechanical engineering section of the Board of Economic Warfare. After the war, he developed the mathematics for a science of geodesics. His most interesting product was the geodesic dome, an efficient building shape that is also remarkably strong.

In 1959, Southern Illinois University offered Fuller a position as a research professor, a testament to his recognized genius, despite his failure at formal education. Fuller accepted. He then moved in 1972 to Philadelphia to become a World Fellow for the University City Science Center. Fuller also wrote books on his design philosophy and applied it to other areas of human endeavor, such as language, education, and even thought. In the

1960s and 1970s, the American counterculture movement adopted Fuller as an icon for
his ability to think thoughts unconstrained by
convention. The environmental movement
found inspiration in his ideas that more could
be done with less if only effective design principles were applied. Fuller died in 1983, the
holder of over 2,000 patents, having received
numerous awards and honorary degrees, and
with a reputation for original thinking. The
1996 Nobel Prize in Chemistry was awarded
to three chemists for their artificial creation
of carbon balls that are formed in shapes reminiscent of Fuller's geodesics. These molecules are called fullerenes, or buckyballs.

See also Environmental Movement
References
The Buckminster Fuller Institute. http://www
    .bfi.org/ (accessed February 13, 2004).
Zung, Thomas T., editor. *Buckminster Fuller: An
    Anthology for the New Millennium.* New York: St.
    Martin's, 2000.

G

Gaia Hypothesis

See Ecology; Environmental Movement; Lovelock, James; Margulis, Lynn

Gamow, George (1904–1968)

George Gamow made significant contributions to the theory of radioactivity, the theory of the Big Bang, and to genetics. Born Georgy Antonovich Gamov in Odessa, Russia, Gamow attended Leningrad University, where he earned a Ph.D. in physics in 1928. He then traveled abroad to study at the University of Göttingen, University of Copenhagen, and Cambridge University, where he worked with the luminaries in the field of quantum physics. He developed a quantum theory to explain radioactivity, and other physicists took notice. In 1933, he traveled to the Solvay International Congress on Physics in Brussels; there he decided not to return to the Soviet Union. Though not a strongly political man, Gamow sensed the changing atmosphere as the tyranny of the Soviet dictator Josef Stalin (1879–1953) grew, reaching into every aspect of Soviet life.

In 1934, Gamow found a post as a professor of theoretical physics at the George Washington University in Washington, D.C. During World War II, his Russian origins meant that he could not work on the Manhattan Project, even though he became a U.S. citizen in 1940, so he worked on high explosives for the U.S. Navy. In 1948, Gamow and his student Ralph Alpher wrote a paper titled "The Origin of Chemical Elements," one of many contributions to the Big Bang theory that Gamow made. Gamow included Hans A. Bethe as an author so that the published paper would appear as authored by Alpher, Bethe, and Gamow, close to alpha, beta, and gamma, the first three letters of the Greek alphabet. Bethe did not object to this famous joke.

During the 1950s, Gamow was the chief champion of the Big Bang theory, and Fred Hoyle (1915–2001) was the chief champion of the steady state theory. Gamow and his team of researchers even hypothesized that the Big Bang had left a faint background radiation in the universe, though they did not realize that radio astronomers might be able to detect it. The Big Bang theory triumphed in the 1960s because of discoveries that the steady state theory could not explain but the Big Bang theory could. The major discovery was the background radiation of the universe, though the explanation for the radiation was developed independently of Gamow's work a decade earlier. In 1954, Gamow also made important proposals about the genetic code embedded within the newly discovered double helix of DNA. In 1956, Gamow moved to

George Gamow, physicist (Bettmann / Corbis)

the University of Colorado. Beginning in the 1930s, Gamow wrote popular articles and books on science for the average person, filled with humor and his own line drawings. He eventually produced some twenty books, including his famous *Mr. Tomkins* series.

> **See also** Astronomy; Big Bang Theory and Steady State Theory; Genetics; Hoyle, Fred; Media and Popular Culture; Physics; Soviet Union
> **References**
> Gamow, George. *Mr. Tompkins in Paperback.* Cambridge: Cambridge University Press, 1965.
> ———. *My World Line: An Informal Autobiography.* New York: Viking, 1970.
> Kragh, Helge. *Cosmology and Controversy: The Historical Development of Two Theories of the Universe.* Princeton: Princeton University Press, 1996.

Gell-Mann, Murray (1929–)

The prominent physicist Murray Gell-Mann devised theories to categorize subatomic particles as well as describe an even more funda-

mental type of matter, which he playfully called the quark. Murray Gell-Mann was born in New York City to Jewish Austrian immigrants. A scholarship allowed him to attend private school, and at age fifteen he entered Yale University. Gell-Mann wanted to study archaeology or linguistics, and his father wanted something more practical, such as engineering. They compromised on physics. Gell-Mann graduated from Yale in 1948, then earned a Ph.D. in physics from the Massachusetts Institute of Technology in 1951. After a year at the Institute for Advanced Study in Princeton, he accepted a position at the Institute for Nuclear Studies at the University of Chicago. In 1955, he moved to the California Institute of Technology, becoming a full professor the following year and remaining there until his retirement in 1993. Gell-Mann married a British archaeologist in 1955, and they produced a son and daughter. His wife died in 1981, and he remarried in 1992.

After World War II, nuclear physicists used cloud chambers and then particle accelerators to smash atoms, tracking the subatomic particles that briefly flared to life before decaying. Dozens of particles had been identified by the early 1950s, with little theoretical understanding of how to organize them. Gell-Mann brought order to this confusion in 1953 by proposing a new property for subatomic particles called strangeness and creating a new theory. In 1961, he found a way to categorize the subatomic particles into groups, creating a system similar to the periodic table of the elements. He predicted the properties of a new particle that was subsequently discovered in 1964. An Israeli physicist, Yuval Ne'eman (1925–), independently devised a similar system, and the two physicists collaborated to write a 1964 book, *The Eightfold Way.* The name for their system is a reference to Buddhism, another example of Gell-Mann's penchant for whimsical names.

Also in 1964, Gell-Mann proposed a new type of fundamental matter that formed subatomic particles. He called the constituents of this fundamental matter quarks and antiquarks, and suggested that three "flavors," or types, existed. Since that time the number of types has risen to six, and quark theory is now part of the standard model that attempts to combine the four basic forces (the weak nuclear force, the strong nuclear force, electromagnetism, and gravity) into a grand unified theory. Gell-Mann received the 1969 Nobel Prize in Physics for his work on strangeness and the eightfold way. In the 1970s, Gell-Mann expanded his theory of quarks into what he called quantum chromodynamics (QCD).

A man of diverse interests, Gell-Mann was drawn in the early 1980s to the new science of complexity, which sought to find order in seemingly random events. Chaos theory emerged from this new science, and Gell-Mann enjoyed speculation in this area, eventually writing a book, *The Quark and the Jaguar: Adventures in the Simple and the Complex* (1994). Gell-Mann and others founded the Santa Fe Institute in New Mexico in 1984 to promote the study of chaos theory and other implications of complexity. Gell-Mann remains active in the study of complexity.

See also Grand Unified Theory; Nobel Prizes; Particle Physics

References

Gell-Mann, Murray. *The Quark and the Jaguar: Adventures in the Simple and the Complex.* New York: W. H. Freeman, 1994.

Johnson, George. *Strange Beauty: Murray Gell-Mann and the Revolution in Twentieth-Century Physics.* New York: Knopf, 1999.

Genetics

The rediscovery of Mendelian genetics at the turn of the century created the science of genetics. Through painstaking research, geneticists developed laws of heredity and developed the theory that genes exist and have a definite chemical nature, but exactly how genes work eluded them. The discovery of the molecular structure of DNA by the zoologist James D. Watson (1928–) and the biologist Francis Crick (1916–2004) in 1953 revolutionized the study of genetics. Later research showed how DNA replication occurs and determined how messenger ribonucleic acid (mRNA) is used to transcribe information from DNA genes to build proteins. The French researchers François Jacob (1920–) and Jacques Monod (1910–1976) were among those who determined how the genetic transcription process is turned on and off by proposing that a repressor molecule binds itself to the DNA at the appropriate point to block transcription. Jacob and Monod also developed the theories of operons and allosteric transitions to help explain how information is transferred from DNA to form enzymes and proteins inside cells. The geneticist Barbara McClintock (1902–1992) contributed her theory of transposable genes, or "jumping genes," developed in the 1940s even before Watson and Crick's discovery, though not appreciated until two decades later.

The central dogma of genetics became that DNA controls everything, with RNA acting as the messenger to create proteins. Molecular biology became the dominant discipline in the biological sciences, claiming that all the really interesting biological processes occurred at the level of molecules. Cellular biologists have countered that the obsession with DNA has neglected the role that the other proteins and enzymes in the cell play as the coded information from genes is expressed. The microbiologist Lynn Margulis (1938–) examined cytoplasmic genes, those genes found outside the cell nucleus in organelles like mitochondria and chloroplasts, and created a theory that organelles were really former bacteria that had fused with other cells. Her 1967 theory of serial endosymbiosis theory (SET) became an important way to understand that bacteria exchanged genes in a manner that accelerated the process of natural selection. The existence of prions (proteinaceous infectious particles) also challenged conventional notions of life and the dominance of DNA, because, though prions act like viruses and cause neurological diseases, they are too small to be viruses or to contain DNA.

James Watson (left) and Francis Crick with their model of part of a DNA molecule in 1953 (A. Barrington Brown / Photo Researchers, Inc.)

The biochemists Herbert Boyer (1936–) and Stanley Cohen (1935–) developed the techniques of genetic engineering in the early 1970s, leading to a large biotechnology industry. Their techniques allowed scientists to cut away strands of DNA from one plasmid and insert it into another plasmid. These new plasmids were then injected into bacteria to create a new organism. The molecular biologists Walter Gilbert (1932–) at Harvard and Frederick Sanger (1918–) at Cambridge University both developed techniques to se-

quence genes by chemically describing individual genes. These technologies laid the foundation for the Human Genome Project, begun in 1990 and completed in 2003. Because people are genetically different from each other, only 99.99 percent of the human genome could be mapped, yielding 3.1 billion pairs making up 35,000 to 40,000 genes.

Mendelian genetics in the first half of the twentieth century combined with Social Darwinism to create the intellectual foundation for the powerful political force of eugenics.

Extrapolating from the small advances in genetics, geneticists and other scientists argued for biological determinism, a theory according to which nature reigned over nurture; they believed that human races contained intrinsic differences and that hybrid races were inferior. Eugenics in the United States resulted in anti-immigration laws, as well as sporadic efforts to sterilize the mentally feeble and poor. Eugenics in Germany eventually led to the Nazi extermination campaigns during World War II. The horrors of Nazi eugenics ended the Western eugenics movements. The Human Genome Project set aside some of its funding for the study of bioethics as it applies to genetics. Some critics objected to the project because the knowledge might lead back to eugenics, allowing people to tailor the DNA of their unborn children to conform to modern images of attractiveness, athleticism, and intellectual ability.

Geneticists have often searched for the gene or small set of genes that cause certain effects in biological organisms or certain diseases. The German geneticist Christiane Nüsslein-Volhard (1942–) and the American biologist Eric Wieschaus (1947–) bred 40,000 fruit fly families, each with a single chemically induced genetic defect, then used a microscope to examine thousands of embryos and larvae. The fruit fly contains 20,000 genes, but Nüsslein-Volhard and Wieschaus found that only 139 are essential to embryonic development. The Chinese-born molecular geneticist Lap-Chee Tsui (1950–) led one of the teams that identified the gene that causes the disease of cystic fibrosis in 1989.

Advances in genetics also made possible cloning, the creation of a genetic duplicate of a previous organism. Starting with frog eggs in 1952, embryologists and geneticists worked to clone simple animals. Eventually, in 1996, Dolly the sheep was born, a clone from an adult somatic cell. Even though Dolly was euthanized at the age of six in 2003, suffering from serious medical problems of the kind that are common in cloned

animals, many scientists and laypeople worry that cloning of human beings lies close in the future.

Other successes in genetics came from more traditional techniques of crossbreeding plants. The Green Revolution of the 1950s through 1970s fed the world's burgeoning population through the development of new strains of wheat and rice. Genetic engineering has been used since the 1970s to engineer transgenic cereal crops that are pest-resistant, fungus-resistant, and virus-resistant, and that can grow in marginal soils and yield even larger returns for each acre.

See also Bioethics; Biotechnology; Chemistry; Cloning; Crick, Francis; Franklin, Rosiland; Green Revolution; Human Genome Project; Margulis, Lynn; McClintock, Barbara; Monod, Jacques; Nüsslein-Volhard, Christiane; Prions; Sanger, Frederick; Symbiosis; Tsui, Lap-Chee; Watson, James D.

References

Chadarevian, Soraya de. *Designs for Life: Molecular Biology after World War II.* Cambridge: Cambridge University Press, 2002.

Crow, James F., and William F. Dove, editors. *Perspectives on Genetics: Anecdotal, Historical, and Critical Commentaries, 1987–1998.* Madison: University of Wisconsin Press, 2000.

Judson, Horace Freeland. *The Eighth Day of Creation: Makers of the Revolution in Biology.* Expanded edition. Woodbury, NY: Cold Spring Harbor Laboratory, 1996.

Kevles, Daniel J. *In the Name of Eugenics: Genetics and the Uses of Human Heredity.* New York: Knopf, 1986.

Wright, Susan. *Molecular Politics: Developing American and British Regulatory Policy for Genetic Engineering, 1972–1982.* Chicago: University of Chicago Press, 1994.

Geology

Geology and its many subdisciplines, including geodesy, geophysics, crystallography, mineralogy, paleogeology, seismology, and petrology, have flourished because of the practical impact of the discipline. Mining and petroleum extraction activities are important foundations of our industrial economy. The major change in geology during the last half

Geothermal activity at Rotorua, North Island, New Zealand (Corel Corporation)

of the twentieth century occurred in the 1960s, when the work of Harry Hess (1906–1969), John Tuzo Wilson (1908–1993), and others led the geological community to accept that plate tectonics existed. The theory of plate tectonics posits that the continents actually move around on a mantle of magma, pushing up mountain ranges where the plates meet and forming rift valleys where plates are pulled apart. The theory has proven to be a powerful explanatory mechanism, which has transformed geology from a science of particulars into a mature science with a body of integrated knowledge. A major impetus for the development of the theory came from oceanographical expeditions that mapped the sea floor, finding midocean mountain ranges as well as magnetic polarity reversals among rocks on the ocean bottom. Studies in oceanography, seismology, and geomagnetism constituted important activities during the International Geophysical Year (1957–1958) that contributed

to finding anomalous evidence pointing toward plate tectonics.

Deep-sea exploration also led to the discovery of other geological wonders, including deep-sea hydrothermal vents, discovered in 1977 by the *Alvin,* a deep-sea submersible, at about 2,700 meters in the Galapagos Rift near the Galapagos Islands. Bacteria in the unique ecology surrounding the vents used chemosynthesis to create energy, converting the sulfides into organic carbon. The geologist and oceanographer Wallace S. Broecker (1931–) and others used analysis of deep-sea cores to show that Milankovitch's ice age theory was correct. The Serbian scientist Milutin Milankovitch (1879–1958) posited in 1920 that the ice ages were caused by astronomical variations in Earth's orbit.

Plate tectonics also finally explained earthquakes and volcanoes. Volcanoes and greater earthquake activity exist on the edges of continental plates. Examination of other planets by spacecraft and landers has given geologists examples to compare with Earth. They have found active volcanism on Jupiter's planet Io and the remains of lava flows on all the inner planets and the Moon, but so far no conclusive evidence of plate tectonics on other planets.

Geological evidence that accumulated in the nineteenth century showing that Earth has existed for a long time constituted one of the major challenges that overthrew the previous biblically based belief in a divine creation just six thousand years ago. The possibility of gradual change over eons of time allowed Charles Darwin (1809–1882) to posit the gradual evolution of species through natural selection. Geologists and paleontologists uncovered fossils of dinosaurs and other extinct species, reinforcing the idea that Earth had been considerably different in earlier eons. The paleobotanist Elso Barghoorn (1915–1984) from Harvard University and the Soviet scientist Boris Vasil'evich Timofeev (1916–1982) identified microfossils from the Precambrian era, the remains of the single-celled life that dominated Earth before the Cambrian era began 550 million years ago. Creationists continued to insist that the biblical flood laid down all the geological strata and fossils during the catastrophic year that water covered the entire planet.

The idea of gradualism, in reaction to creationism, became a firm dogma within the geological community, and geologists instinctively rejected any theories that proposed any form of catastrophism. The geologist Walter Alvarez (1940–) and his father, Nobel laureate and physicist Luis Walter Alvarez (1911–1988), discovered high concentrations of iridium in a 65-million-year-old layer of clay in Gubbio, Italy, at the stratigraphical boundary between the Cretaceous era and the Tertiary era, also called the K-T boundary. Because iridium is deposited on Earth from extraterrestrial sources through meteorites and micrometeorites, the father and son proposed that the dinosaurs were driven to extinction by the impact of an asteroid or comet. Other scientists in the 1980s identified an impact crater in Chicxulub, Yucatán,

Mexico, as the source of the K-T boundary iridium. The dramatic impacts of the Shoemaker-Levy comets on Jupiter in 1994 showed that extraterrestrial hazards in the form of comets and asteroids still exist. The pioneering planetary geologist Eugene Shoemaker (1928–1997) had made his reputation earlier by proving that meteorite impact events were more common in earlier Earth history than previously supposed. Some scientists have proposed that all the major extinctions in Earth history were caused by impact events, making even more room for catastrophism in geological theory.

See also Alvarez, Luis W., and Walter Alvarez; Barghoorn, Elso; Broecker, Wallace S.; Creationism; Deep-Sea Hydrothermal Vents; Earthquakes; Hess, Harry; International Geophysical Year; Kuno, Hisashi; Plate Tectonics; Shoemaker, Eugene; Shoemaker-Levy Comet Impact; Space Exploration and Space Science; Volcanoes; Wilson, John Tuzo

References

Gohau, Gabriel. *A History of Geology.* Revised edition. New Brunswick, NJ: Rutgers University Press, 1991.

Hallam, Anthony. *A Revolution in the Earth Sciences: From Continental Drift to Plate Tectonics.* Oxford: Clarendon Press, 1973.

Raup, David M. *The Nemesis Affair: A Story of the Death of Dinosaurs and the Ways of Science.* Revised and expanded edition. New York: Norton, 1999.

Young, Davis A. *Mind over Magma: The Story of Igneous Petrology.* Princeton: Princeton University Press, 2003.

Gilbert, Walter (1932–)

Walter Gilbert trained as a physicist, then turned to molecular biology and discovered a process to sequence genes. Walter Gilbert ("Wally") was born in Boston, Massachusetts, where his father was an economist at Harvard University and his mother was a child psychologist. His father moved the family to Washington, D.C., during World War II. Though lackluster in his studies, Gilbert was obviously bright, and he entered Harvard in 1949. He graduated summa cum laude in 1953 with an A.B. in physics. He earned an

M.S. in physics from Harvard a year later. For his doctorate, he moved to England to attend Cambridge University, where he worked on mathematical formulas to predict the scattering of elementary particles. He received a Ph.D. in mathematics in 1957. Gilbert married in 1953 and fathered a son and daughter.

Gilbert met Francis Crick (1916–2004) and James D. Watson (1928–) at Cambridge, who had just become famous for their discovery of the double-helix structure of deoxyribonucleic acid (DNA). Gilbert returned to Harvard for a yearlong fellowship, then worked as a research assistant to the prominent physicist Julian Schwinger (1918–1994). In 1959, he became an assistant professor in physics. In 1960, Watson moved to Harvard, and Gilbert renewed their friendship. Watson was working on the problem of messenger RNA (mRNA), and Gilbert agreed to help him. Gilbert found that he enjoyed molecular biology so much that he switched fields, becoming an associate professor in the biophysics department in 1964.

Messenger RNA is used to transcribe information from DNA genes to build proteins. The French researchers François Jacob (1920–) and Jacques Monod (1910–1976) asked the question why proteins were sometimes created in a cell and sometimes not. What turned the transcription process on and off? They proposed that a repressor molecule bound itself to the DNA at the appropriate point to block transcription. Gilbert chose to examine the genes that created the proteins needed to digest the milk sugar lactose. Working with *Escherichia coli (E. coli),* in 1966 Gilbert and his colleagues identified the repressor molecule used to stop proteins from being produced.

Gilbert continued to work on repressor molecules. Building on earlier efforts at sequencing genes, a process of chemically describing genes, Gilbert and his students created a better way to sequence genes and in 1977 successfully sequenced every base sequence used in a protein. Frederick Sanger (1918–) at Cambridge University also devel-

oped a similar technique. The 1980 Nobel Prize in Chemistry was shared, with half being awarded to Gilbert and Sanger, and the other half being awarded to Paul Berg (1926–) for other genetics research.

Approached by venture capitalists, Gilbert and others formed, in 1978, a biotechnology company called Biogen. Gilbert served as chairman of the scientific board of directors and eventually became chief executive officer. In 1982, Gilbert decided that the company required more of his time and he resigned from Harvard. Gilbert led Biogen for three and a half years, before resigning because he found that his commitment to scientific research was not always compatible with the profit motives of a corporation. Biogen later became quite successful, based on the patents filed during the years of Gilbert's tenure.

Gilbert returned to Harvard in the Department of Cellular and Developmental Biology. He became an early passionate advocate of the Human Genome Project in the mid-1980s, giving the idea credence because he had developed one of the main techniques for sequencing genes. At one point, frustrated by the attempt to get the project going, he tried to find funding to create a company to do the project. He failed to find funding; the Human Genome Project officially began in 1990. Gilbert believed that decoding the blueprint of the human body would do more than help scientific understanding. He expected the project to change how human beings view themselves, though he also warned that what makes human beings unique individuals is more than our genes.

See also Biotechnology; Crick, Francis; Genetics; Human Genome Project; Nobel Prizes; Watson, James D.

References

Gilbert, Walter. "A Vision of the Grail." Pp. 83–97 in *The Code of Codes: Scientific and Social Issues in the Human Genome Project.* Edited by Daniel J. Kevles and Leroy Hood. Cambridge: Harvard University Press, 1992.

Hall, Stephen S. *Invisible Frontiers: The Race to Synthesize a Human Gene.* New York: Atlantic Monthly, 1987.

Glashow, Sheldon L. (1932–)

The American theoretical physicist Sheldon L. Glashow developed the electroweak theory, combining the weak nuclear force and the electromagnetic force, two of the basic four physical forces, and invented the property of charm for quark theory. Sheldon Lee Glashow was born in New York City, where his parents were Jewish Russian immigrants and his father owned a plumbing business. His father had changed the family name from Gluchovski to Glashow. Glashow went to the Bronx High School of Science, where he met Steven Weinberg (1933–). He graduated from Cornell University in 1954 with a B.S. in physics and received a Ph.D. in physics from Harvard University in 1959. His major professor and mentor was Julian Schwinger (1918–1994), who shared the 1965 Nobel Prize in Physics for his contribution to quantum electrodynamics (QED).

Glashow studied as a postdoctoral fellow at the University of Copenhagen from 1958 to 1960. In 1960, using gauge symmetry, he published a theory unifying the weak nuclear force and the electromagnetic force into an electroweak force. This theory became an important step on the road to the elusive grand unified theory to combine into one standard model the four basic forces of electromagnetism, gravity, the weak nuclear force, and the strong nuclear force. Weinberg and the Indian-born, Cambridge-educated Abdus Salam (1926–1996) also worked on elaborating this theory as it developed during the next two decades.

Glashow taught at the University of California at Berkeley until 1967, when he returned to Harvard. He continued to work on his electroweak theory, predicting the existence of several subatomic particles, later confirmed by experiments in particle accelerators. After Murray Gell-Mann predicted the existence of three particles more basic than any known subatomic particle, which he called quarks, Glashow further developed the theory in the mid-1960s to include a fourth quark. Glashow also proposed a new property for quarks, which he called "charm." Quark theory now includes six quarks. Glashow, Weinberg, and Salam shared the 1979 Nobel Prize in Physics. Glashow married in 1972 and fathered four children.

See also Gell-Mann, Murray; Grand Unified Theory; Nobel Prizes; Particle Physics; Schwinger, Julian
References
Glashow, Sheldon L. The Charm of Physics. New York: American Institute of Physics, 1991.
Glashow, Sheldon L., with Ben Bova. Interactions: A Journey through the Mind of a Particle Physicist and the Matter of This World. New York: Warner, 1988.

Global Positioning System

See Satellites

Global Warming

In the 1970s, scientists noticed a slight drop in world temperatures from 1940 to 1970, and some alarmed scientists feared that a new ice age might be coming. But by the 1980s, scientists had determined that the average temperature on Earth was rising, on the order of approximately 0.6 degree Celsius (1 degree Fahrenheit) during the course of the twentieth century, with about a third of that change coming in the last twenty-five years. Whether this global warming was part of a long-term natural pattern or an effect of human industrial activity has become a heated issue, starting in the 1990s. Water vapor, carbon dioxide, and other gases in the atmosphere reflect back heat radiating from the Earth's surface to create a greenhouse effect. Without the greenhouse effect, the average temperature on Earth would be about –18 degrees Celsius (0 degree Fahrenheit). By the end of the century, global warming, not global ice, had become an important rallying point for environmental groups.

Accurate records of past temperatures kept by human beings are scant before the twentieth century. Deep-core drilling of the Greenland ice cap and other ice deposits has

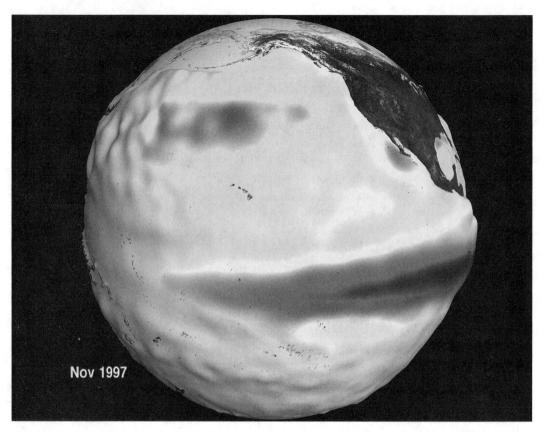

NASA-created image illustrating elevated sea surface temperatures caused by El Niño (NASA/Goddard Space Flight Center Scientific Visualization Center)

revealed clues to past temperatures and carbon dioxide levels. Scientists have been surprised to find how much past temperatures have varied, even when the Earth was not in the midst of an ice age. The Serbian scientist Milutin Milankovitch (1879–1958) proposed in 1920 that orbital variations cause the coming and receding of ice ages. Milankovitch's theory is now accepted, with the addition that minor variations in solar output could also be a factor.

Many scientists argue that the greenhouse effect has been enhanced by human industrial activities during the past two centuries. Current studies find that the amount of carbon dioxide in the atmosphere has risen from 280 parts per million to 370 parts per million in the last 200 years, as the Western world has industrialized and increased the

use of fossil fuels (Woods Hole Research Center). Current forecasts predict that if left unchecked, concentrations of carbon dioxide in the atmosphere at the end of the twenty-first century could reach 490 to 1,260 parts per million.

The scientific effort to understand global warming has encouraged paleoclimatological studies to determine past climate fluctuations, as well as the creation of computer programs to model the Earth's atmosphere. These models require sophisticated algorithms and large computing resources. The first effort to create a numerical model of Earth's atmosphere was published in 1922 by the English mathematician Lewis Fry Richardson (1881–1953). Although his model was limited to the speed at which manual calculations could be made, his effort showed promise. Using supercomputers, scientists in the 1980s and 1990s created ever more sophisticated climate models. Studies of the atmospheres of other planets and the larger moons in the solar system are also used in these models.

Although global warming does not cause the El Niño weather phenomenon, its discovery has contributed to a general sense of concern about the climate. The explanation by geologist Wallace S. Broecker (1931–) of the thermohaline circulation system in the northern Atlantic, which delivers heat from the tropics up to the water off of northwest Europe, indicates how important the oceans are in regulating the Earth's temperatures.

Global warming became a scientific issue with serious political and economic implications in the 1980s. The United Nations established the Intergovernmental Panel on Climate Change (IPCC) in 1988, and the international team of scientists published its first report in 1990, followed by later updates. The United Nations Conference on Environment and Development in Rio de Janeiro in 1992 was known as the Earth Summit, and included global warming among the many other environmental and economic development issues. The United Nations Conference on Climate Change in 1997 in Kyoto, Japan, led to the Kyoto Protocol, an international treaty that required nations to reduce their carbon dioxide emissions. The treaty was not ratified by the U.S. Senate, and the United States withdrew from the treaty.

The possible consequences of global warming include dramatic shifts in agricultural productivity and increased desertification. Founded in 1990, the Hadley Centre for Climate Prediction and Research in the United Kingdom was one of the first of many organizations devoted to the study of global warming, and it has issued forecasts of the likely consequences of further global warming. The direst forecasts by some scientists predict the possible melting of the polar ice caps and the Greenland ice cap. Smaller island nations might disappear, and depending on how much of the ice melts, many coastal areas would be inundated.

See also Broecker, Wallace S.; Computers; Deep-Core Drilling; El Niño; Environmental Movement; International Geophysical Year; Meteorology; Oceanography

References

Houghton, John. *Global Warming: The Complete Briefing.* Second edition. Cambridge; New York: Cambridge University Press, 1997.

Mayewski, Paul Andrew, and Frank White. *The Ice Chronicles: The Quest to Understand Global Climate Change.* Hanover, NH: University Press of New England, 2002.

Michaels, Patrick J., and Robert C. Balling, Jr. *The Satanic Gases: Clearing the Air about Global Warming.* Washington, DC: Cato Institute, 2000.

Woods Hole Research Center. "Scientific Evidence." http://whrc.org/resources/online_publications/warming_earth/scientific_evidence.htm (accessed November 2004).

Goodall, Jane (1934–)

Jane Goodall became the world's leading expert on chimpanzee behavior through patience and empathy. Goodall was born in London, where her father was an engineer and her mother was a novelist. She grew up in a country home and showed a continuous fascination with the outdoors. After working as a secretary and a film production assistant, Goodall accepted the invitation of a childhood friend to visit Africa in 1957. She met Louis S. B. Leakey (1903–1972) in Kenya and went to work for the paleontologist as an assistant secretary. Leakey encouraged her to undertake a study of chimpanzees in the hope that understanding modern primates would lead to insights into how humanity's own ancestors behaved.

In 1960, Goodall set up camp in the Gombe Stream Game Reserve near Lake Tanganyika. Her mother accompanied her because the government authorities would not allow a single Englishwoman to be alone in the bush. Chimpanzees in the wild normally flee humans, so Goodall found a high clearing where she could watch the chimpanzees on

Jane Goodall, primatologist, 1972 (Bettmann/Corbis)

the surrounding terrain through binoculars. After a time, the local chimpanzee troop became used to her and came closer. She began to follow the chimpanzees around the forest and while this initially provoked aggressive displays that she feigned to ignore, the troop eventually came to accept her and even to wait for her to catch up in their travels.

Goodall made several astounding initial discoveries. Up until then, scientists had thought that chimpanzees were vegetarians, but Goodall witnessed them eating meat. She also observed a chimpanzee fishing for termites by stripping a twig of leaves and using the twig to draw the termites out of their mound. This was the first evidence of tool creation and tool use by an animal other than humans. National Geographic television specials on her work turned Goodall into a celebrity scientist and brought funding to her project. In 1964 she

married Hugo Van Lawick, a Dutch wildlife photographer. Their son, nicknamed Grub, became a common sight on the wildlife films they produced. Goodall divorced in 1974, and her second marriage in 1975 ended with the death of her husband in 1980.

At Leakey's suggestion, Goodall entered Cambridge University and received a Ph.D. in ethology in 1965, a most unusual achievement for someone who had never received a bachelor's degree. Goodall's unorthodox techniques have not always been accepted by other scientists. They objected to her naming the chimpanzees in her study, instead of using numbers; they objected to her feeding the troop so that they would stay closer to her camp for study; they objected to her anthropomorphizing the chimpanzees and reporting anecdotes instead of more abstract forms of knowledge; and they objected to her insistence that chimpanzees experience emotions in the same way humans do. Research assistants in the 1970s helped Goodall with her work, and they observed the troop under study break into two troops. A four-year-long war ensued between the two troops until every member of one troop was killed. Goodall's research assistants also witnessed the killing of an infant and cannibalization of the body. These events shocked Goodall. In the 1980s, as her research grew more accepted, Goodall began to use her fame in a relentless effort to preserve chimpanzee habitat in Africa and improve conditions for captive chimpanzees around the world, especially those chimpanzees used in biomedical research.

See also Anthropology; Environmental Movement; Fossey, Dian; Goodall, Jane; Leakey Family; National Geographic Society; Zoology

References
Goodall, Jane. *Reason for Hope: A Spiritual Journey.* New York: Warner, 1999.
———. *Through a Window: My Thirty Years with the Chimpanzees of Gombe.* Boston: Houghton Mifflin, 1990.
Montgomery, Sy. *Walking with the Great Apes: Jane Goodall, Dian Fossey, Biruté Galdikas.* Boston: Houghton Mifflin, 1991.

Gould, Stephen Jay (1941–2002)

The paleontologist Stephen Jay Gould was an important theorist and popularizer of the theory of evolution. Gould was born in New York City, where his father worked as a court stenographer. At the age of five, he was taken by his father to the Museum of Natural History, sparking Gould's interest in paleontology. His parents encouraged his intellectual interests and in 1963 Gould graduated from Antioch College with a B.A. in geology. He received his Ph.D. in paleontology from Columbia University in 1967, where he studied land snail fossils from Bermuda. After teaching for a year at Antioch, Gould accepted an appointment at Harvard University, where he became a full professor of geology in 1973 and also served as curator of invertebrate paleontology at Harvard's Museum of Comparative Zoology.

Gould's original scientific research concentrated mostly on land snails of the West Indies and evolution. In 1972, Niles Eldredge (1943–) and Gould published "Punctuated Equilibria: An Alternative to Phyletic Gradualism," which argued that evolution did not always occur in gradual steps, taking long periods of time. The lack of intermediate fossils for many species indicated that evolution instead often occurred in spurts, with rapid speciation in a short period of time, even within a couple of thousand years. The theory was not widely accepted but did provoke considerable intellectual ferment, and Gould later argued that a pluralism of ideas within the evolutionary paradigm led to better scientific theories.

Gould's monthly columns for *Natural History* magazine, written from 1974 to 2001, were widely praised for their readability and insight into evolution and the history of science. Gould once said that if he had not been so interested in paleontology, he would have become a historian. The columns at times wandered into curious asides, reflecting Gould's multifaceted interests. Many of Gould's twenty-three books were compilations of these columns. A strong believer in

human potential, Gould rejected sociobiology, promoted by his fellow Harvard faculty member, Edward O. Wilson (1929–), as too determinist. Gould also objected to the evolutionary psychology promoted by the zoologist Richard Dawkins (1941–), with its emphasis on genes as the focus of evolution.

Gould's 1981 book *The Mismeasure of Man* examined the ways that psychological intelligence tests have been used to limit human potential and have often promoted racial and cultural assumptions. Using the social prominence that his popular writings gave him, Gould campaigned relentlessly against the rise of creationist science, which he saw as a contradiction in terms. As for his own religious life, Gould considered himself to be an agnostic. The culmination of Gould's scientific thought, the massive *Structure of Evolutionary Theory* (2002), was published two months before his death. Gould won many awards for his scientific work and his writing, including an American Book Award. Gould married in 1965 and fathered two children. Divorced in 1995, he remarried and moved to New York City. Gould survived a diagnosis of mesothelioma in 1982 and died of another form of cancer in 2002.

See also Creationism; Dawkins, Richard; Evolution; History of Science; Media and Popular Culture; Psychology; Sociobiology and Evolutionary Psychology; Wilson, Edward O.

References
Gould, Stephen Jay. *Full House: The Spread of Excellence from Plato to Darwin.* New York: Harmony, 1996.
————. *Wonderful Life: The Burgess Shale and the Nature of History.* New York: Norton, 1989.

Grand Unified Theory

The search for a grand unified theory (GUTS), combining the four known fundamental physical forces of nature into one mathematical model, became an important quest in the late twentieth century. The great English scientist Isaac Newton (1642–1727) mathematically explained the force of gravity in the late seventeenth century. In the nineteenth century the Scottish scientist James Clerk Maxwell (1831–1879) mathematically explained the force of electromagnetism. In the early half of the twentieth century, Albert Einstein (1879–1955) refined the work of Newton with his general theory of relativity. Einstein spent the last quarter century of his life trying to create a theory that combined his own theory with Maxwell's electromagnetism, but failed. With the development of quantum mechanics, physicists discovered two more forces of nature: the strong nuclear force, which holds nuclear particles together, and the weak nuclear force, which causes radiation.

Beginning in 1960, the American theoretical physicists Sheldon L. Glashow (1932–) and Steven Weinberg (1933–), as well as the Indian-born, Cambridge-educated Abdus Salam (1926–1996), developed the electroweak theory, combining the weak nuclear force and the electromagnetic force. Tests in particle accelerators confirmed their predictions of a weak neutral current. Their theory also predicted the existence of three vector bosons that, along with the photon, mediate the interactions within the electroweak theory. Experimental physicists using the particle accelerators at the Conseil Européen pour la Recherche Nucléaire (CERN) discovered the three bosons in 1983.

The American physicist Murray Gell-Mann (1929–) proposed the existence of quarks in 1964, later expanding this insight into a theory called quantum chromodynamics (QCD) to explain the strong nuclear force. This theory proposed gluons as the force particles to mediate the strong force. Gravitons have been proposed as the mediating particle for the force of gravity, but have not been found and are not widely accepted.

The currently accepted and experimentally proven theories are called the standard model. In the 1960s, cosmologists, used to thinking about the vast expanses of the universe, accepted the Big Bang model to explain the origin of the universe and all matter within the universe. The first few minutes of the Big Bang, as matter formed from energy, replicated many of the conditions that the particle physicists studied in their quest for a grand unified theory. Supersymmetry theories emerged from this marriage of the physics of the very large and the physics of the very small, which seemed reasonable to expect, since any grand unified theory had to explain everything from the Big Bang to the smallest interaction between subatomic particles.

The Japanese-born physicist Yoichiro Nambu (1921–) realized in 1970 that quarks act as if they are connected by strings. Further work by Nambu and others led to the development of string theory in the 1980s. String theory deals with hypothetical strings that are near the Planck length in size (about 10^{-33} centimeters or 10^{-34} inches), so small that no instruments can measure them. These strings manifest themselves as the larger known elementary particles, such as quarks, bosons, and gluons. String theory demanded complex mathematics, using up to twenty-six multiple dimensions, and provided a promising avenue to the development of a grand unified theory. String theory later combined with supersymmetry to create superstring theory.

Multiple grand unified theories have been proposed, many of them based on string theory, but all the theories require experimental proof that can only come from particle accelerators far beyond current technology, or extraordinarily precise measurements to detect gravitons, or measurements of the natural abundance of other esoteric particles thought to exist in certain quantities after the Big Bang. Occasionally another fundamental force is proposed, a fifth to add to the known four, but none have ever been accepted by the physics community. The notion of parallel universes, a consequence of quantum theory, is now seriously entertained among sober-minded theorists.

See also Gell-Mann, Murray; Glashow, Sheldon L.; Nambu, Yoichiro; Particle Accelerators; Particle Physics; Physics

References

Crease, Robert P., and Charles C. Mann. *The Second Creation: Makers of the Revolution in Twentieth-century Physics.* New York: Macmillan, 1986.

Greene, Brian. *The Elegant Universe: Superstrings, Hidden Dimensions, and the Quest for the Ultimate Theory.* New York: Norton, 1999.

Kaku, Michio. *Hyperspace: A Scientific Odyssey through Parallel Universes, Time Warps, and the 10th Dimension.* New York: Oxford University Press, 1994.

Kaku, Michio, and Jennifer Thompson. *Beyond Einstein: The Cosmic Quest for the Theory of the Universe.* Revised edition. New York: Anchor, 1995.

The Official String Theory Web Site. http://superstringtheory.com (accessed February 13, 2004).

Weinberg, Steven. *Dreams of a Final Theory: The Search for the Fundamental Laws of Nature.* New York: Hutchinson Radius, 1993.

A new kind of engineered wheat, grown at Colombia's Tibaitita Agriculture Station, as part of the Green Revolution (Ted Spiegel/Corbis)

Green Revolution

The Green Revolution of the 1950s through 1970s came from the combined use of nitrogenous fertilizers and new strains of staple crops to dramatically increase food production in Third World countries. Nitrogenous fertilizers came from the development of the Haber-Bosch ammonia synthesis prior to World War I. Fritz Haber (1868–1934), a German Jew, received the 1918 Nobel Prize in Chemistry for an accomplishment that led to billions of people being fed. It is estimated that almost a third of the current world population of over six billion people is fed only because the use of nitrogenous fertilizers has increased crop yields.

The new crop strains came from the efforts of American scientists sponsored by private foundations. American plant specialist Norman E. Borlaug (1914–) went to Mexico in 1944 to work for the Rockefeller Foundation on a project to create new strains of wheat. The resulting strains produced higher yields and resisted diseases better, though they also required more water and depleted the soil more quickly, requiring fertilizers to help them grow. Many developing nations adopted these wheat strains, and Mexico's wheat harvest allowed that nation to become self-sufficient in wheat in 1956. Nations who adopted the new strains often doubled their wheat harvest within only a couple of years. India and Pakistan staved off starvation in the 1960s through this technology. In 1962, the Ford Foundation joined the Rockefeller Foundation to found the International Rice Research Institute in the Philippines at Los Banos. Robert Chandler and his team at Los Banos quickly created new strains of rice that proved as successful as Borlaug's wheat strains. The work of both Borlaug and Chandler benefited from earlier strains of dwarf wheat and rice developed by Japanese plant breeders.

Borlaug received the Nobel Peace Prize in 1970 in recognition of his important achievement. Borlaug did not believe, however, that the Green Revolution solved the population crisis; rather, the increased food production bought time to find more permanent solutions to the problem. Largely as a result of the Green Revolution, worldwide grain production increased from 692 million tons in 1950 to 1.9 billion tons in 1992, outpacing the growth in population. In order to meet the growing need for food, biotechnology advocates argue that more refined techniques from biotechnology will be necessary to engineer transgenic cereal crops that are pest-resistant, fungus-resistant, and virus-resistant, and can grow in marginal soils and yield even larger returns for each acre.

See also Agriculture; Biology and the Life Sciences; Biotechnology; Foundations; Nobel Prizes; Population Studies

References
Brown, Lester R. *Seeds of Change: The Green Revolution and Development in the 1970s.* New York: Praeger, 1970.

H

Harlow, Harry F. (1905–1981)

The experimental psychologist Harry F. Harlow changed the way modern psychologists look at the need for affection and the intricacies of mother-child relationships. Harry Frederick Israel was born in Fairfield, Iowa, to parents who owned a general store. His father had attended medical school before dropping out and dabbled in many pursuits. After a year at Reed College in Oregon, Harlow transferred to Stanford University in 1924. Originally an English major, he changed to psychology and graduated with a Ph.D. in 1930. In graduate school, the young man changed his last name from Israel to Harlow. His family was not Jewish, but the Jewish sound of the name could have been a serious career obstacle in those anti-Semitic times.

Harlow accepted an appointment at the University of Wisconsin at Madison and began behavioral experiments with rats, cats, and frogs, but found that primates were enough like humans to be more useful. He founded a primate research center, and long hours of work led to considerable recognition as a primate researcher. Harlow's first graduate student was Abraham Maslow (1908–1970), who later gained fame as a psychologist.

Harlow valued experiment over grand theories. The current theories of the time emphasized the physical drives for food, water, and procreation as primary motives, with cognitive motives like the need for loving affection and curiosity relegated to secondary roles. Prevailing Freudian and behaviorist wisdom even advocated that parents not give their children too much affection. Using rhesus monkeys, Harlow engaged in maternal deprivation research; newborn monkeys were taken from their mothers and isolated. Two surrogate mothers were offered, one cloth-covered and cuddly and the other made of wire. Even when only the wire mother offered milk from a nipple, the monkeys preferred the touch and warmth of the cloth-covered mothers. He also noted that monkeys raised without their own mothers grew up to be neglectful or abusive mothers to their own offspring, and that monkeys raised without affectionate mothers manifested behavioral disorders while still newborns and also as mature monkeys. Numerous variations on these experiments led Harlow to conclude that affection was a primary need. Among the supporters of his research were the National Institutes of Health and the Ford Foundation. Harlow served as the president of the American Psychological Association in 1958–1959. His 1958 presidential address, "The Nature of Love," became a classic paper; in it he summarized his work and argued that love was a primary drive for primates and humans.

Harlow married Clara Mears in 1932, and they had two sons. They divorced in 1946. He married a fellow researcher in 1948, and they had a son and a daughter. After his second wife died of cancer in 1971, Harlow renewed contact with his first wife, and they remarried in 1972. Both of his wives were partners in his research and writing. Years of smoking, alcoholism, emotional isolation, long work hours, and Parkinson's disease took their toll on Harlow, and his health declined. His wife helped him complete a final collection of his work, *The Human Model: Primate Perspectives* (1979), and he died in 1981. Harlow received many awards, including a National Medal of Science in 1967. Harlow's research methods, condemned nowadays as inhumane, would not be approved by contemporary ethics review boards, but his results are part of the accepted wisdom of contemporary psychology.

See also Ethics; Foundations; National Institutes of Health; Psychology

References

Blum, Deborah. *Love at Goon Park: Harry Harlow and the Science of Affection.* Cambridge, MA: Perseus, 2002.

Harlow, Harry F., and Clara Mears Harlow. *From Learning to Love: The Selected Papers of H. F. Harlow.* New York: Praeger, 1987.

Hawking, Stephen (1942–)

The theoretical physicist Stephen Hawking became a celebrity as a result of overcoming his physical disabilities to produce important work on black holes and the Big Bang theory of the creation of the universe. Stephen William Hawking was born in Oxford, England, 300 years to the day after the death of Galileo Galilei (1564–1642), a coincidence that appealed to his mischievous sense of humor. Hawking grew up in London, where his father, a medical doctor, directed the division of parasitology at the National Institute for Medical Research. Hawking first studied mathematics at University College at Oxford University, but soon turned to physics. After

graduating with his B.A. in 1962, he moved to Cambridge University for graduate studies. He preferred theory to experiment and found cosmology to offer the grandest theories. Early in his studies he was diagnosed with amyotrophic lateral sclerosis (ALS), commonly called Lou Gehrig's disease, a neuromuscular degenerative disorder with no cure. After the disease stabilized, Hawking attacked his studies with renewed vigor. He married in 1965 and credited his wife with giving him the determination to succeed professionally and in his personal life. After he earned his Ph.D. in 1966, Hawking remained at Cambridge on the staff of the Institute of Theoretical Astronomy and remains as a faculty member in the applied mathematics department.

Dennis Sciama (1926–1999), Hawking's thesis advisor, introduced the young man to the mathematician Roger Penrose (1931–), and Penrose's interest in singularities, later called black holes, intrigued Hawking. Singularities were thought to represent what happened to certain stars at the end of their life cycle and what the primordial point at the beginning of the universe looked like before the Big Bang. Penrose and Hawking began to work on the problem and in 1970 published a theory that argued that while black holes could not be observed, relativistic radiation from near the event horizon of black holes should be detectable. The event horizon is the boundary of a black hole, the limit past which incoming mass can no longer escape the intense gravity of the black hole.

The study of singularities is not only about the eventual fate of large stars, but also finds application in the theory of the Big Bang. The *Big Bang* is the name given to the explosion of a massive singularity to create the universe. In the 1970s, Hawking proposed that in the immediate aftermath of the Big Bang, numerous miniature black holes were formed, each no larger than a proton. Combining quantum mechanics with relativity theory showed Hawking, to his astonishment, that these miniature black holes would emit radiation, before eventually exploding, scattering en-was forced to abandon his cane and became confined to a wheelchair. Eventually he lost the ability to talk or even move most of his body. Hawking has two sons and a daughter, though his first marriage eventually resulted in divorce from the strain of his disease. He remarried in 1995. Hawking also turned to popularizing cosmology, giving public lectures, appearing in television programs, and writing several best-selling books, including *A Brief History of Time* (1988). The combination of his intellect, his ideas, and his disability fascinated the public. Hawking seemed to enjoy his iconic status, as when he appeared as himself on the science fiction television series *Star Trek: The Next Generation*.

See also Astronomy; Big Bang Theory and Steady State Theory; Black Holes; Penrose, Roger
References
Hawking, Stephen. *Black Holes and Baby Universes and Other Essays.* New York: Bantam, 1993.
White, Michael, and John Gribbin. *Stephen Hawking: A Life in Science.* New York: Dutton, 1992.

Stephen Hawking, physicist, 1982 (M. Manni / Bettmann / Corbis)

ergy and particles. This revolutionary idea, running counter to all current theory on black holes, became known as Hawking Radiation. According to his theory, these miniature black holes were an important part of the sequence of events following the Big Bang, but no miniature black holes now remain.

Hawking's brilliance was apparent to all, and he moved up the academic ranks. He became a professor of gravitational physics in 1977 and the Lucasian Professor of Mathematics in 1980. The same position had been held by the great scientist Isaac Newton (1642–1727). His disability relieved Hawking of teaching duties, and he devoted his time to research, remaining on the cutting edge of cosmological thought. As of 2004, Hawking remained active in his field. Penrose and Hawking shared the prestigious Wolf Prize for Physics in 1988, just one of the many honors that Hawking has received. As his disease progressed, destroying his ability to control his voluntary muscles, his brain and senses remained unaffected. In 1969 he

Hess, Harry (1906–1969)

Harry Hess did not initially support the heretical idea of continental drift, but when he worked out his explanation of how sea floor spreading in midoceanic ridges occurred, he came to support that idea, and proposed a theory of how plate tectonics worked. Harry Hammond Hess was born in New York City. He earned his B.A. in geology at Yale University in 1927. He graduated with a Ph.D. in geology from Princeton University in 1932, and two years later he began teaching there. He became a professor in 1948 and remained at Princeton until his retirement in 1966. In the 1930s, Hess worked with the navy on submarine gravity studies and became an officer in the naval reserve. During World War II, Hess served as the captain of the assault transport USS *Cape Johnson*. As his ship traveled the Pacific Ocean, Hess used its sonar to map the ocean floor, searching for seamounts.

After the war and his return to Princeton,

Hess continued to be involved in the exploration of the ocean bottom. Oceanographers were astonished to find that midoceanic ridges existed in all the major oceans. Studies of these ridges revealed that they were geologically young, no more than 400 million years old, and had a central rift valley, high heat flows, a lack of sediment deposits, and unusual magnetic readings. In 1960, Hess wrote an essay, "History of Ocean Basins," which he called geopoetry, as if the notion of writing such a history was a fanciful one. He circulated the essay among the geological community. In the essay, he proposed magma as the lubricant that allowed continental plates to move and argued that the midoceanic ridges were the result of sea floor spreading. The ideas in this widely circulated manuscript supported the heretical theory of continental drift first proposed by the German meteorologist Alfred Wegener (1880 –1930) in 1912. Most geologists rejected that theory.

In 1962, Hess published his essay as a scholarly article. Frederick Vine (1939–) and Drummond Matthews (1931–1997) in Cambridge, England, expanded the work of Hess by proposing the hypothesis that magnetic strips on the ocean floor were a recording of reversals in Earth's magnetic field. Lawrence Morley (1920–) of Canada independently proposed a similar hypothesis, and it became known as the Vine-Matthews-Morley hypothesis. By the mid-1960s the hypothesis was accepted, and Hess was vindicated. At his death, Hess was involved with the Apollo project as one of the geologists who intended to examine the rock samples brought back by the *Apollo 11* astronauts from the first Moon landing.

See also Apollo Project; Geology; Oceanography; Plate Tectonics

References
Frankel, Henry. "Hess's Development of His Seafloor Spreading Hypothesis." Pp. 345–366 in *Scientific Discovery: Case Studies.* Edited by Thomas Nickles. Dordrecht, the Netherlands: D. Reidel, 1980.

"Hess, Harry Hammond." *A Princeton Companion.* http://etc.princeton.edu/CampusWWW/Companion/hess_harry.html (accessed February 13, 2004).

History of Science

The history of science was a small field in 1950, with not even twenty professors on the subject in the world. Few U.S. universities offered graduate training. By the late 1960s, there were some 150 historians of science in just the United States. The history of science is now a vibrant discipline in Western nations, with numerous doctoral programs, academic journals, scholarly conferences, and dedicated archives sustaining a vigorous worldwide program of study. With over 3,000 members, the American-based History of Science Society, founded in 1924, is the world's largest scholarly society studying the subject. The History of Science Society publishes *Isis,* the leading journal in the field. Many other such societies exist, usually in a national context, such as the British Society for the History of Science.

The last half of the twentieth century saw history of science studies becoming ever more specialized. At the same time, books and articles on the subject turned from their usual focus on narrow technical or conceptual themes (such as the relationship of science to nature) to more emphasis on the social or cultural context in which scientific change occurred (the relationship of science to social structures). This change could also be described as a shift from an "internalist" approach, concentrating on issues within the discipline, to an "externalist" approach, concentrating on the discipline in relation to the external world. Perspectives of other disciplines, such as sociology, philosophy, economics, literary criticism, and anthropology, became more common in externalist studies, enriching the analysis and fostering interdisciplinary studies. The new field of science and technology studies (STS), combining history, sociology, and philosophy, emerged. In the

Soviet Union and other communist countries, the study of the history of science remained a slave to Marxist ideology until communism crumbled and allowed historians to engage in unfettered analysis.

The 1962 book *The Structure of Scientific Revolutions,* by the physicist-turned-historian Thomas S. Kuhn (1922–1996), is the single most important book in the field. In the book, Kuhn adopted the term *paradigm* to describe the common worldview, shared practices, and accepted theories of a group of scientists. Using Ptolemaic geocentric astronomy and Copernican heliocentric astronomy as examples, Kuhn argued that scientists have usually practiced "normal science," in the sense that they have accepted the current paradigm in their field and their experiments have either confirmed the paradigm or resulted in minor changes. Repeatedly, however, as time has gone on, anomalies have occurred in observations and experiments that could not be explained by the current paradigm. When enough of these anomalies accumulate, the old paradigm becomes ripe for revolution, and a new paradigm emerges that explains the anomalies and incorporates the old paradigm inside it.

The English historian Herbert Butterfield (1900–1979) condemned the presentism in contemporary history books in his work *The Whig Interpretation of History* (1931). Like writers of political or cultural history, the writers of the history of science are particularly vulnerable to ideological advocacy of one scientific vision over another, and to the extent that they are presentists (or Whigs, to use Butterfield's analogy), they advocate the scientific vision of the present as the true one. Many argue that the best historical writing comes from solidly placing past events in their social, cultural, and intellectual contexts, as opposed to viewing history as a triumphal march of obvious decisions and developments to the present. Many of the best historians of science have taken Butterfield to heart and set their narratives and analysis within solid historical contexts.

Social constructionism, postmodernism, and certain developments in the philosophy of science sparked in the 1980s what became known as the "science wars." The work of historians of science was influenced by the perspectives of postmodernist theorists who emphasized the human fallibility inherent in scientific endeavors. Social constructionists, present on the cutting edge of sociology, anthropology, and history in the 1980s and 1990s, advocated a position of cultural and cognitive relativism. The agenda of social constructionists was not welcomed by practicing scientists, who vigorously defended the scientific method as a way of objectively obtaining knowledge and defended the body of scientific knowledge as being a correct view of the universe.

How do case studies and narratives from the history of science help the practitioners of actual science today? To take one example, the story of the cold fusion controversy, as documented by journalists and historians, provides a reminder of the importance of properly documenting and announcing scientific breakthroughs. Many scientists are very aware of the history of their disciplines, but are impatient with studies that lean toward social constructionism or postmodernism and find them not at all helpful. From the perspective of the practice of history, many historians have seen the history of science and its sister disciplines of the history of technology and the history of medicine as examples of truly global history: historical research and writing that can transcend national and ideological boundaries.

A sampling of some classics in the history of science from the second half of the twentieth century would have to include Alexander Koyré, *Galileo Studies* (1978, originally published in French in 1939); Herbert Butterfield, *The Origins of Modern Science, 1300–1800* (revised edition, 1957); Joseph Needham, *Science and Civilization in China* (seven volumes, 1954–1984); Thomas S. Kuhn, *The Copernican Revolution: Planetary Astronomy in the Development of Western Thought*

(1957); Charles E. Rosenberg, *The Cholera Years: The United States in 1832, 1849, and 1866* (1968); Charles Coulston Gillispie, editor, *Dictionary of Scientific Biography* (18 volumes, 1970–1980); Daniel J. Kevles, *The Physicists: The History of a Scientific Community in Modern America* (1977); Stillman Drake, *Galileo at Work: His Scientific Biography* (1978); Michael Ruse, *The Darwinian Revolution* (1979); I. Bernard Cohen, *The Newtonian Revolution: With Illustrations of the Transformation of Scientific Ideas* (1980); Stephen Jay Gould, *The Mismeasure of Man* (1981); Jack Morrell and Arnold Thackray, *Gentlemen of Science: Early Years of the British Association for the Advancement of Science* (1981); Margaret W. Rossiter, *Women Scientists in America: Struggles and Strategies to 1940* (1982); Martin J. S. Rudwick, *The Great Devonian Controversy: The Shaping of Scientific Knowledge among Gentlemanly Specialists* (1985); Daniel J. Kevles, *In the Name of Eugenics: Genetics and the Uses of Human Heredity* (1985); Steven Shapin and Simon Schaffer, *Leviathan and the Air Pump: Hobbes, Boyle, and the Experimental Life* (1985); David C. Lindberg and Ronald L. Numbers, editors, *God and Nature: Historical Essays on the Encounter between Christianity and Science* (1986); Richard Rhodes, *The Making of the Atomic Bomb* (1987); Crosbie Smith and M. Norton Wise, *Energy and Empire: A Biographical Study of Lord Kelvin* (1989); David Lindberg, *The Beginnings of Western Science* (1992); Donald Worster, *Nature's Economy: A History of Ecological Ideas* (second edition, 1994); and Richard S. Westfall, *The Life of Isaac Newton* (1994). As the above list shows, the history of science hit its stride in the 1980s in producing outstanding works of scholarship. Many other books written in the 1990s and the opening years of the twenty-first century will certainly qualify as classics after a measure of time has passed.

See also Cold Fusion; Kuhn, Thomas S.; Philosophy of Science; Social Constructionism; Sociology of Science

References

Durbin, Paul T., editor. *The Culture of Science, Technology, and Medicine.* New York: Free Press, 1980.

Gillispie, Charles C. "Recent Trends in the Historiography of Science." *Bulletin for the History of Chemistry* 15–16 (1994): 19–26.

Hessenbruch, Arne. *Reader's Guide to the History of Science.* Chicago: Fitzroy Dearborn, 2000.

Jasanoff, Sheila, et al., editors. *Handbook of Science and Technology Studies.* Thousand Oaks, CA: Sage, 1995.

Kragh, Helge. *An Introduction to the Historiography of Science.* New York: Cambridge University Press, 1987.

Olby, R. C., et al., editors. *Companion to the History of Modern Science.* New York: Routledge, 1990.

Rossiter, Margaret W., editor. "Catching Up with the Vision: Essays on the Occasion of the 75th Anniversary of the Founding of the History of Science Society." Supplement to *Isis* 90 (1999).

Söderqvist, Thomas, editor. *Historiography of Contemporary Science and Technology.* New York: Harwood Academic, 1997.

Hodgkin, Dorothy Crowfoot (1910–1994)

Dorothy Crowfoot Hodgkin determined the chemical structure and shape of penicillin, the vitamin B_{12}, and insulin. She was born Dorothy Crowfoot in Cairo, Egypt, where her father was an archaeologist. As a child, she was fascinated by chemistry and crystals, a preoccupation that English society of her time approved for women. She attended Somerville College, the women's college at Oxford University, where she decided between her twin interests in archaeology and chemistry, graduating in 1932 with a degree in chemistry. She learned crystallography at Oxford and continued to work in that area with the noted chemist John D. Bernal (1901–1971) at Cambridge University. In 1934, Somerville invited her back to teach there. She taught at Oxford while earning her Ph.D. in chemistry from Cambridge in 1937, and in that same year married Thomas Hodgkin, a historian who specialized in African studies.

Hodgkin remained at Oxford, pursuing her studies of crystallography while raising a family. She bore two sons and a daughter, and the family often maintained two households,

Dorothy Crowfoot Hodgkin, chemist, 1964 (Hulton-Deutsch Collection / Corbis)

as her husband's work took him to Africa and her own kept her at Oxford. Hodgkin moved up the academic ranks, becoming a university lecturer in 1946, a university reader in 1957, and finally the Wolfson Research Professor of the Royal Society in 1960.

Using x-ray crystallography photographs, obtained by a process in which x-rays reflected through crystals resulted in photographs, Hodgkin determined the structure of cholesterol iodide in 1943. Knowing the structure of a molecule allows chemists to better understand how that molecule behaves in chemical reactions. In 1949, she cracked the structure of penicillin. Beginning in 1948, her group took more than 2,500 x-ray photographs of the vitamin B_{12}, and she helped pioneer the use of computers to analyze molecules. The eventual chemical formula for B_{12} that she announced in 1957 was $C_{63}H_{88}N_{14}O_{14}PCo$. She then turned to an even greater challenge, the structure of the insulin molecule, which contained 777 atoms. She succeeded in 1969, helped by advances in computers that allowed more complex calculations on measurements taken from x-ray photographs.

Hodgkin was elected to the Royal Society in 1947, only the third woman to receive this honor. She considered this honor to be greater than the Nobel Prize in Chemistry that she received in 1964. Nobel prizes are often shared between recipients, but her award was a solo honor. In 1965, the Queen of England awarded her the Order of Merit. The only other woman to receive that honor up until then had been Florence Nightingale (1820–1910). Hodgkin served as chancellor of Bristol University from 1970 to 1988.

Hodgkin was renowned for her patient work and her dexterous fingers. Her skills are all the more remarkable because in 1934 she contracted rheumatoid arthritis, which painfully inflamed her hands and feet and eventually crippled her. A committed socialist and pacifist, Hodgkin remained politically active her entire life, though she maintained that her science and her politics were separate endeavors. A former student of hers, Margaret Thatcher (1925–), turned from chemistry to become an influential conservative prime minister of the United Kingdom.

See also Chemistry; Computers; Nobel Prizes
References

Ferry, Georgina. *Dorothy Hodgkin: A Life.* London: Granta, 1998.

McGrayne, Sharon Bertsch. *Nobel Prize Women in Science.* Second edition. Secaucus, NJ: Carol, 1998.

Hoyle, Fred (1915–2001)

Sir Fred Hoyle developed and championed the steady state theory of the universe. Fred Hoyle was born in Bingley, Yorkshire, England, to parents in the wool business; he attended Cambridge University, winning prizes and recognition as a mathematician. He married in 1939 and fathered a son and a daughter. During World War II, Hoyle worked on radar for the British Admiralty, where two coworkers, Hermann Bondi (1919–) and Thomas Gold (1920–), were also interested in astrophysics. The three of them developed the steady state theory of cosmology, which they published in 1948, with Hoyle contributing the mathematical work. After the war, Hoyle returned to Cambridge, becoming the Plumian Professor of Astronomy in 1958.

Hoyle became the leading proponent of the steady state theory and vigorously defended the idea that the universe had always existed and that matter was constantly being created to fuel its expansion. The opposing theory proposed a moment of creation, a primordial egg that exploded to create the universe. Hoyle dubbed the opposing theory the "Big Bang," and the name stuck. Debate over the two competing theories raged among astronomers during the 1950s. The discovery of background radiation in the universe in 1964, at the temperature predicted by the Big Bang theorists, resolved the dispute for most astronomers.

Because George Gamow had proposed that all the heavier elements were formed during the Big Bang, Hoyle needed to find a way that heavy elements could have been created in a steady-state universe. Hoyle and his collaborators developed the theory of nucleosynthesis, arguing that the heavier elements are created by nuclear fusion as part of the life cycle of stars. This breakthrough is now a standard part of astronomy and earned Hoyle's collaborator, William A. Fowler (1911–1995), a share of the 1983 Nobel Prize in Physics. Hoyle also made significant contributions to the theory of stellar evolution, specifically on the theory of accretion by which stars are formed, and worked on quasars and the physics of supermassive objects.

Hoyle raised the funds to found the Institute of Theoretical Astronomy at Cambridge in 1966. A bitter dispute with fellow faculty at the university led to his retirement from Cambridge in 1972, the same year that he was knighted. He resumed his career at the

Cardiff University of Wales. While at Cardiff, he branched out into other areas of interest, including publishing a theory that the ice ages were caused by comet impacts rather than permutations of Earth's orbit. Working with a former student, Sri Lankan–born Chandra Wickramasinghe (1939–), he created a theory of panspermia, which argued that life did not spontaneously originate on Earth, but came from biological material contained in interstellar dust clouds. The Cardiff Center for Astrobiology was founded to further investigate this theory. Neither the theory on ice ages nor panspermia has been generally accepted by the scientific community.

Hoyle also wrote science fiction novels, plays, short stories, two works of autobiography, and a book on Stonehenge and archeoastronomy, and was well-known for his numerous popular science books. His most noted novels were *The Black Cloud* (1957) and *A for Andromeda* (1962). The year before he died, Hoyle published a book proposing a quasi–steady state theory, yet another effort to promote his favorite theory. His penchant for thinking outside of the mainstream of the scientific community enabled Hoyle to make major contributions, but also meant that he left behind many rejected ideas. Hoyle received many important awards, including the 1997 Crafoord Prize from the Royal Swedish Academy of Sciences. The Crafoord prize is awarded by the Royal Swedish Academy of Sciences in those areas not covered by the Nobel prizes. Perhaps his penchant for controversy and odd theories explains why Hoyle failed to receive a Nobel prize.

See also Astronomy; Bell Burnell, Jocelyn; Big Bang Theory and Steady State Theory; Gamow, George; Media and Popular Culture; Physics; Quasars

References
Hoyle, Fred. *Home Is Where the Wind Blows: Chapters from a Cosmologist's Life.* Mill Valley, CA: University Science, 1994.
Rees, Martin. "Fred Hoyle: Obituary." *Physics Today* 54, no. 11 (November 2001): 75.

Human Genome Project

The Human Genome Project created a complete map of all the genes in the human body, including every base pair. In the mid-1980s, the advent of new gene sequencing techniques encouraged visionaries to begin to promote the idea of sequencing the entire human genome. Individuals at the United States Department of Energy (DOE) and National Institutes of Health (NIH) both started to seek funding for such a project. The DOE justified its interest as a continuation of its long-term studies of the effects of radiation on humans and pointed to its experience with multibillion-dollar big science projects. The molecular biologist and Nobel laureate Walter Gilbert (1932–) became an early proponent of the idea and persuaded his fellow Nobel laureate James D. Watson (1928–) of its feasibility.

The Human Genome Project officially began in 1990 as a joint NIH/DOE project, federally funded at $200 million a year for fifteen years. Watson became the first director of the organization, serving until 1992, when the physician Francis Collins took over the job. The project began by sequencing the genomes of simpler organisms, such as *Escherichia coli,* in order to gain experience and develop better gene-sequencing technologies. Work on the genome of the mouse accompanied intensified work on the human genome. Britain's Wellcome Trust joined the project in 1993, and many other international collaborators joined, with organizations and individuals from over twenty countries contributing. The Internet allowed the collaborators to share and publish their data.

When the project began, scientists knew that the human genome contained about 3 billion base pairs, making up an estimated 50,000 to 100,000 genes. Many of the base pairs are not parts of genes, but just debris connecting the double-helix-shaped deoxyribonucleic acid (DNA) molecule together. As the project progressed, scientists were astonished to find that the human genome

James D. Watson, zoologist (National Human Genome Research Institute)

contained only about 30,000 to 35,000 genes. The project intended to sequence every base pair. The physiologist J. Craig Venter of the National Institute for Neurological Disorders and Stroke disagreed with the conservative sequencing approach of the project, arguing that they should seek out the genes and only sequence them, not everything else. In 1991, a venture capital firm lured Venter away from the NIH to form The Institute for Genomic Research (TIGR).

In 1995, Venter and TIGR announced that they had sequenced the genome of the bacterium *Haemophilus influenzae,* which contained 1.8 million base pairs. They used a shotgun technique, cutting the entire genome into smaller pieces and using powerful computers to match the base pairs at the ends of the fragments. In 1996, the NIH succeeded in sequencing the yeast genome, using their more methodical and more accurate methods. In 1998, Venter raised the stakes by announcing a partnership with the Perkin-Elmer Corporation, which made a sophisticated automated sequencing ma-

chine. Venter's new company, Celera Genomics, intended to use 300 of the machines to sequence the entire human genome in just three years for only $300 million. They planned to publish their results through their own Web site and not the usual GenBank public database, and also sought to make a profit from commercial users of the data.

The two projects feuded as their race intensified. A temporary truce allowed the two to agree to the simultaneous publication of rough drafts of the genome in 2000. In April 2003, the Human Genome Project announced that it had mapped 99.99 percent of the human genome (100 percent was not reachable because people are genetically different from each other), finishing the project in thirteen years, instead of the originally projected fifteen, and coming in under budget at $2.7 billion. The final genome contained 3.1 billion pairs, making up 35,000 to 40,000 genes. The Sanger Institute in England had done 30 percent of the work, and the political leaders of the United States, United Kingdom, France, Germany, Japan, and China signed the statement announcing final success. Celera chose not to complete their effort and continued to offer their research via subscription.

Up to 5 percent of the funding for the Human Genome Project was devoted to studies of the ethical, legal, and social issues involved, a unique innovation for a big science project. It is expected that the human genome map will help scientists create new medical treatments. It may also enable the genetic engineering of human beings, stoking fears of a return of eugenics. Besides curing diseases, will people start to modify their own bodies and the bodies of their children?

See also Big Science; Bioethics; Genetics; Gilbert, Walter; Internet; National Institutes of Health; Watson, James D.

References
Bodmer, Walter, and Robin McKie. *The Book of Man: The Human Genome Project and the Quest to Discover Our Genetic Heritage.* Boston: Little, Brown, 1994.

Davies, Kevin. *Cracking the Genome: Inside the Race to Unlock Human DNA*. New York: Free Press, 2001.

Kay, Lily E. *Who Wrote the Book of Life? A History of the Genetic Code*. Stanford, CA: Stanford University Press, 2000.

Kevles, Daniel J., and Leroy Hood, editors. *The Code of Codes: Scientific and Social Issues in the Human Genome Project*. Cambridge: Harvard University Press, 1992.

Roberts, Leslie. "Controversial from the Start." *Science* 291 (February 16, 2001): 1182–1188.

Sloan, Phillip R., editor. *Controlling Our Destinies: Historical, Philosophical, Ethical, and Theological Perspectives on the Human Genome Project*. Notre Dame, IN: University of Notre Dame Press, 2000.

Hydrogen Bomb

The hydrogen bomb, a thermonuclear device, uses the process of fusion to create a much larger explosion than is possible with the fission-based atomic bomb. The first U.S. atomic bombs were exploded in 1945, the culmination of the expensive and secret Manhattan Project. J. Robert Oppenheimer (1904–1967) led the American researchers at the Los Alamos Laboratory in New Mexico who designed the atomic bomb. Atomic bombs are based on fission, a process in which a stream of neutrons splits the nucleus of an atom, breaking it apart, releasing further neutrons and energy in the process. These neutrons then split other atoms, creating a chain reaction. Isotopes of uranium, U-235 or U-238, or the plutonium isotope P-239 are used as the atomic fuel.

A more powerful type of bomb was thought possible even before the fission bombs were successfully developed. The first theoretical suggestion of such a bomb was put forward by a physicist at the University of Kyoto, Tokutaro Hagiwara (1897–1971), in May 1941, six months before the Pearl Harbor attack brought America into World War II. Japan never developed a serious atomic weapons program during World War II. In September 1941, conversations between Enrico Fermi (1901–1954) and Edward Teller (1908–2003) led to speculation that a more powerful type of bomb, a superbomb based on fusion instead of fission, was possible. Such a bomb, using isotopes of hydrogen, deuterium, and tritium, would force hydrogen nuclei together to fuse into larger nuclei of helium, releasing excess energy in the process. This conversion of mass into energy followed the prediction of Albert Einstein's famous equation: $E=mc^2$. Fusion is what powers the Sun, turning hydrogen into helium, and helium into heavier elements. Current cosmological theory holds that all atoms in the universe, other than hydrogen, were created in the solar furnaces of stars.

Teller advocated research into fusion bombs as well as fission bombs during the Manhattan Project, but Oppenheimer chose to concentrate on the technically less challenging fission bomb. The 1949 explosion of the Soviet Union's first atomic bomb shocked the Americans. Teller and Atomic Energy Commission (AEC) chairman Lewis L. Strauss (1896–1974), a former investment banker, argued that the United States needed to accelerate development of the superbomb. The general advisory committee of the AEC, which formulated military and civilian nuclear policy for the United States, debated the decision. Oppenheimer, chair of the committee, and others opposed pursuing the superbomb on moral and technical grounds. They saw the superbomb as being so powerful that its only possible targets were cities, whereas atomic bombs might be used to destroy many types of military targets. The committee also questioned the difficulty of actually building the device. This opposition eventually led to a 1954 AEC hearing that stripped Oppenheimer of his security clearance, effectively exiling him from nuclear weapons work.

President Harry S. Truman (1884–1972) decided to build the superbomb, and Teller led this effort. Flawed calculations by Teller initially delayed the superbomb project.

The first hydrogen bomb tested in 1952 by the United States, which obliterated an island in the Eniwetok Atoll (Corbis)

Teller wanted to build a megaton-capable weapon immediately, and later observers have noted that if the Americans had scaled back their ambitions to about half a megaton, still much larger than kiloton-capable atomic bombs, the first hydrogen bomb could have been exploded in 1949. Eventually Teller's errors were corrected. In 1951, the mathematician Stanislaw Ulam (1909–1984) and Teller solved the major technical obstacle by electing to use x-rays generated by an atomic bomb trigger to start the fusion reaction in the hydrogen bomb. The first hydrogen bomb was exploded on November 1, 1952, on the small island that formed part of Eniwetok Atoll. The island disappeared, leaving a mile-wide crater 175 feet deep. Teller became the "father of the hydrogen bomb."

The brilliant Andrei Sakharov (1921–1989) designed the first Soviet hydrogen bomb, which was exploded on August 12, 1953. A leading scientist on the American project, the German-born Klaus Fuchs (1911–1988), had fed detailed intelligence to the Soviet Union. The intelligence from Fuchs and others certainly helped the Soviet atomic bomb effort, but how much it helped the Soviet hydrogen bomb effort is disputed by historians. A nuclear arms race followed the development of atomic and hydrogen bombs by the Soviets, turning the cold war into an ideological conflict with the potential to devastate much of the planet. Ever larger bombs were detonated, with at least one Soviet explosion in excess of 50 megatons in yield. These open-air tests led to measurable poisoning of the atmosphere until open-air testing was banned in 1963 by a treaty between the United States and Soviet Union. The United Kingdom (first public test in 1952), China (first public test in 1964), France (first public test in 1961), Israel (no public test), South Africa (no public test), India (first public test in 1998), and Pakistan (first public test in 1998) have all developed atomic or hydrogen bombs. After the end of the cold war, a Comprehensive Nuclear Test Ban Treaty was negotiated in the 1990s, but not all nuclear-capable nations have signed or ratified it.

See also Big Science; Cold War; Ethics; Nuclear Physics; Oppenheimer, J. Robert; Sakharov, Andrei; Teller, Edward

References

Bailey, George. *Galileo's Children: Science, Sakharov, and the Power of the State.* New York: Arcade, 1990.

Goncharov, German A. "Thermonuclear Milestones." *Physics Today,* November 1996, 44–61.

Herken, Gregg. *Brotherhood of the Bomb: The Tangled Lives and Loyalties of Robert Oppenheimer, Ernest Lawrence, and Edward Teller.* New York: Henry Holt, 2002.

Holloway, David. *Stalin and the Bomb: The Soviet Union and Atomic Energy 1939–1956.* New Haven: Yale University Press, 1994.

Rhoades, Richard. *Dark Sun: The Making of the Hydrogen Bomb.* New York: Simon and Schuster, 1995.

Schweber, Silvan S. *In the Shadow of the Bomb: Bethe, Oppenheimer, and the Moral Responsibility of the Scientist.* Princeton, NJ: Princeton University Press, 2000.

I

In Vitro Fertilization

In vitro fertilization (IVF) is a medical technique for removing a mature egg from a woman, fertilizing it with donor sperm in the laboratory, allowing it to divide several times until it becomes a blastocyst, and then implanting the blastocyst in a uterus. Studies have estimated that one in ten married couples experience some form of infertility problem. A common cause of these problems is defective fallopian tubes. Normally, an egg is fertilized in the ovary, divides into a blastocyst, then descends through the fallopian tubes to implant itself on the wall of the uterus, where it grows into a baby.

IVF was initially developed for breeding animals, but the Cambridge University physiologist Robert Edwards (1925–) saw the promise of applying this technique to humans. He teamed up with Patrick Steptoe (1913–1988), a gynecologist in Oldham, England, who had pioneered the technique of laparoscopy. The laparoscope, a narrow tube with a fiber-optic light at the end, allowed a surgeon to peer into the abdominal cavity. Steptoe used his skill to extract mature eggs from the ovaries of volunteers, and Edwards fertilized them in the laboratory. In 1972, Steptoe implanted a fertilized human egg into a uterus for the first time. None of the approximately thirty women in the test group who became pregnant carried the babies for more than a trimester. Funding for their experiments was cut amidst a general outcry from Members of Parliament, religious leaders, and other scientists. Among the objections was fear that a successful technique would not stop at helping infertile couples, but would lead to genetically engineered babies and surrogate mothers. Steptoe financed the continuation of their research with profits from his practice, which included legal abortions, and the two scientists moved their work to a smaller hospital in Oldham, Dr. Kershaw's Cottage Hospital.

Continued criticism from their colleagues caused Steptoe and Edwards to suspend reporting on their research. They persisted despite many failures until late 1977, when they decided to stop using a fertility drug with their patients and started to implant blastocysts that were only eight cells large. Up until then, they had waited for the blastocyst to become much larger, as it would in nature before descending from the ovary to the uterus. The patient Lesley Brown was impregnated with her own egg fertilized by her husband's sperm. On July 25, 1978, a daughter named Louise Joy was delivered by Caesarian section. The world press proclaimed the world's first test-tube baby.

By the year 2000, over half a million children had been born worldwide through the use of IVF since 1978, and the number of clinics specializing in the procedure is in the hundreds. The implantation process has continued to be the weakest point in the IVF process. Although other mammals have high rates of success, the rate of success in humans has been as low as one in five tries. Much of the research in IVF has focused on bettering these implantation statistics, resulting in slow improvements. Laparoscopy has also been replaced by using a vaginal ultrasound probe to retrieve eggs.

Normally, multiple embryos are created as insurance, because of the high incidence of failure in the implantation process. Excess embryos are frozen in liquid nitrogen and show no side effects when unfrozen and later implanted, even if years have passed. What to do with these frozen embryos after a successful pregnancy and the couple do not want to have more children? Who owns the embryos if the couple divorces before pregnancy? Should embryos be used for further scientific research? These legal and ethical questions have been answered differently by legislatures, courts, and public opinion in different countries. Attitudes toward abortion are a major indicator of how people respond to these issues. The destruction of frozen embryos was one of the issues that prompted the Catholic Church to issue a statement in 1987 opposing IVF.

In vitro fertilization has been joined by other forms of assisted reproductive technology (ART), such as intracytoplasmic sperm injection to help men with low sperm counts and operations on fetuses in the womb to correct abnormalities. Many of these ART technologies bring ethical concerns. There is no biological reason why a blastocyst must be reimplanted in the same woman who donated the egg, and this fact made possible the emergence of surrogate mothers who carry and bear children for other women.

The development of IVF laid the essential groundwork for the possible future genetic engineering of children. Techniques for avoiding certain genetic diseases in vitro have been developed in the 1990s, and effective techniques for sex selection are close to being developed. IVF and other forms of ART allow individuals to overcome problems with infertility, though from the larger perspective of world population growth, fertility rates within the human population are more than adequate to sustain the population.

See also Edwards, Robert; Ethics; Medicine; Population Studies; Steptoe, Patrick

References

Bonnicksen, Andrea L. *In Vitro Fertilization: Building Policy from Laboratories to Legislatures.* New York: Columbia University Press, 1989.

Edwards, Robert G. "The Bumpy Road to Human *In Vitro* Fertilization." *Nature Medicine* 7, no. 10 (October 2001): 1091–1094.

Gosden, Roger. *Designing Babies: The Brave New World of Reproductive Technology.* San Francisco: W. H. Freeman, 1999.

Sher, Geoffrey, Virginia Davis, and Jean Stoess. *In Vitro Fertilization: The A.R.T. of Making Babies.* New York: Facts on File, 1995.

Silver, Lee M. *Remaking Eden: How Genetic Engineering and Cloning Will Transform the American Family.* New York: Avon, 1997.

Integrated Circuits

Besides creating an enormous consumer electronics industry, integrated circuits had become pervasive in all areas of scientific endeavor by the late 1970s. Integrated circuits are made of wafers of silicon with electronic components, such as transistors, diodes, capacitors, and resistors, etched into the silicon by chemicals or light. The semiconducting nature of silicon is used to direct the flow of electrons through the electronic components, using binary values and the rules of Boolean algebra, a mathematical curiosity created a century earlier by the self-taught British mathematician George Boole (1815–1864).

The earlier innovation necessary to create integrated circuits was the invention of the transistor in 1947 at Bell Telephone Laboratories in Murray Hill, New Jersey, by William Shockley (1910–1989), Walter H. Brattain (1902–1987), and John Bardeen (1908–

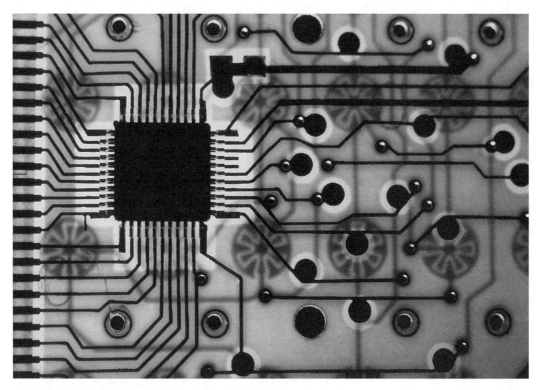

Integrated circuit (Kelly Harriger / Corbis)

1991), for which the three men shared the 1956 Nobel Prize in Physics. Integrated circuits were independently invented by Jack S. Kilby (1923–) of Texas Instruments and Robert Noyce (1927–1990) at Fairchild Semiconductor in 1958 and 1959. Kilby obtained the first patent, and Noyce developed the planar process to place metal connections on the semiconductor wafer; royalties were paid to both men for their inventions. Kilby received half of the 2000 Nobel Prize in Physics after Noyce had been dead for ten years.

Fairchild Semiconductor brought the first integrated circuit to market in 1961. The Apollo project became the first big customer for integrated circuits and proved a key driver to accelerate the growth of the semiconductor industry. Quite quickly, integrated circuits became the main technology of the computer industry, and integrated circuits made possible later space exploration. The in-

tegrated circuit led to the invention of the microprocessor in the early 1970s at Intel Corporation. A microprocessor combines all the central logic necessary for a computer onto a single chip.

In 1964, Gordon E. Moore (1929–) noticed that the density of transistors and other components on integrated chips doubled every year. He charted this trend and predicted that it would continue until about 1980, when the density of integrated circuits would decline to doubling every two years. Variations of this idea became known as Moore's Law. Since the early 1970s, chip density on integrated circuits, both microprocessors and memory chips, has doubled about every eighteen months. About fifty electronic components per chip was the density in 1965; five million electronic components were placed on an individual chip in 2000. An individual transistor on a chip is now about 90 nanometers in size. At times, different commentators have predicted that this trend would hit an obstacle that engineers could not overcome and slow down, but that has not yet happened. The electronic components on integrated circuits are being packed so close that by the second decade of the twenty-first century it is believed that quantum effects will begin to make it impossible to maintain the rate of increasing density predicted by Moore's Law.

In the late 1990s, the technology used to manufacture integrated circuits was adapted by replacing transistors on the silicon with deposits of DNA. These DNA microarrays, or DNA chips, are used to quickly and simultaneously test a biological substance against thousands of pieces of different DNA, leading to a thousandfold increase in productivity by experimental biologists.

See also Apollo Project; Bardeen, John; Computers; Nobel Prizes

References
Ceruzzi, Paul E. *A History of Modern Computing.* Cambridge: MIT Press, 1998.
Queisser, Hans. *The Conquest of the Microchip: Science and Business in the Silicon Age.* Cambridge: Harvard University Press, 1988.
Reid, T. R. *The Chip: How Two Americans Invented the Microchip and Launched a Revolution.* New York: Random House, 2001.
Riordan, Michael, and Lillian Hoddeson. *Crystal Fire: The Birth of the Information Age.* New York: Norton, 1997.
Slater, Robert. *Portraits in Silicon.* Cambridge: MIT Press, 1987.

International Geophysical Year (1957–1958)

The International Geophysical Year (IGY) lasted for eighteen months, from July 1, 1957, to December 31, 1958, bringing together most of the globe's nations in a concerted scientific effort to learn more about the planet. Its origins go back to 1950, when several prominent geophysicists met at the home of the American physicist James Van Allen (1914–), and Lloyd Berkner (1905–1967) suggested that another international polar year be organized. (The original

International Polar Year [1882–1883] had investigated the Arctic region. The Second International Polar Year, held fifty years later in 1932–1933, also investigated the Arctic and parts of Antarctica.) Within two years, Berkner's idea had became a proposal for an International Geophysical Year in 1957–1958 under the auspices of the International Council of Scientific Unions (ICSU), a body founded in 1931 to foster international cooperation in science. The United Nations Educational, Scientific and Cultural Organization (UNESCO) began to partially sponsor the ICSU in 1947 and lent its efforts to the IGY. In the United States, the National Science Foundation helped fund the U.S. efforts, with considerable cooperation from the U.S. military, under the auspices of the National Academy of Sciences. The IGY was timed to coincide with the peak of the solar activity cycle, when sunspots and solar flares are more frequent. The Englishman Sydney Chapman (1888– 1979) headed the global effort.

The advances in technology during World War II gave scientists radar, sonar, rockets, aircraft, and computers to assist in their more peaceful efforts. The main subjects and tools of the IGY included meteorology, geomagnetism, the aurora and air glow, the ionosphere, solar activity, cosmic rays, longitude and latitude studies, glaciology, oceanography, rockets and satellites, seismology, gravimetry, and nuclear radiation. An IGY committee also coordinated the idea of world days, specific times set aside when scientists at numerous places around the globe make simultaneous measurements, for example during the total solar eclipse on October 12, 1958.

In 1954, an IGY committee agreed to an American plan to launch artificial satellites to orbit Earth as sensor platforms during the IGY. Sergey Pavlovich Korolev (1907–1966), the "Chief Designer" of the Soviet ballistic missile program, used the impending IGY to persuade the Soviet government to build a launcher and satellite to preempt the United States. The project was kept secret, and the

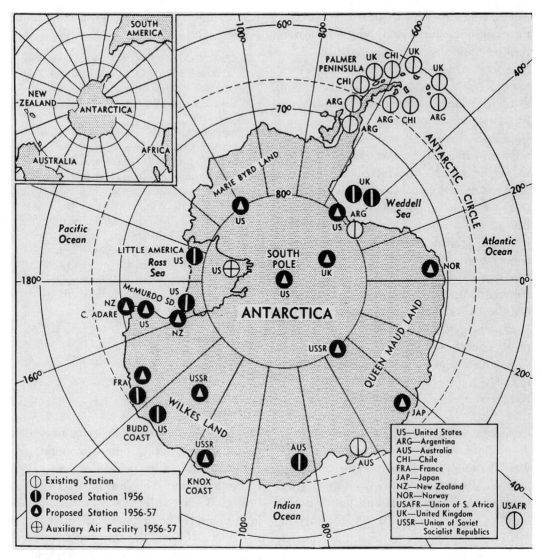

Map of proposed research stations in Antarctica during the 1957–1958 International Geophysical Year (Bettmann/Corbis)

United States did not even know that a competition was going on until the Soviets surprised the world by launching *Sputnik I* on October 4, 1957. In their haste to beat the Americans, the Soviets made the first satellite quite simple, with few scientific instruments.

The first successful American satellite, *Explorer 1,* was launched on January 31, 1958. The satellite discovered zones of radiation around Earth, later named the Van Allen Radiation Belts in honor of the creator of the onboard scientific instrument that detected the radiation. Scientists later learned that Earth's magnetic field created these belts of radiation and that the belts helped protect Earth from some of the more harmful effects of solar radiation.

In the run-up to the IGY, explorers swarmed onto Antarctica, setting up dozens of camps and monitoring stations. The naval and military forces of many nations participated in establishing these bases, some of which have remained manned and useful. The IGY also prompted the negotiation of a 1961

Antarctic Treaty, which declared Antarctica the possession of no nation and set the frozen continent aside for scientific study.

A total of sixty-seven nations participated in the IGY, with the notable major exception of China, which withdrew from its initial efforts. John Tuzo Wilson (1908–1993), the Canadian geophysicist who later played an important role in the theory of plate tectonics, was president of the International Union on Geodesy and Geophysics during the IGY and played an important leadership role. China invited him to visit them during the IGY, and he gained some limited, informal cooperation. Measurements of geomagnetism on the ocean floors during the IGY contributed to the development of the theory of plate tectonics in the 1960s. Eighty-six vessels participated in measurements of the planet's oceans and seas.

Scientists began the systematic monitoring of the planet's ozone layer during the IGY, creating an important base of information when chlorofluorocarbons became an important environmental issue in the 1970s. The fallout from open-air nuclear testing was also more carefully monitored during the IGY, leading to alarm over the higher than expected levels, and contributing to the creation of an open-air test ban treaty in 1963 between the United States and Soviet Union.

The IGY was followed by the International Geophysical Co-operation of 1959, which allowed the organized tapering off of IGY activities. Many organizations and efforts, such as those in Antarctica, continued after IGY on a reduced scale, and the organizational effects of the IGY are still felt today. Both the Americans and Soviets embarked on vigorous space programs as part of their cold war ideological struggle. The IGY also inspired other, smaller-scale efforts, such as the International Years of the Quiet Sun (1964–1965) and the International Decade of Ocean Exploration (1970–1980). Among the efforts directly sponsored by the ICSU

was the International Biological Programme (1964–1974).

See also Cold War; National Science Foundation; Ozone Layer and Chlorofluorocarbons; Plate Tectonics; Satellites; Space Exploration and Space Science; Wilson, John Tuzo

References

Chapman, Sydney. *IGY: Year of Discovery.* Ann Arbor: University of Michigan Press, 1959.

Eklund, Carl R., and Joan Beckman. *Antarctica: Polar Research and Discovery during the International Geophysical Year.* New York: Holt, Rinehart and Winston, 1963.

Siddiqi, Asif A. "Korolev, Sputnik, and the International Geophysical Year." http://www.hq.nasa.gov/office/pao/History/sputnik/siddiqi.html (accessed February 13, 2004).

Sullivan, Walter. *Assault on the Unknown: The International Geophysical Year.* New York: McGraw-Hill, 1961.

Wilson, John Tuzo. *IGY: Year of the New Moons.* New York: Knopf, 1961.

Internet

The Internet forms a global computer network interconnecting other computer networks, providing an unparalleled means of communication. The Advanced Research Projects Agency (ARPA) was formed in 1958 in response to the first artificial satellite, the Soviet *Sputnik I,* and the emerging space race. ARPA was an agency of the Pentagon, and the researchers at ARPA were given a generous mandate to develop innovative technologies. In 1962, a psychologist from the Massachusetts Institute of Technology's Lincoln Lab, J. C. R. Licklider (1915–1990), joined ARPA to take charge of the Information Processing Techniques Office (IPTO). In 1963 Licklider wrote a memo proposing an interactive network allowing people to communicate via computer. This project did not materialize. In 1966, Robert Taylor (1932–), then head of the IPTO, noted that he needed three different computer terminals to connect to three different machines in different locations around the nation. Taylor also recognized that universities working with the IPTO needed more computing resources. Instead of the government buying machines for each university, why not share machines? Taylor revitalized Licklider's idea, secured $1 million in funding, and hired twenty-nine-year-old Larry Roberts (1937–) to direct the creation of ARPAnet.

Universities were reluctant to share their precious computing resources and concerned about the processing load of a network. Wes Clark of Washington University of St. Louis proposed an Interface Message Processor (IMP), a separate smaller computer for each main computer on the network, that would handle the network communication. Another important idea came from Paul Baran (1926–) of the RAND Corporation, who had been concerned about the vulnerability of the U.S. telephone communication system since 1960, but had yet to convince the telephone monopoly AT&T of the virtues of his ideas on distributed communications. He devised a scheme of breaking signals into blocks of information to be reassembled after reaching their destination. These blocks of information traveled through a "distributed network," where each "node," or communication point, could independently decide which path the block of information took to the next node. This allowed data to automatically flow around blockages in the network. Donald Davies (1924–2000) at the British National Physical Laboratory (NPL) independently developed a similar concept in 1965, which he termed packet-switching, each packet being a block of data. Although Baran was interested in a communications system that could continue to function during a nuclear war, ARPAnet was purely a research tool, not a command-and-control system.

A small consulting firm in Cambridge, Massachusetts, Bolt, Beranek and Newman (BBN), got the contract to construct the IMP in December 1968. They decided that the IMP would only handle the routing, not the transmitted data content. As an analogy, the IMP looked only at the address on the envelope, not at the letter inside. Faculty and graduate students at the host universities created host-to-host protocols and software to enable the computers to understand each other. Because the machines did not know how to talk to each other as peers, the researchers wrote programs that fooled the computers into thinking they were talking to preexisting dumb terminals.

ARPAnet began with the installation of the first 900-pound IMP in the fall of 1969 at UCLA, followed by three more nodes at the Stanford Research Institute (SRI), University of California at Santa Barbara, and University of Utah. The first message transmitted between UCLA and SRI was "L," "O," "G," the first three letters of "LOGIN," but then the system crashed. Initial bugs were overcome, and ARPAnet added an extra node every month in 1970. Improvements and new protocols quickly followed. Electronic mail, file transfer, and remote login became the dominant applications.

In 1981 the National Science Foundation (NSF) created the Computer Science Network (CSNET) to provide universities that did not have access to ARPAnet with their own network. In 1986, the NSF sponsored the NSFNET "backbone" to connect five supercomputing centers together. The backbone also connected ARPAnet and CSNET together. The idea of the Internet, a network of networks, became firmly entrenched. The open technical architecture of the Internet allowed numerous innovations to easily be grafted onto the whole. When ARPAnet was dismantled in 1990, the Internet was thriving at universities and technology-oriented companies. The NSF backbone was later dismantled when the NSF realized that commercial entities could keep the Internet running and growing on their own. The introduction in 1991 of the World Wide Web, developed by Tim Berners-Lee (1955–), took advantage of the modular architecture of the Internet and became widely successful. What began with four nodes in 1969 as a creation of the cold war became a worldwide network of networks, forming a single whole. In early 2001, an estimated 120 million computers were connected to the Internet in every country of the world. The Internet has proved to be an extraordinary tool for scientists to use in communicating with each other and for laypersons to learn about science. Unfortunately, the Internet has also been used to spread pseudoscientific theories and scientific misinformation.

See also Berners-Lee, Timothy; Big Science; Cold War; Computers; Engelbart, Douglas; National Science Foundation

References

Abbate, Janet. *Inventing the Internet.* Cambridge: MIT Press, 2000.

Hafner, Katie, and Matthew Lyon. *Where Wizards Stay Up Late: The Origins of the Internet.* New York: Simon and Schuster, 1996.

Internet Society (ISOC). "Histories of the Internet." http://www.isoc.org/internet/history/ (accessed February 13, 2004).

Moschovitis, Christos J. P., Hilary Poole, Tami Schuyler, and Theresa M. Senft. *History of the Internet: A Chronology, 1943 to the Present.* Santa Barbara, CA: ABC-CLIO, 1999.

Segaller, Stephen. *Nerds 2.0.1: A Brief History of the Internet.* New York: TV Books, 1988.

J

Jet Propulsion Laboratory

The Jet Propulsion Laboratory (JPL) in Pasadena, California, is jointly operated by the California Institute of Technology (Caltech) and the National Aeronautics and Space Administration (NASA). In the 1930s, Theodore von Kármán, head of Caltech's Guggenheim Aeronautical Laboratory, began working on rocket propulsion. During World War II, Kármán and his Caltech group developed jet-assisted take-off (JATO) rockets for the U.S. Army Air Corps. This effort led to a project to create an American answer to the German V-2 rocket, and the U.S. Army Ordnance department contracted with Caltech to create the Jet Propulsion Laboratory. In 1947, two years after Germany was defeated, JPL launched the first Corporal rocket.

After NASA was created in 1958, JPL was transferred from the control of the army to the new civilian space agency. JPL had already built *Explorer 1,* the first satellite launched by the U.S. Army earlier on January 31, 1958. JPL came to specialize in lunar and interplanetary spacecraft, designing and operating all such spacecraft for NASA. Numerous Ranger, Surveyor, and Lunar Orbiter spacecraft were flown to scout the way for the Apollo Moon landings. *Mariner 2,* launched in 1962, was the first spacecraft to fly by another planet when it passed Venus. Other Mariner spacecraft explored Mercury, Venus, and Mars. Pioneer and Voyager spacecraft visited the outer planets and then left the solar system. The two Viking landers tried to find life on Mars in 1976, but failed. The Galileo spacecraft began orbiting Jupiter in 1995 and sent a probe into that gas giant. Magellan mapped the surface of cloud-shrouded Venus with radar from 1990 to 1994. The joint NASA–European Space Agency spacecraft Ulysses, launched in 1990, used a gravitational assist past Jupiter to enter a solar orbit that flew over the poles of the Sun, exploring sections of space outside the ecliptic plane. In the 1990s, in an attempt to stretch funding to cover more missions, JPL and NASA began to build cheaper spacecraft with fewer redundant systems and smaller loads of instruments. Two missions to Mars in the late 1990s failed, leading to questions about this approach. Numerous other missions have already been launched or are planned to continue to explore the solar system, including robotic visits to comets, asteroids, and Pluto.

Having developed sophisticated sensors for interplanetary exploration, JPL also created other spacecraft to concentrate on Earth exploration. Among these efforts were Seasat in 1978, which examined the oceans with a variety of sensors, and four space shuttle missions, in 1981, 1984, 1994, and 2000, that

The Galileo Jupiter probe undergoes testing before its launch. (Roger Ressmeyer / Corbis)

used radar to image the Earth. JPL also operated the Deep Space Network (DSN) for NASA, which communicated with distant spacecraft, such as the Pioneer and Voyager spacecraft, that have left the solar system. The three main DSN sites are in the Mojave Desert in California; at Tidbinbilla, outside of Canberra, Australia; and near Madrid, Spain. With a budget that has reached over a billion dollars a year, JPL is an example of Big Science. JPL engineers have been pioneers in robotics, microelectronics, image processing, telepresence, and numerous other technologies. JPL has also undertaken contracts with the United States Department of Defense and other federal agencies.

See also Apollo Project; Big Science; European Space Agency; National Aeronautics and Space Administration; Space Exploration and Space Science; Telescopes; Voyager Spacecraft

References

Jet Propulsion Laboratory. http://www.jpl.nasa. gov/ (accessed February 13, 2004).

Koppes, Clayton R. *JPL and the American Space Program: A History of the Jet Propulsion Laboratory.* New Haven, CT: Yale University Press, 1982.

K

Kinsey, Alfred C. (1894–1956)

The zoologist-turned-sociologist Alfred C. Kinsey revolutionized the study of human sexuality in the United States and helped lay the sociological and psychological groundwork for the Sexual Revolution, the sea change in American sexual mores that occurred in the 1960s. Alfred Charles Kinsey was born in Hoboken, New Jersey, where his father taught technical subjects at the Stevens Institute of Technology. The unhappiness of his sickly and miserable childhood was relieved by his love of nature, and he graduated in 1916 with a B.S. in biology from Bowdoin College. In 1920, he earned an Sc.D. degree from Harvard University and accepted a position as an assistant professor of zoology at Indiana University. Kinsey married a graduate student in chemistry at Indiana University in 1921, and they eventually had two sons and two daughters. Kinsey established himself as a prominent entomologist, traveling to Mexico and Central America as part of a long-term study of the gall wasp. His extensive collection of data showed a man who harnessed his obsessive nature to intellectual pursuits. He published articles and books on the gall wasp, and wrote several high school biology textbooks. In 1929 he became a full professor.

In 1938, Kinsey offered to teach a class on marriage, with the intention of surveying students on their sexual behavior. This led to a larger project to create better data on American sexual habits. Researchers in Europe had established the academic discipline of sexology, but American efforts in this area were stymied by the lack of funds and a general sense that sociology did not belong in the bedroom. Kinsey and his associates started with detailed questionnaires about people's sexual practices and moved to personal interviews. In 1940, Kinsey found a steady source of research funding with the National Research Council and the Rockefeller Foundation, and founded the Institute of Sex Research in 1942. After compiling the results from thousands of interviews, Kinsey and his associates published *Sexual Behavior in the Human Male* in 1948. This technical book became an instant best-seller, with the profits being returned to the institute. The book showed that sexual activity varied greatly among men, with a strong correlation to social and economic classes. The book showed that what was considered sexually deviant behavior, such as masturbation and homosexuality, was much more widespread than commonly assumed.

Kinsey's 1953 book, *Sexual Behavior in the Human Female*, was also very popular; it explained that women enjoyed different types of sexual behavior. Contemporary critics condemned Kinsey for attacking the moral

Alfred Kinsey, zoologist-turned-sociologist, 1956 (Library of Congress)

foundations of society and insulting women. Critics feared that if a given sexual practice was shown to be common, then it would be considered morally acceptable. Later critics concentrated on Kinsey's science, complaining that Kinsey's voyeuristic interview techniques were biased and that his method of selecting research subjects severely compromised his statistical results, overstating the incidence of different sexual practices.

Kinsey's interest in sexuality derived from his personal situation and beliefs. He believed strongly that Americans were sexually repressed, and he believed in having as much sexual variety in his own life as possible. A conflicted man, he struggled to reconcile youthful homosexual yearnings with his religious upbringing and later engaged in many types of compulsive sexual behavior, including sadomasochism. Kinsey's two books revolutionized how Americans thought about sex, helping to create an intellectual environment that only needed the invention of an ef-

fective birth control pill to create the Sexual Revolution of the 1960s.

See also Birth Control Pill; Foundations; Masters, William H., and Virginia Johnson; Psychology; Sociology
References
Bullough, Vern L. *Science in the Bedroom: A History of Sex Research.* New York: Basic, 1994.
Jones, James H. *Alfred C. Kinsey: A Public / Private Life.* New York: Norton, 1997.
Robinson, Paul. *The Modernization of Sex: Havelock Ellis, Alfred Kinsey, William Masters and Virginia Johnson.* Ithaca, NY: Cornell University Press, 1989.

Kübler-Ross, Elisabeth (1926–2004)

A controversial psychiatrist, Elisabeth Kübler-Ross changed how medical professionals and laypeople viewed the psychology of the terminally ill. Born in Switzerland, Kübler-Ross was horrified by the savagery of Nazi Germany during World War II. Her father and brother once witnessed Nazi guards machine-gunning Jewish refugees trying to escape across the border from Germany into Switzerland. During the war, she worked as a volunteer in a hospital. After Germany's defeat, she hitchhiked through war-torn Europe, working at odd jobs and at clinics. She decided to become a medical doctor, returned to Switzerland, and graduated in 1957 from the University of Zurich with her medical degree. The following year she married an American, Emanuel (Manny) Robert Ross, whom she met when they were both medical students. They moved to New York City, where Kübler-Ross served as an intern and resident at several area hospitals. She decided to specialize in psychiatry, and in 1962, she moved to Colorado, where she served as a fellow and then an instructor in psychiatry at two Denver hospitals. She bore a son and daughter during this time.

In 1965, she moved to Illinois and became an assistant professor of psychiatry at the University of Chicago Medical School. Her husband worked as a neuropathologist. A teaching seminar that she organized for medical personnel, chaplains, and theology students from the Chicago Theological Seminary led to a series of conversations with terminally ill students. Kübler-Ross and her associates usually interviewed the patients about their feelings while the other seminar participants watched from behind a one-way glass barrier. Kübler-Ross found that other doctors did not want to talk to their terminally ill patients about death, but that the patients desperately wanted to talk about their impending death.

Kübler-Ross's best-selling 1969 book, *On Death and Dying,* presented her conclusions from the seminar. She found that terminally ill patients went through a series of five stages: denial and isolation, anger, bargaining, depression, and then acceptance. Not all patients followed the stages in sequence and some even skipped stages, but the model proved to be a useful psychological theory.

These stages were later extended to apply to the emotional cycle that the family members of terminally ill patients went through, as well as to the emotional cycle of the bereaved after a person they cared for had died.

Critics accused Kübler-Ross of not publishing a research methodology and being subjective, rather than objective, in her approach to interviews and clinical study. Her studies have not been repeated, but they are nevertheless generally accepted. Through *On Death and Dying* and later books, Kübler-Ross gained worldwide fame. She became instrumental in promoting the hospice movement, the creation of places near hospitals to care for the physical and emotional needs of terminally ill patients. She also contributed much to thanatology, the study of death, and championed the Death Awareness movement.

Struck by the stories of some patients who reported near-death experiences, Kübler-Ross came to believe in a joyous life after death. Seeking further information, she turned to mediums and charlatans, and also interviewed thousands of people about their near-death experiences. Never one to shrink in the face of controversy, Kübler-Ross publicly proclaimed her belief in life after death. In later books, she described her own out-of-body experience, her belief in spirit guides, and her eager desire to see what lay after death. She became an icon to seekers in New Age alternative religious movements. Discouraged by this change in his wife's beliefs, Kübler-Ross's husband divorced her after twenty-one years of marriage. Kübler-Ross eventually retired to a retreat in Arizona, but remained active in advocacy and writing until her death in 2004.

Elisabeth Kübler-Ross, psychiatrist (Bettmann/Corbis)

See also Bioethics; Medicine

References
Bennetts, Leslie. "Elisabeth Kübler-Ross's Final Passage." *Vanity Fair,* June 1997, 70–89.
Chaban, Michèle. *The LifeWork of Dr. Elisabeth Kübler-Ross and Its Impact on the Death Awareness Movement.* Lewiston, NY: Mellen, 2000.
Kübler-Ross, Elisabeth. *TheWheel of Life: A Memoir of Living and Dying.* New York: Scribner, 1997.

Kuhn, Thomas S. (1922–1996)

The physicist Thomas S. Kuhn became the most influential historian of science of his time. Thomas Samuel Kuhn was born in Cincinnati, Ohio, where his parents were secular Jews and his father worked as a consulting engineer. His parents moved to Manhattan and then Croton-on-Hudson, a small town near New York City. They placed Kuhn in progressive private schools where the ability to think critically was emphasized. He entered his father's alma mater, Harvard University, in 1940 to study physics. Because of the coming of World War II, his studies were redirected to electronics; he graduated summa cum laude with a bachelor's degree in 1943 and then worked on radar countermeasures in Europe. After the war, he returned to theoretical physics at Harvard and earned his master's degree in 1946. The philosophy of science increasingly attracted his interest, but he did finish his doctorate in physics in 1949. The president of Harvard asked Kuhn to work as an assistant to a history of science course that he taught, preparing a case study on physics from Aristotle to Galileo. Kuhn was not initially drawn to history, but found that he needed to understand history in order to further develop his understanding of the philosophy of science.

On a summer day, while reading the physical theories of the ancient Greek philosopher Aristotle (384–322 B.C.), Kuhn had what he called an epiphany. It was not that Aristotle was wrong because he had not understood the principles that Isaac Newton (1642–1727) would develop two millennia later, but that Aristotle's science made sense in light of the information the Greek philosopher had and the worldview of Greek philosophy. Suddenly he realized how differently Aristotle had viewed reality and why his physics made sense. He later acknowledged that this ability to see something from another person's intellectual point of view was a common trait in intellectual historians, but not a trait that he had acquired as a physicist.

Kuhn taught at Harvard from 1951 to 1956, then moved to the University of California at Berkeley, where he held a joint appointment in the history and philosophy departments. He published his first book, *The Copernican Revolution: Planetary Astronomy in the Development of Western Thought,* in 1957. In 1962, he published his seminal work, *The Structure of Scientific Revolutions.* In the book, Kuhn adopted the term *paradigm* to describe the joint worldview, shared practices, and accepted theories of a group of scientists. He used Ptolemaic geocentric astronomy and Copernican heliocentric astronomy as examples. Kuhn argued that scientists usually practiced "normal science," which meant that they accepted the current paradigm in their field and that their experiments either confirmed the paradigm or resulted in minor changes. As time went by, anomalies occurred in observations and experiments that could not be explained by the current paradigm. When enough of these anomalies accumulated, the old paradigm was ripe for revolution. A new paradigm emerged that explained the anomalies and incorporated the old paradigm inside it.

The Structure of Scientific Revolutions was phenomenally successful, selling 750,000 copies in the next three decades and being translated into more than forty languages. Kuhn's work was very influential in the practice of the history of science, and widely applied to other academic areas, influencing sociology, economics, political science, and anthropology. The word *paradigm* even became a buzzword among business consultants.

Kuhn was later criticized for having up to twenty-one different meanings for the word *paradigm* in his book. Many scholars in various disciplines interpreted Kuhn's work as demonstrating that the practice of science was irrational, subjective rather than objective, and that perhaps science was really just a form of rhetoric. The later rise of social constructionism drew on Kuhn, though he disagreed with their epistemological relativism. Socially awkward his entire life, as well as annoyed at how his work was misinterpreted, Kuhn tended to be surprised at expressions of respect from his colleagues.

In 1961, Kuhn began a project to collect an archive of primary materials in twentieth-century physics, including collecting taped interviews with prominent figures. This important effort eventually led to a major study, *Black-Body Radiation and the Quantum Discontinuity* (1978), in which he concluded that Max Planck (1858–1947) did not discover quantum theory in 1900. Traditional accounts had credited Planck with this discovery and argued that the physicist had just not been very good at expressing himself in writing. Kuhn took those writings at their word and decided that Planck was too confused about what he found to be credited with the discovery.

In 1964, Kuhn moved to Princeton University. In 1979, he moved to the Massachusetts Institute of Technology to become the Laurance S. Rockefeller Professor of Philosophy; he retired in 1991. Kuhn married in 1948, having three children. The marriage dissolved in divorce in 1978, and he remarried in 1982. When he died in 1996, he had completed two-thirds of an expansion on his ideas entitled "The Plurality of Worlds: An Evolutionary Theory of Scientific Discovery."

See also History of Science; Popper, Karl; Social Constructionism

References
Fullmer, June Z. "Thomas S. Kuhn (1922–1996)." *Technology and Culture* 39, no. 2 (1998): 372–377.
Heilbron, J. L. "Thomas Samuel Kuhn, 18 July 1922–17 June 1996." *Isis* 89, no. 3 (September 1998): 505–515.

Kuno, Hisashi (1910–1969)

The petrologist Hisashi Kuno became a world authority on the formation of magma and basalts. Kuno was born in Tokyo, Japan, where his father was an artist. He enjoyed mountain climbing and skiing, and when he entered Tokyo Imperial University in 1929, he gravitated to geology. The 1923 Tokyo earthquake had killed tens of thousands of people and focused national attention on the geological sciences. Kuno began a much admired study of Hakone Volcano in 1931. As part of his study, he became unusually skilled in the use of the petrographic microscope. He made important contributions to understanding the process by which pyroxenes are formed, which are crystalline mineral silicates found in metamorphic and igneous rocks.

Kuno became an assistant professor of petrology at the university, but in 1941, with the coming of World War II, he was drafted into the Japanese Army and sent to Manchuria. Captured by Soviet troops at the end of the war, he spent a year in captivity before returning home in September 1946. He resumed his scientific work and finished his study of the Hakone Volcano as his dissertation in 1948, obtaining his doctorate of science degree. The publication of parts of his dissertation in English brought him to the attention of Harry Hess (1906–1969) at Princeton University, and in 1951 Kuno was invited to spend a year studying with Hess in America. Even with an international reputation, Kuno remained an active fieldworker. His studies of Hawaii made important contributions to understanding how different types of magma were formed.

In 1954, Kuno received the prestigious prize of the Japan Academy, and in 1955, he became a professor of petrology at the University of Tokyo (the new name of Tokyo Imperial University). Kuno mentored many students who became important geologists. One of his techniques was to hold special seminars in English and have his students make presentations in English, so that his students could acquire the skills to participate in the larger global community of geologists. He also became active in organizing international conferences and building up scientific societies in Japan and internationally. Kuno was chosen as a principal investigator to study the rocks being brought back from the Moon by *Apollo 11,* though he died of stomach cancer before receiving the rocks. Kuno was married and was survived by a son and a daughter.

> *See also* Earthquakes; Geology; Hess, Harry; Volcanoes

References
Foster, Helen L. "Memorial to Hisashi Kuno." *The Geological Society of America* 1 (1973): 27–37.
Yagi, Kenzo. "Memorial of Hisashi Kuno." *The American Mineralogist* 55 (March–April 1970): 573–583.

L

Lasers

Invented in the 1950s, lasers are indispensable tools in scientific investigations and high-technology communications equipment. The laser began with Charles H. Townes (1915–), born in Greenville, South Carolina, who received his Ph.D. in physics from the California Institute of Technology in 1939. Townes worked at Bell Telephone Laboratories from 1933 to 1947; he joined the faculty of Columbia University in 1948, where he pursued his interest in applying the technology developed during World War II for microwave radar to using microwaves in spectroscopy. In 1951, Townes conceived the idea of the maser (microwave amplification by stimulated emission of radiation), a machine that stimulated the atoms in ammonia gas to emit coherent microwave radiation. The famous German-born Albert Einstein (1879–1955) recognized in 1916 that stimulated emission was possible. In 1954, Townes and his graduate students built the first functional maser.

The second individual in the story of the laser, Arthur L. Schawlow (1921–1999), grew up in Toronto, Ontario, after being born in Mount Vernon, New York. Schawlow attended the University of Toronto, learning optical spectroscopy as part of graduate studies that led to a Ph.D. in 1949. A Carbide and Carbon Chemicals postdoctoral fellowship brought him to Columbia University to study with Townes. Townes introduced Schawlow to his younger sister, and the couple married. In 1951, Schawlow moved to Bell Telephone Laboratories, but he continued to collaborate with Townes on weekends. They wrote a book, *Microwave Spectroscopy,* in 1955, and in 1958 published an article describing how to create an optical maser, or laser (light amplification by stimulated emission of radiation). In 1960, drawing on the work of Townes and Schawlow, the physicist Theodore H. Maiman (1927–), working at Hughes Research Laboratories in California, used a rod of synthetic ruby as the medium to build the first laser. Maiman received the 1987 Japan Prize for his laser work.

A physics graduate student at Columbia University, Gordon Gould (1920–), also thought of the laser in 1957. He kept a notebook filled with dates and his emerging ideas, intending to patent them, though he assumed that he had to make the idea practicable before filing for a patent. He filed for a patent in 1959, months after Townes and Schawlow had filed for their own patents. Gould lost his patent case before the United States Court of Customs and Patent Appeals in 1965, but the issue was revived in 1977, when Gould received a more narrow patent on the optical pumping of lasers; the patent was finally

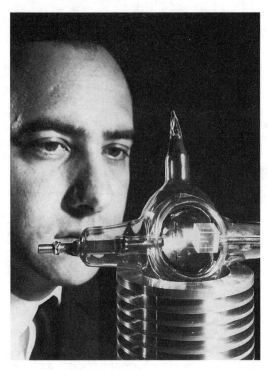

Theodore H. Maiman studies a ruby crystal in the shape of a cube in the first laser. (Bettmann/Corbis)

awarded some eighteen years after he applied. The conflict between Gould and the Townes-Schawlow patents over invention priority is sometimes referred to as the laser wars.

Townes moved to the Massachusetts Institute of Technology in 1961, and moved on to the University of California at Berkeley in 1967, where he retired in 1986. Townes received half of the 1964 Nobel Prize in Physics, with Aleksandr M. Prokhorov (1916–2002) and Nikolay G. Basov (1922–2001), both of the Lebedev Institute for Physics, Akademija Nauk, in Moscow, sharing the other half. The two Soviets had independently developed a maser-laser.

Schawlow became a professor of physics at Stanford University in 1961, where he continued to help develop the technology of lasers while applying lasers to analyzing atoms and molecules through laser spectroscopy. This technique allowed scientists to detect trace elements in larger samples be-

cause the lasers were so precise. Schawlow shared half of the 1981 Nobel Prize in Physics with the Dutch-born Nicolaas Bloembergen (1920–) for their contribution to the development of laser spectroscopy. The other half of the 1981 prize went to the Swedish physicist Kai M. Siegbahn (1918–) for his contribution to the development of high-resolution electron spectroscopy. Schawlow did not receive the prize for his contribution to the invention of the laser but for laser spectroscopy. Schawlow remained at Stanford until his retirement in 1991.

The United States military poured money into laser research during the 1960s, and though initial dreams of laser weapons were not realized, laser range finders revolutionized tank warfare in the 1970s. The considerable effort put into developing laser or laser-like weapons found limited success because of the enormous power requirements. Lasers did lead to the development of optical fibers, which came to be the transmission mechanism of choice for efficient, no-heat, high-capacity data communications. The Apollo astronauts left a laser reflector on the Moon so that a laser could precisely measure the distance from the Earth to the Moon. Lasers have also been applied to laser eye surgery, microsurgery, optical scanners, compact discs for data as well as audio and visual media, holography, welding, and precise measurements of Earth movement to monitor volcanoes and earthquakes.

See also Apollo Project; Medicine; Physics

References

Harbison, James P., and Robert E. Nahory. *Lasers: Harnessing the Atom's Light.* New York: Scientific American Library, 1998.

Hecht, Jeff. *Laser Pioneers.* Revised edition. San Diego: Academic, 1992.

———. "Winning the Laser-Patent War." *Laser Focus World,* December 1994, 49–51.

Taylor, Nick. *Laser: The Inventor, The Nobel Laureate, and the Thirty-Year Patent War.* New York: Simon and Schuster, 2000.

Townes, Charles H. *How the Laser Happened: Adventures of a Scientist.* New York: Oxford University Press, 1999.

Leakey Family

Beginning with Louis S. B. Leakey (1903–1972), the Leakey family and their associates have profoundly changed contemporary views of humanity's evolutionary ancestors through their paleoanthropological discoveries and their controversial theories. Louis Seymour Bazett Leakey was born in Kabete, Kenya, where his parents were Christian missionaries to the Kikuyu people. He grew up with the Kikuyu; learned their language, folklore, and culture, as well as how to stalk game; and was initiated into their tribe at the age of thirteen. His 1966 autobiography was entitled *White African*. Leakey's parents tutored him until they sent him to England at the age of sixteen, where he attended public school before entering Cambridge University in 1922. A head injury received while playing rugby led to a leave of absence from the school. He joined an archaeological expedition sponsored by the British Museum to what is now the country of Tanzania, where he learned field techniques. This sparked his lifelong preoccupation with the archaeology and anthropology of Africa.

Leakey returned to England to earn undergraduate degrees in archaeology and anthropology from Cambridge in 1926 and his doctorate in 1929. He began to lead his own expeditions to Africa while holding various minor academic appointments at Cambridge. Evolutionists at that time believed that early humans had evolved into modern humans in Asia, as evidenced by the discovery of Peking Man and Java Man. Leakey argued that Africa was the birthplace of humanity and intended to prove it.

Leakey married H. Wilfrida Avern in 1928, and they had a son and daughter. In 1933 he met Mary Douglas Nicol (1913–1996), an amateur archaeologist with the ability to draw, and she agreed to draw the illustrations for his book *Adam's Ancestors* (1934). Louis and Mary began an affair and, after his divorce, married in 1936.

Mary was born in London as the only child of a landscape painter and his wife. The family moved frequently, and she received only erratic schooling. After her father's death when she was thirteen years old, Mary was sent to convent schools. Her mother gave up on schooling after Mary was twice expelled. Mary was fortunate to have met archaeologists in France and England through her father and was intrigued by excavations and cave paintings. From 1930 to 1934, though quite young, Mary served as an assistant to Dorothy Liddell, one of the few women archaeologists then working, on a Neolithic site at Hembury, England. Her illustration skills opened opportunities for her, including her meeting with Louis.

The couple moved to Africa, where Louis studied the Kikuyu for two years for the private Rhodes Trust. Mary proved to be an enthusiastic and skilled archaeologist, continuing excavations even during World War II, while Louis worked for British Intelligence. The couple had three sons and a daughter, beginning in 1940, though the daughter died as an infant.

After the war, Louis became the full-time curator of the Coryndon Museum in Nairobi, Kenya. In 1948, Mary discovered on an island in Lake Victoria the partial skull and teeth of a new primate species named *Proconsul africanus,* which lived at least 20 million years ago. Though the couple excavated in many places in Kenya, working on more recent modern human sites, they often searched for much older material. Olduvai Gorge, a 35-mile-long valley cut by a stream into the sediments of an ancient lake bed, is located in northern Tanzania. Louis always thought that Olduvai Gorge held considerable promise and began fieldwork there in 1931. Mary continued this work with him, taking advantage of the geological strata exposed along the Olduvai Gorge walls.

In 1959, Mary found hundreds of fragments of a hominid skull in Olduvai Gorge; these fragments she eventually reconstructed. Louis called the hominid thus discovered *Zinjanthropus bosei* and pointed out that it was found in strata dated to 1.75

Louis S. B. Leakey, anthropologist (Bettmann / Corbis)

an important indicator of how ancestors to modern humans might have evolved tool use. Louis also recruited the Canadian Biruté Galdikas (1946–) to study orangutans in Borneo at Tanjung Puting, but she had less success because of the solitary nature of that animal.

Louis maintained a vigorous schedule traveling around the world, raising funds, and promoting his scientific views to professional and lay audiences. Mary remained in Africa, concentrating on fieldwork, and the couple drifted apart. Louis died in London in 1972. Louis published regularly throughout his life, including over a dozen popular and professional books, and received honorary degrees and other awards for his work, including the Founder's Medal of the Royal Geographic Society in 1964.

Mary continued their work, and in 1978, at Laetoli, she found the fossilized footprints of three hominids, which were dated at 3.6 million years old. The three are thought to be the footprints of a man, a woman or smaller man, and a child. The gait of the footprints clearly showed that these hominids walked upright, much sooner than anyone had suspected. Mary remained active as she grew older, taking up the fund-raising and public promotion roles that her husband had played. She died in Nairobi, Kenya, in 1996. Despite her lack of formal education, Mary received numerous honorary degrees and awards, and wrote several books. Her oldest son became a herpetologist, and her youngest son became a Kenyan politician.

Mary and Louis's middle son, Richard Erskine Frere Leakey (1944–), was born in Nairobi. He did not want to follow his parents' career, and while still a teenager he started his own safari business. On a 1967 flight, he spied sedimentary rock out the airplane window and thought that it would be a good place to find fossils. Despite little formal schooling and no university education, Richard decided to apply to be director of the National Museums of Kenya. Though only twenty-three years old, he was a Kenyan citi-

million years ago. Louis enthusiastically promoted the significance of this find, and the National Geographic Society became a major sponsor of the Leakeys' research in 1960, allowing Louis to resign his museum position the following year. The skull was thought for a time to be a direct ancestor of modern humans, but further discoveries led even Louis to agree that it was a separate line of hominids, and the skull was reclassified as *Australopithecus boisei*.

Louis thought that the study of the great apes might lead to insights into the behavior of humanity's own ancestors. He encouraged three women to undertake this task, believing that women had the patience to do what male scientists had earlier failed to do. He selected women who were unschooled in fieldwork and found funding for Jane Goodall (1934–) to study the chimpanzees and Dian Fossey (1932–1985) to study the mountain gorillas. Both of them succeeded in making the animals accept them so that they could study their behavior in the wild. Goodall found that chimpanzees used primitive tools, zen and he was selected for the post. Richard thought of himself as a naturalist rather than an anthropologist or archaeologist. Funded by the National Geographic Society, Richard led an expedition to the shores of Lake Turkana to excavate the rocks that he had seen from the air. In the following years, his team excavated fossils from perhaps 200 individual hominids, including the 1.6-million-year-old Turkana Boy, an almost complete skeleton of *Homo erectus*. The discoveries at Lake Turkana showed that at least three different species of hominids lived together in close proximity millions of years ago, though controversy remains over which line of hominids led to modern humans.

Richard developed the National Museums into a large and successful organization before he resigned in 1989 to become director of the Kenya Wildlife Service (KWS). He won international acclaim for his vigorous efforts to root out corruption and his struggle to stem the losses of African wildlife to poachers. With others, he championed an international ban on ivory sales, which reduced the ongoing slaughter of elephants for their tusks. Richard lost the lower parts of both legs in a 1993 airplane crash. He resigned from the KWS in 1994. A year later he helped found a political party and he remains active in politics and conservation. Richard's second marriage, to Mauve Epps (1942–), a zoologist student of his father's, produced three daughters. The middle daughter, Louise N. Leakey (1972–), earned her Ph.D. in paleontology from the University of London and has continued in the family tradition by working with her mother at the Lake Turkana site.

See also Anthropology; Evolution; Fossey, Dian; Foundations; Goodall, Jane; National Geographic Society; Zoology

References

Cole, Sonia. *Leakey's Luck: The Life of Louis Seymour Bazett Leakey, 1903–1972*. London: Collins, 1975.
The Leakey Foundation. http://www.leakeyfoundation.org (accessed February 13, 2004).
Leakey, Mary. *Disclosing the Past*. New York: Doubleday, 1984.
Leakey, Richard E., and Roger Lewin. *Origins Reconsidered: In Search of What Makes Us Human*. New York: Doubleday, 1992.
Leakey, Richard E., and Virginia Morrell. *Wildlife Wars: My Fight to Save Africa's Natural Treasures*. St. Martin's, 2001.
Morrell, Virginia. *Ancestral Passions: The Leakey Family and the Quest for Humankind's Beginnings*. Simon and Schuster, 1995.

Levi-Montalcini, Rita (1909–)

The neuroembryologist Rita Levi-Montalcini and her colleague, Stanley Cohen (1922–), discovered nerve-growth factor (NGF) in 1954, the agent that activates nerve growth. Levi-Montalcini was born to an upper-class secular Jewish family in Turin, Italy, where her father managed a factory. Her last name is a combination of her father's and mother's last names. Her dogmatic father insisted on controlling the household, and Levi-Montalcini remembers herself as a submissive child on the surface with resentment and strength inside. She excelled in school, though few educational opportunities were available to girls. When a former governess contracted cancer and died, Levi-Montalcini resolved that she wanted to become a doctor. She realized that she did not want to be a wife and did not feel strong maternal instincts toward children. Her father reluctantly agreed when she told him that she wanted to attend medical school. Levi-Montalcini and a cousin hired tutors in order to engage in remedial studies to prepare her in the sciences.

In 1930, Levi-Montalcini entered the Turin School of Medicine. A professor of human anatomy and histology, Giuseppe Levi (1872–1965), became her mentor, and she began to work in his laboratory. After she graduated in 1936, she remained as an assistant to Levi. When the fascist government of Italy decreed that all Jews be fired from their university jobs and forbidden to practice medicine, she was dismissed. Levi and Levi-Montalcini moved to Belgium for a time to continue their neurological research there. When Levi-Montalcini returned to Turin at the end of 1939, she secretly practiced medicine for a time before setting up a rudimentary laboratory in her bedroom to continue her research on nerves using chicken embryos. Levi later returned from Belgium and worked with her.

When the Italian government surrendered to the Allies in 1943, the German army occupied the country. Levi-Montalcini and her family tried to flee, but found themselves trapped in Florence. They hid from the Nazis until Allied troops liberated the city in September 1944. Levi-Montalcini worked as a doctor for the Allied health service until the war ended, then she returned to working with Levi as a research assistant at the University of Turin. An article by Levi and Levi-Montalcini on their wartime research was noticed by the prominent neurological researcher Viktor Hamburger (1900–2001) of the zoology department at Washington University in St. Louis, Missouri. He invited her to visit, and she arrived in 1946, remaining at Washington University at first as a research associate, before becoming a member of the zoology faculty in 1951.

Levi-Montalcini experienced an epiphany while peering through her microscope in 1947. She realized that many nerve cells die as nerve tissue develops and that there must be a signal of some sort to keep nerve cells developing. A student of Hamburger's found in 1950 that grafting a mouse tumor onto a chick embryo would excite nerve cell growth. Levi-Montalcini replicated the experiment and concluded that some type of growth factor must be produced by the tumor. In 1953, Levi-Montalcini traveled to Rio de Janeiro to work with a friend from medical school. In Brazil, she succeeded in getting nerve cells to grow in vitro in a petri dish by dropping a piece of tumor into the dish, causing the nerve cells to blossom quickly, creating a halo of new nerve cell fibers.

Returning to St. Louis, she found that Hamburger had brought in the biochemist Stanley Cohen to work with her. Cohen was able to purify the NGF, finding the salivary glands of male mice particularly rich in the protein. For six years the productive team expanded on the implications of NGF before Cohen moved to another university in 1959. Eventually it became apparent that NGF is what keeps young nerve cells from dying and allows them to mature. Other researchers eventually isolated the NGF protein molecule, then sequenced its amino acids in 1971. In 1983, researchers learned how to synthetically manufacture human NGF.

Levi-Montalcini became an American citizen in 1956 while retaining her Italian citizenship. In 1968, the National Academy of Sciences elected her as a member, only the tenth woman so honored up until then. In 1969, she began to divide her time between Washington University and Italy, where she became director of the Laboratory of Cell Biology at the Council of National Research in 1969. She retired in 1979. Levi-Montalcini and Cohen shared the 1986 Nobel Prize in Physiology or Medicine for their discovery of NGF.

See also Medicine; Neuroscience; Nobel Prizes
References
Levi-Montalcini, Rita. *In Praise of Imperfection: My Life and Work.* New York: Basic, 1988.
McGrayne, Sharon Bertsch. *Nobel Prize Women in Science.* Second edition. Secaucus, NJ: Carol, 1998.

Lévi-Strauss, Claude (1908–)

The theoretical anthropologist Claude Lévi-Strauss developed structuralism, a systematic way of understanding culture. Lévi-Strauss was born in Brussels, Belgium, where his father worked as an academic painter. He earned degrees in law and philosophy at the University of Paris in 1929 and 1931, then taught philosophy in secondary schools for three years, while profiting from association with the intellectuals who orbited the existentialist philoso-

*Claude Lévi-Strauss, theoretical anthropologist, 1981
(Sophie Bassouls / Corbis Sygma)*

pher Jean-Paul Sartre (1905–1980). In 1934, Lévi-Strauss became a professor of sociology at the University of São Paulo in Brazil. In his five years in Brazil, he studied the native Indians of the interior of the country, having developed a fascination with anthropology.

With the coming of World War II, Lévi-Strauss served in the French army as a liaison officer to their British allies. After France fell to the Nazis, he escaped to the United States because of the danger from being a Jew. He obtained a visiting professorship at the New School for Social Research in New York City for the duration of the war. While there, he met the linguist Roman Jakobson (1896–1982) and adopted Jakobson's ideas on structural linguistics, a structured way of viewing language. Lévi-Strauss returned to Paris, earned a doctorate in letters from the Sor-

bonne in 1948, and began to publish in anthropology. He became the director of a laboratory of social anthropology at the University of Paris in 1950. Though he did occasional fieldwork, his reputation came from his theoretical work. His *Elementary Structures of Kinship,* published in 1949, and his *Structural Anthropology,* published in 1961, helped establish him as a major thinker in French society.

In these books and others, all written in challenging and complex prose, Lévi-Strauss tried to reduce human behavior and culture to its essentials and to illustrate the structured relationships between those essentials. His greatest success came from describing the incest taboo and kinship systems as together constituting a foundation for culture. In religion, he argued that myths are attempts to resolve contradictions in human behavior, rather than to create justifications for human behavior, as others had previously argued. Lévi-Strauss sought to explain human behavior and culture in a way uninfluenced by the historical flow of time, looking for truths that applied to all people at all times. His structuralism was not only applied to anthropology, but was just as influential in philosophy, literary criticism, and the study of religions.

In 1959, Lévi-Strauss was appointed to the chair of social anthropology at the College of France. Beginning in the mid-1960s, translation of his books into English spread the influence of structuralism, making structuralism an important school of thought in global anthropology. Lévi-Strauss married three times, fathering one child in each of the second and third marriages.

See also Anthropology
References
Clarke, Simon. "The Origins of Levi-Strauss's Structuralism." *Journal of the British Sociological Association* 12 (1978): 405–439.
Hénaff, Marcel. Translated by Mary Baker. *Claude Lévi-Strauss and the Making of Structural Anthropology.* Minneapolis: University of Minnesota Press, 1998.
Pace, David. *Claude Lévi-Strauss: The Bearer of Ashes.* New York: Routledge, 1983.

Libby, Willard F. (1908–1980)

The chemist Willard F. Libby developed the process of carbon-14 dating, which is used extensively in archaeology. Willard Frank Libby was born in Grand Valley, Colorado, to a poorly educated farmer. The family moved to an apple ranch in California, where his parents encouraged him to go to college. In 1926, Libby enrolled in the University of California at Berkeley, where he intended to become a mining engineer, but found chemistry more interesting. He earned his bachelor's in 1931 and his doctorate in 1933, and he specialized in detecting extremely low amounts of radioactivity.

Libby remained at Berkeley, first as an instructor before moving up to associate professor. In 1941, while spending a year as a Guggenheim Fellow at Princeton University, Libby was recruited into the Manhattan Project. He spent World War II at Columbia University working on ways to separate uranium isotopes. After the war, he became a professor of chemistry at the University of Chicago, where he conducted research at the Institute for Nuclear Studies.

At Chicago, Libby hypothesized that a rare isotope of carbon called carbon-14 was absorbed by plants through photosynthesis. After a plant died, no more carbon-14 would be absorbed, and the plant remains would gradually lose the carbon-14 that it had accumulated through radioactive decay. By measuring the amount of carbon-14 in the dead plant material, one could determine when it died. Another scientist had already determined in 1940 that the half-life of carbon-14, which is mildly radioactive, is 5,730 years. Libby developed a technique to convert a sample into gas (thereby destroying the sample) and measure the carbon-14 content.

Libby tested his technique on tree samples, where tree rings gave an independent measure of the age of the sample. He also tested historical artifacts with known ages. By testing samples from different latitudes, he found that his technique applied to all locations. This process revolutionized the practice of archaeology. Now archaeologists only need something organic—wood, cloth, plant remains, or charcoal from a fire–to positively date their finds. The technique becomes more inaccurate the further back one tries to measure, but measurements for the first ten thousand years are quite accurate.

Libby served as a member of the United States Atomic Energy Commission (AEC) from 1954 to 1959, before returning to academe at the University of California at Los Angeles. Libby received the 1960 Nobel Prize in Chemistry for his carbon-14 work. A student of Libby's at Chicago, F. Sherwood Rowland (1927–), later shared the 1995 Nobel Prize in Chemistry for his work on chlorofluorocarbon gases. Libby remained active in research, directing the Institute of Geophysics and Planetary Physics from 1962 until his retirement in 1976. Libby married in 1940, fathering twin daughters. The marriage dissolved in divorce in 1966, and he later remarried.

See also Archaeology; Chemistry; Nobel Prizes; Nuclear Physics; Rowland, F. Sherwood

References
Burleigh, Richard. "W. F. Libby and the Development of Radiocarbon Dating." *Antiquity* 55 (1981): 96–98.
Nobel *e*-Museum. "Barbara McClintock—Autobiography." http://nobelprize.org/medicine/ (accessed February 13, 2004).
———. "Willard F. Libby—Biography." http://nobelprize.org/chemistry/ (accessed February 13, 2004).

Lorenz, Edward N. (1917–)

The meteorologist Edward N. Lorenz applied computers to modeling the atmosphere and made important contributions to meteorology and chaos theory. Edward Norton Lorenz was born in West Hartford, Connecticut. He earned an A.B. degree in mathematics at Dartmouth College in 1938 and an A.M. in mathematics at Harvard University in 1940. During World War II, Lorenz worked for the Army Air Corps as a meteorologist. Lorenz earned an S.M. in meteorology at the Massachusetts Institute of Technology (MIT) in 1943, and an Sc.D. in meteorology from MIT in 1948. He began his academic career on the staff of MIT and joined the faculty of MIT in 1955.

In 1963, Lorenz published an article in the *Journal of Atmospheric Sciences* entitled "Deterministic Nonperiodic Flow," a seminal article in the development of chaos theory. He later introduced the term, the *butterfly effect,* to describe how a small change in initial conditions can have a dramatic effect on later events. Even the simple mathematical models of the atmosphere of that time illustrated the butterfly effect. This idea later became one of the foundations of chaos theory. Because of the basic nature of the complex systems described by chaos theorists, even the best knowledge of current weather conditions limits the ability of meteorologists to predict weather too far into the future. Lorenz used computer modeling to illustrate his ideas and developed a set of well-known equations used in chaos theory. The title of a famous 1972 paper, "Predictability: Does the Flap of a Butterfly's Wings in Brazil Set off a Tornado in Texas?", succinctly summarized the relationship of chaos theory and meteorology, as well as explaining the origin of the term *butterfly effect.*

Lorenz retired from MIT in 1987, but remained active in research. Among his many honors, Lorenz received the Crafoord Prize in 1983 and the Kyoto Prize in 1991. Lorenz married in 1948 and had three children.

See also Chaos Theory; Computers; Meteorology
References
Lorenz, Edward N. "Deterministic Nonperiodic Flow." *Journal of Atmospheric Sciences* 20 (March 1963): 130–141.
————. *The Essence of Chaos.* Seattle: University of Washington Press, 1993.

Lorenz, Konrad (1903–1989)

The zoologist Konrad Lorenz founded modern ethology and later applied what he learned from the study of animal behavior to controversial ideas about human behavior.

Konrad Lorenz was born in Vienna, Austria, the son of an orthopedic surgeon. His self-made father owned an estate with a large garden and marshes, and the young boy rapidly showed an obsession with nature and animals. As a child, he experienced a duck imprinting on him, a process in which newborn animals can mistakenly adopt a person as their surrogate parent and learn their behavior from that person. His father wanted him to study medicine, so Lorenz attended Columbia University in New York City for a short time before returning to Vienna. At the University of Vienna he studied comparative anatomy and zoology, earning his medical degree in 1928 and a Ph.D. in zoology in 1933. He went on to hold various academic positions in anatomy and ethology.

In the late 1920s and 1930s, Lorenz became active in researching the behavior of colonies of birds that he raised. He published ethological articles and became best known for developing his theory of imprinting. He argued that instincts were innate but did not manifest themselves until a threshold of stimuli was crossed. As an ethologist, Lorenz emphasized biological determinism in the form of instinctual drives, as opposed to the behavioral and environmental theories espoused by American psychologists, such as James D. Watson (1878–1958) and B. F. Skinner (1904–1990).

In 1940, after the Nazis had absorbed Austria into Germany, Lorenz wrote an article that heavily used Nazi terms in describing behavior. He regretted this article after the true magnitude of Nazi crimes later became apparent. In 1941, he was drafted into the German army as a doctor, though he had never practiced medicine. A year later he was captured by Soviet troops and spent the next six years as a camp doctor to fellow prisoners of war. He wrote a manuscript on epistemology, using cement sacks for paper and potassium permanganate as ink. The Soviets allowed him to take the manuscript with him when he was repatriated with other Austrians in 1948, and he later published it as a book.

Lorenz's friends found the liberated scientist an academic position, and he resumed his research. Up until this point in his life, human behavior and human culture had held little interest for him, but he began to apply theories derived from animal behavior to human behavior and became known for books such as *King Solomon's Ring* (1952) and *On Aggression* (1966). In the latter book he argued that aggression in animals served useful purposes in protecting territory and preserving the species. He argued that aggression in humans is instinctual but that the development of artificial weapons has made this drive much more deadly. Most animal species have behavioral mechanisms for limiting the effects of aggression, allowing the loser to flee a fight or submit to the winner. Lorenz thought that humans lacked many of these behavior limitations. Some critics felt that Lorenz was justifying war, though he denied this.

In 1973, Lorenz shared the Nobel Prize in Physiology or Medicine with two fellow ethologists, Karl von Frisch (1886–1982) and Nikolaas Tinbergen (1907–1988). Lorenz also retired in 1973 and returned to Austria, but remained active in research and writing. Lorenz married a childhood friend and medical doctor in 1927, who practiced as a gynecologist, and they had two daughters and a son. He died in Altenburg, Austria, in 1989.

See also Nobel Prizes; Psychology; Skinner, B. F.; Zoology

References

Evans, Richard I., editor. *Konrad Lorenz: The Man and His Ideas.* New York: Harcourt Brace, 1975.

"Konrad Lorenz—Autobiography." *Nobelprize.org.* http://nobelprize.org/medicine/ (accessed February 13, 2004).

Nisbett, Alec. *Konrad Lorenz: A Biography.* New York: Harcourt Brace, 1977.

Lovelock, James (1919–)

The biophysicist and inventor James Lovelock gained fame (and in some quarters notoriety) for his "Gaia hypothesis": the hypothesis that proposed that the Earth was like a living organism, maintaining surface and atmospheric conditions so that the chemistry and temperature would sustain life. James Ephraim Lovelock was born in Letchworth, Hertfordshire, England, where his father was an art dealer and his mother worked as a local government official. He loved nature, science, and technology as a child, and his father encouraged this interest by supplying him with material to build things. Lovelock disliked school, preferring to learn on his own, but chose to go to college. He attended the Birkbeck College at the University of London for a year before transferring to the University of Manchester in 1939. He earned a B.Sc. in chemistry in 1941 and went to work at the National Institute for Medical Research (NIMR) in London. He became known for his ability to invent and construct scientific instruments. He also earned a Ph.D. in medicine in 1949 and a D.Sc. in biophysics in 1959, both from University of London.

In the early 1950s, Lovelock developed a machine to successfully freeze and thaw live hamsters. In 1957, he invented a small electron capture detector. In the late 1960s, Lovelock combined his electron capture detector with a gas chromatograph, which Lovelock and other scientists used to demonstrate the levels of pesticide pollution in the environment. Lovelock also detected the high levels of chlorofluorocarbons that prompted the discovery by the chemists Mario Molina (1943–) and F. Sherwood Rowland (1927–) that those man-made chemicals were eroding the ozone layer. Lovelock originally disagreed with their conclusions, thinking that Molina and Rowland were unduly alarmist, but later changed his mind and supported the efforts to ban the chemicals.

In 1961, the National Aeronautics and Space Administration (NASA) invited Lovelock to come to the United States to develop scientific instruments for spacecraft. Lovelock took the opportunity to become an independent scientist, supporting himself from revenue from his inventions and consulting contracts. He also held various temporary academic positions while converting his country home in Bowerchalke into a scientific laboratory. He later moved his family and laboratory to another country home at Saint Giles on the Heath.

In 1965, while a consultant at NASA's Jet Propulsion Laboratory, Lovelock learned from a conversation with the astronomer Carl Sagan (1934–1996) that the sun had been a quarter less luminous in the past. How had Earth managed to sustain life? This prompted Lovelock to develop the theory that Earth was like a living organism, maintaining surface and atmospheric conditions so that the chemistry and temperature would sustain life. In 1967, Lovelock presented this idea to other NASA scientists and engineers, who were familiar with the notion of feedback systems and found his idea plausible. Other scientists were not so enthusiastic.

A neighbor of Lovelock's, the novelist William Golding, suggested that he name the hypothesis Gaia after the Greek goddess of the Earth. In 1970, Lovelock began to work with the microbiologist Lynn Margulis (1938–) on the idea. She found the idea compatible with her own promotion of symbiosis and vigorously championed his idea. Many biologists were suspicious of such a holistic notion. The metaphor of Gaia made Earth seem alive. Activists in the environmental movement and advocates of New Age spirituality eagerly adopted the notion of Gaia, further alienating scientists who were uncomfortable with the idea's mystical overtones. Lovelock has remained active as a scientist and speaking out on issues that concern him; for instance, he advocates the use of clean nuclear power.

In 1942, Lovelock married and became the father of two sons and two daughters. After the death of his first wife, he remarried in 1991. Among his many awards and honorary degrees are citations from NASA for his contributions to space exploration and being elected as a Fellow of the Royal Society.

See also Ecology; Environmental Movement; Jet Propulsion Laboratory; Margulis, Lynn; Molina, Mario; National Aeronautics and Space Administration; Ozone Layer and Chlorofluorocarbons; Rowland, F. Sherwood; Sagan, Carl

References
Bingham, Roger. "James Lovelock." Pp. 159–166 in *A Passion to Know: 20 Profiles in Science.* Edited by Allen L. Hammond. New York: Scribner's, 1984.
Lovelock, James. *Gaia: A New Look at Life on Earth.* Oxford: Oxford University Press, 1979.
———. *Homage to Gaia: The Life of an Independent Scientist.* Oxford: Oxford University Press, 2000.

Lysenko, Trofim (1898–1976)

The biologist Trofim Lysenko promoted a theory of non-Mendelian genetics that found favor in the Stalinist Soviet Union and distorted the science of biology in that country for decades. Trofim Denisovich Lysenko was born to peasants in Karlovka, Ukraine, in the old Russian Empire. He earned a doctorate of agricultural science in 1925 from the Kiev Agricultural Institute, then worked at the Gyandzha Experimental Station in the Caucasus until 1929, when he moved to the Ukrainian All-Union Institute of Selection and Genetics in Odessa.

A well-known figure in Soviet agriculture, the plant breeder Ivan V. Michurin (died 1935), argued that plants adapt to their environmental conditions by acquiring the characteristics that they need. This revival of Lamarckianism was more compatible with Marxist ideology, which held that environmental influences were paramount, than with the theory of Mendelian genetics. Lysenko adopted the ideas of Michurin as his own and promoted what he called vernalization, a technique that he claimed would convert winter wheat seeds to spring wheat seeds by simply freezing the seeds. He claimed that his new seeds would dramatically increase wheat yields.

Supported by the Soviet leader Josef Stalin

(1879–1953), Lysenko became a powerful Soviet bureaucrat. From 1935 to 1938, he was the scientific director, then director, at the Odessa institute. From 1940 to 1965, Lysenko directed the Institute of Genetics of the Academy of Science of the USSR and served as president of the powerful V. I. Lenin All-Union Academy of Agricultural Sciences.

As the effective dictator of Soviet agricultural science, Lysenko wielded his power ruthlessly. Soviet scientists who did not support his ideas found themselves unemployed. Some Soviet scientists who disagreed with Lysenko died from either execution or starvation after being arrested and taken into custody. At one point, at the height of his power, Lysenko claimed that wheat planted in a hostile environment would convert to rye.

Lysenko survived even though widespread adoption of his ideas led to agricultural failures. After Stalin's death in 1953, Lysenko's power declined. The new Soviet leader, Nikita S. Khrushchev (1894–1971), believed in Lysenko's Marxist science, but allowed scientific criticism of Lysenko to go unpunished. In 1964, after changes in the Soviet leadership, Lysenko was finally completely discredited and exiled to a remote agricultural station, though it took even more time for the effects of his regime to fade. An interesting contrast to Lysenko's pseudoscientific claims is the Green Revolution, begun in the 1940s, which transformed agriculture in the rest of the world, dramatically increasing crop yields through the use of Mendelian genetics.

See also Agriculture; Evolution; Green Revolution; Soviet Union

References
Joravsky, David. *The Lysenko Affair.* Cambridge: Harvard University Press, 1970.
Medvedev, Zhores A. *The Rise and Fall of T. D. Lysenko.* New York: Columbia University Press, 1969.
Soyfer, Valery N. *Lysenko and the Tragedy of Soviet Science.* New Brunswick, NJ: Rutgers University Press, 1994.

M

Mandelbrot, Benoit (1924–)

Benoit Mandelbrot developed the mathematics of fractals, which came to be an important way to visualize chaos theory. Benoit Mandelbrot was born in Warsaw, Poland, where his father manufactured textiles and his mother was a physician. His uncle, a mathematician with a strong analytical bent, influenced Mandelbrot's choice of a vocation, but was disappointed with the young man's preference for geometry over analysis. After World War II started, the family moved to France, where Mandelbrot waited out the subsequent German occupation. In 1944, after the liberation of Paris by Allied troops, Mandelbrot scored high in university entrance exams. Mandelbrot's mathematical talent centered on seeing equations as geometric shapes. After only a few days at the École Normale Supérieure, he chose to move to the École Polytechnique, where he graduated in 1947. Mandelbrot moved to the United States, earning a master's in aeronautics in 1948 from the California Institute of Technology. After returning to Europe, he earned his doctorate in mathematics from the University of Paris in 1952. He married in 1955 and fathered two sons.

Mandelbrot was never comfortable studying a single subject and became an academic wanderer, regularly moving from one temporary position to another in Europe and the United States. International Business Machines Corporation (IBM) became the home that he could not find at a university, and there he regularly worked on research and as a consultant. Mandelbrot studied linguistics, Brownian motion, economics, random noise on telephone lines, and natural phenomena. He consistently sought the patterns and shapes in things, and early in the 1960s he found that his varied interests all came together in the idea of fractals.

Mandelbrot invented the word *fractal* by deriving it from the Latin word *fractus,* which means "fragmented." Fractals often describe irregular shapes and often are self-similar, in the sense that the closer one looks at the fractal, the more similar fractals appear, just on a smaller scale. Computer scientists worked with Mandelbrot, making complex multicolor graphic images of fractals. They noticed that the images resembled shapes in nature, such as the curve of coastlines, and the patterns in turbulent liquids, snowflakes, blood vessels, and other natural systems that exhibit mathematically chaotic behavior. In 1979, he developed his Mandelbrot Set, which supported his theory of fractals. Mandelbrot's 1982 book, *The Fractal Geometry of Nature,* with its gorgeous computer-generated pictures, spread knowledge of his insights.

Mandelbrot fractal image generated by computer (Alfred Pasieka/Photo Researchers, Inc.)

After the pairing of fractals with the chaos theory pioneered by the meteorologist Edward N. Lorenz (1917–), Mandelbrot received recognition for his contribution. Among his many awards are a Barnard Medal for Meritorious Service to Science (1985) and a Wolf Foundation Prize in Physics (1993). Mandelbrot eventually became a professor at Yale University in 1987.

See also Chaos Theory; Computers; Lorenz, Edward N.; Mathematics

References

Barcellos, Anthony. "Benoit Mandelbrot." Pp. 205–225 in *Mathematical People: Profiles and Interviews.* Edited by Anthony Barcellos. Cambridge, MA: Birkhauser Boston, 1985.

Margulis, Lynn (1938–)

The microbiologist Lynn Margulis revived the old theory of microbiological symbiosis and convinced her colleagues of its validity. She was born Lynn Alexander in Chicago, where her father was a lawyer and businessman. An avid reader, she wanted to be an explorer and a writer. She entered the University of Chicago at the age of fifteen, where she intended to follow the humanities, but found science more intriguing. She graduated with an A.B. in liberal arts in 1957 and promptly married a graduate student she had met, the astronomer Carl Sagan (1934–1996). She earned a master's degree in zoology and genetics from the University of Wisconsin in

1960, then followed her husband to the University of California, where she earned her Ph.D. in genetics in 1965 at Berkeley. She and Sagan divorced in 1963, and she took their two sons to Massachusetts, where she conducted research at Brandeis University before finding a position at Boston University in 1966. A second marriage (1967–1980) produced a son and daughter and left her with the last name of Margulis after her divorce. She found she enjoyed being a mother more than being a wife. Her extraordinary energy and strong capacity for work enabled her to raise four children and at the same time make significant contributions to science.

In her research, Margulis found her interest drawn to cytoplasmic genes, those genes found outside the cell nucleus in organelles such as mitochondria and chloroplasts. Geneticists of the time completely ignored cytoplasmic genes and their relationship to biological evolution. Margulis found in her research that earlier scientists had shown that microbes could transmit induced characteristics, which led her to the idea that organelles were really former bacteria that had fused with other cells. After numerous rejections, she published her first article on what became known as the serial endosymbiosis theory (SET) in 1967. A book on the *Origin of Eukaryotic Cells* followed in 1970. Eukaryotic cells contain nuclei, and she argued that they evolved when non-nucleated bacteria fused together billions of years ago. She also argued that chloroplasts, which convert sunlight into energy via photosynthesis, were originally free-living microbes that symbiotically merged with other cells. As evidence accumulated in the 1970s, SET came to be accepted, though Margulis's contention that most evolutionary change is the result of symbiogenesis is not widely accepted. Always central in Margulis's thinking is the idea that microbes constitute the dominant form of life on Earth. Margulis expanded on symbiosis by proposing that cilia and other locomotion appendages of microbes originally evolved when spirochetes merged with archaebacteria. This hypothesis is not widely accepted.

In 1970, Margulis began to work with the biophysicist James Lovelock (1919–) on his idea of Gaia, which is the hypothesis that Earth is like a living organism, maintaining surface and atmospheric conditions so that the chemistry and temperature will sustain life. She found the idea compatible with her view of the planet and vigorously championed his work. Many biologists were suspicious of such a holistic notion. The metaphor of Gaia, the ancient Greek goddess of Earth, made Earth seem alive, and activists in the environmental movement and advocates of New Age spirituality eagerly adopted the notion of Gaia, further alienating scientists who were uncomfortable with the idea's mystical overtones. Margulis did not initially see a connection between SET and Gaia, but she came to see Gaia as symbiosis on a larger scale.

A literary agent persuaded Margulis to add popular articles and books to her academic output. Margulis has become a well-known and prolific writer, promoting symbiosis, evolution, and Gaia. Many of her books are coauthored with her son, Dorion Sagan. Margulis moved to the University of Massachusetts at Amherst in 1988 as a distinguished professor in botany. She changed departments in 1993 to biology and then to geosciences in 1997. These moves reflect her strong interdisciplinary focus and abilities. She became a member of the National Academy of Sciences in 1983 and received the National Medal of Science in 1999. Among her international awards are membership in the Russian Academy of Natural Sciences.

See also Ecology; Environmental Movement; Lovelock, James; Sagan, Carl; Symbiosis
References
Margulis, Lynn. *Symbiotic Planet: A New Look at Evolution.* New York: Basic, 1998.
Margulis, Lynn, and Dorion Sagan. *Slanted Truths: Essays on Gaia, Symbiosis, and Evolution.* New York: Springer-Verlag, 1997.

Masters, William H. (1915–2001), and Virginia Johnson (1925–)

The physician William H. Masters and psychologist Virginia Johnson, commonly referred to as Masters and Johnson, pioneered scientific research on human sexual anatomy and physiology. William Howell Masters was born in Cleveland, Ohio, to wealthy parents. He attended Hamilton College in Clinton, New York, where he excelled in athletics and developed an interest in scientific research. Earning a B.S. from Hamilton in 1938, he then earned an M.D. from the University of Rochester in 1943. Masters married in 1942 and fathered a son and daughter.

Masters wanted to conduct research in human sexuality, but on the advice of an older doctor, he determined to first establish his credentials as a legitimate researcher in another field. After multiple residencies and internships in obstetrics and gynecology, pathology, and internal medicine, Masters joined the medical school faculty at Washington University in St. Louis. He conducted research in obstetrics and gynecology, publishing articles on hormones and the problems of aging until 1954, when Masters started to conduct physiological research into human sexuality. He felt that he needed a female assistant to better communicate with female subjects, so he hired Virginia Eshelman Johnson in 1957.

Virginia Eshelman was born in Springfield, Missouri, where her father was a farmer. She graduated from high school in 1941 at the age of sixteen and attended the local Drury College. After a year of school, she went to work and eventually became a local singer. She also attended at times the University of Missouri and Kansas City Conservatory of Music. Her first marriage ended after two days, the second ended in divorce, and her third marriage was to a dance-band leader named George V. Johnson. She bore a son and daughter, and then that marriage ended in divorce in 1956. She retained her married name and applied for a job at Washington University, where she came to work

for Masters. Though she never completed any academic degrees, Johnson rapidly became a full partner in the sexuality research.

Where Alfred Kinsey's (1894–1956) sexual research was riddled with biases and poor scientific technique, relying mostly on secondary research in the form of interviews, Masters and Johnson applied solid techniques in primary research. They used movie cameras, electrocardiographs, electroencephalographs, and other advanced tools to study subjects masturbating or engaged in coitus. Masters and Johnson started by hiring prostitutes, but found that their sexual responses were not normal, so they solicited couples for their study, though they realized that their subjects were atypical in allowing observation of such intimate situations.

In 1959, Masters and Johnson began to cautiously publish the results of their research on hundreds of human subjects in scientific journals. When a 1964 article criticized their work as dehumanizing the sexual experience, they became bolder and published an academic book, *Human Sexual Response,* in 1966. Intended for physicians and other researchers, the book became a surprise best-seller, just as Kinsey's books had been. Masters and Johnson concluded that sexual enjoyment did not have to decline with age, and discovered a chemical present in some women that functioned as a contraceptive. The pair of researchers continued their research while appearing on television shows, occasionally writing for *Playboy* magazine, and writing more books: *Human Sexual Inadequacy* (1970), *The Pleasure Bond: A New Look at Sexuality and Commitment* (1975), and *Homosexuality in Perspective* (1979).

In 1959, Masters and Johnson began to apply their research to clinical sex therapy. Couples came to their clinic for two or three weeks of intensive therapy followed by periodic consultations. Masters and Johnson later offered postgraduate training at Washington University in sex therapy.

Masters and Johnson were explicit about their feminist focus, and their work found favor with the feminist movement. Some critics felt that their emphasis on technique took the passion out of sex and failed to account for the importance of psychological factors. Their emphasis on achieving orgasm was also criticized as reductionistic. Other critics decried the bias of Masters and Johnson toward monogamy as the best setting for sexual satisfaction.

In 1964, after federal funding for their research ended, Masters and Johnson founded the Reproductive Biology Research Foundation to obtain funding from private donors and private research foundations. They later founded the Masters and Johnson Institute in 1973. Masters and Johnson married in 1971 and divorced in 1993, and after he retired in 1994, their Masters and Johnson Institute closed.

> *See also* Birth Control Pill; Feminism;
> Foundations; Kinsey, Alfred C.; Medicine;
> Psychology; Sociology of Science
> *References*
> Bullough, Vern L. *Science in the Bedroom: A History of Sex Research.* New York: Basic, 1994.
> Robinson, Paul. *The Modernization of Sex: Havelock Ellis, Alfred Kinsey, William Masters and Virginia Johnson.* Ithaca, NY: Cornell University Press, 1989.

Mathematics

As the twentieth century progressed, mathematics became more important to every field of scientific endeavor. Physics had long had an intimate reciprocal relationship with mathematics, and now such diverse fields as evolution, geology, genetics, and economics became more dependent on mathematical models.

In the 1950s, a movement known as the "New Mathematics" began in the United States, emphasizing the teaching of theory and axioms to secondary and even elementary school children, rather than just applied mathematics. Despite its influence on curriculum, ultimately mathematics education moved away from the emphasis on theory, recognizing that students usually learn how

to do mathematics before learning why mathematics works the way it does. Mathematics education also found a new audience. Women mathematicians have been rare in the history of mathematics, but social and cultural changes in the last four decades have opened up the field for women. Approximately a fifth of all doctorates in mathematics granted in the United States by the end of the century were awarded to women.

The invention of electronic digital computers depended on the unusual base system of two (binary) and the development of the rules of Boolean algebra, a mathematical curiosity created in the mid-nineteenth century by the self-taught British mathematician George Boole (1815–1864). The Hungarian-born mathematician John von Neumann (1903–1957) expanded on the concept of stored programs and laid the theoretical foundations of all modern electronic digital computers in a 1945 report and later work. His ideas came to be known as the "von Neumann Architecture." Von Neumann also developed, with the economist Oskar Morgenstern (1902–1976), in 1944 the idea of game theory as a way to understand economics.

The relationship between computers and mathematics has remained strongly reciprocal. Mathematicians still provide important support for new algorithms and methods of computing, especially in the fields of parallel processing and networking, and computers have become ever more important tools for mathematicians. Many fields of mathematics that had lain dormant have been revitalized by using the processing power of computers to do repetitive and tedious work. When Kenneth Appel (1932–) and Wolfgang Haken (1928–) completed their formal proof of the four-color theorem in 1976, they used a computer to reduce the remaining possible configurations, which marked the first time that a computer had been used to construct a formal proof. The four-color theorem showed that any geographical map of countries or other regions can be colored using only four colors in such a way that for any

J. Robert Oppenheimer (left) and John von Neumann stand before an early computer, 1952. (Bettmann/Corbis)

given region all adjacent regions will receive different colors.

The meteorologist Edward N. Lorenz (1917–) applied computers to modeling the atmosphere and made important contributions to meteorology and chaos theory. He introduced his term, the *butterfly effect,* to describe how a small change in initial conditions can have a dramatic effect on later events. Even the simple mathematical models of the atmosphere of that time illustrated the butterfly effect. This idea later became one of the foundations of chaos theory. The French mathematician Benoit Mandelbrot (1924–) developed the mathematics of fractals, which came to be an important way to visualize chaos theory. Fractals often describe irregular shapes and are self-similar, in the sense that the farther one looks into the fractal, the more similar fractals appear, just on a smaller scale.

Perhaps one of the most far-reaching impacts of mathematics on computing has been

in the area of the development of sophisticated encryption schemes that are so secure that their algorithms are published in order to challenge other mathematicians to find any flaws in them. In 1976, Whitfield Diffie (1944–), Ralph Merkle (1952–), and Martin Hellman (1945–) developed asymmetric encryption, also called public key encryption. The technique is based on prime numbers and uses a pair of keys, one considered private and kept secret, and the other considered public and distributed to anyone. Any data encrypted with one key may only be decrypted with the other key of the pair. This has led to secure computer networks, secure e-mail, and digital signatures, and will certainly have an even more profound impact in the future.

The brilliant mathematician John Nash (1928–) expanded on the work of von Neumann and Morgenstern to better understand noncooperative games, cooperative games, two-person bargaining games, and mutually optimal threat strategies in those kinds of games; he also created what became known as Nash equilibria, a way of finding an optimal strategy in games including two or more players. The Hungarian exile John C. Harsanyi (1920–2000) and the German Reinhard Selten (1930–) both expanded on Nash's work, and in the 1960s game theory became an integral part of economic theory. The three men shared the 1994 Nobel Prize in Economics. Another noted event in mathematics occurred when the English-born mathematician Andrew Wiles (1953–) announced a proof of Fermat's last theorem in 1993, exciting considerable public attention. This theorem posits that for the equation xn + yn = zn, there exist no solutions where n is a positive whole number greater than 2.

See also Chaos Theory; Computers; Economics; Lorenz, Edward N.; Mandelbrot, Benoit; Nash, John F.; Neumann, John von; Wiles, Andrew
References
Farmelo, Graham. *It Must Be Beautiful: Great Equations of Modern Science.* London: Granta, 2002.

Katz, Victor J. *A History of Mathematics: An Introduction.* Second edition. Reading, MA: Addison-Wesley, 1998.

Peterson, Ivars. *The Mathematical Tourist: Snapshots of Modern Mathematics.* San Francisco: W. H. Freeman, 1988.

Mayer, Maria Goeppert (1906–1972)

Maria Goeppert Mayer won a Nobel Prize in Physics for her work on the organization of protons and neutrons in orbitals around the atomic nucleus. Maria Goppert (as her surname was originally spelled) was born in Kattowitz, Germany, the child of a professor of medicine and a schoolteacher. Mayer's father was in the sixth straight generation of Gopperts to work as a university professor, and he moved the family to Göttingen when she was four years old. Mayer grew up an academic princess, doted upon as an only child, encouraged by her father to become a scholar, a vivacious attraction at parties, and endowed with a powerful intellect.

Mayer entered the University of Göttingen to study mathematics, but friendships with Max Born (1882–1970) and other influential physicists turned her attention to the emerging field of quantum mechanics. In 1930, she married Joseph E. Mayer (1904–1983), an American chemist boarding with her mother. She earned her doctorate in physics a few months later. Three Nobel prize winners sat on her dissertation committee. Her husband strongly supported Mayer in her scientific pursuits throughout her life, encouraging her to stay involved even when she was inclined to withdraw. Mayer accompanied her husband to America, where he joined the faculty of Johns Hopkins University. Like most American universities, Hopkins had an anti-nepotism policy that prevented her from being hired, so she conducted research, taught, and supervised graduate students as an unpaid associate. In her first nine years, while bearing a son and a daughter, she published ten papers and co-wrote an influential chemistry textbook with her husband. The couple employed a maid so that Mayer might have the time to pursue science, though she confessed to feeling guilty about leaving the children at home with the maid.

After her husband moved to Columbia University, Mayer obtained her first paid position in 1941, teaching part-time at Sarah Lawrence College. During World War II, Mayer and her husband worked on wartime research, and she worked on the Manhattan Project for a time. When the war ended, the couple moved to the University of Chicago, and Mayer returned to working as an unpaid associate. A former student invited her to work part-time at Argonne National Laboratory outside Chicago as a nuclear physicist. While there, she dove into the problem of why some isotopes are more stable than others. When the Italian-born physicist and Nobel laureate Enrico Fermi (1901–1954) asked her if spin-orbit coupling had anything to do with it, the solution presented itself to her. She developed a theory of nuclear shell orbits. Simultaneously, three Germans developed the same theory, and all four published at about the same time. Mayer co-wrote a book with one of the Germans, J. Hans D. Jensen (1907–1973) of the University of Heidelberg, *Elementary Theory of Nuclear Shell Structure* (1955).

The Mayers moved to the University of California at San Diego in 1960, where Mayer was offered a regular full-time professorship in physics and her husband a position as a professor of chemistry. Half of the 1963 Nobel Prize in Physics was awarded to Mayer and Jensen, and the other half to Eugene Wigner (1902–1995) for unrelated work. After years of declining health, Mayer died in San Diego in 1972.

See also Nobel Prizes; Nuclear Physics; Physics
References
McGrayne, Sharon Bertsch. *Nobel Prize Women in Science.* Second edition. Secaucus NJ: Carol, 1998.

Nobel *e*-Museum. "Maria Goeppert Mayer—
Biography." http://nobelprize.org/medicine/
laureates/index.html (accessed February 13,
2004).

Sachs, Robert G. "Maria Goeppert Mayer, June
28, 1906–February 20, 1972." *Biographical
Memoirs of the National Academy of Sciences* 50
(1979): 311–328.

Ernst Mayr, ornithologist (Rick Friedman/Corbis)

Mayr, Ernst (1904–)

The ornithologist Ernst Mayr helped create
the modern version of the theory of evolution by natural selection, often referred to as
the evolutionary synthesis, before turning to
writing important works on the history of biology. Ernst Mayr was born in Kempten,
Germany, where his father was a judge. He
began his higher education as a medical student, but was influenced to change by his love
of ornithology, graduating in 1926 from the
University of Berlin with a doctorate in zoology. He became assistant curator of the university's Zoological Museum for the next six
years. During that time, he undertook an or-
nithological expedition to New Guinea and
remained there for over a year, collecting
specimens and exploring the island with two
expeditions. In 1929, he joined an American
Museum of Natural History expedition to the
Solomon Islands. The American Museum of
Natural History invited Mayr to New York as
a research associate in 1931, and the following year, when he was named associate curator of the museum's Whitney-Rothschild
Collection, he resigned his position in Berlin
and remained in the United States. He later
became a U.S. citizen. Mayr married in 1935
and fathered two daughters.

While writing his *List of New Guinea Birds*
(1941), Mayr immersed himself in taxonomy,
and developed a new definition of species that
rapidly became widely accepted. Instead of
categorizing species by their morphological
differences or genetic differences, he looked
at a species as being a group in nature in
which the individuals interbreed with each
other and are geographically isolated from
other similar species. In 1942, Mayr published *Systematics and the Origin of Species*, the
first of his major contributions to the theory
of evolution. Although many scientists
thought evolution required dramatic mutations, Mayr revived the old idea of geographic
speciation and argued that small mutations in
geographically isolated populations led to
rapid evolution. Mayr also proposed the
founder effect, which was caused when some
geographic quirk or partial die-off caused a
population to be founded by a few individuals. The work of Theodosius Dobzhansky
(1900–1975) on evolution was important to
Mayr's own thoughts. In the late 1940s, Mayr
founded the Society for the Study of Evolution, edited its journal, and helped push forward what became known as the modern
evolutionary synthesis.

Mayr became curator of the Whitney-
Rothschild Collection in 1944, then moved

on to Harvard University in 1953 as the Alexander Agassiz Professor of Zoology. From 1961 to 1970, he served as director of Harvard's Museum of Comparative Zoology. He continued his efforts to create a new systematics of evolution in his 1963 book, *Animal Species and Evolution*.

Mayr turned to the history of his discipline and evolution in the late 1950s. After his retirement in 1975, he pursued historical studies ever more vigorously. He published *The Growth of Biological Thought: Diversity, Evolution, and Inheritance* in 1982, 974 pages long, and *One Long Argument: Charles Darwin and the Genesis of Modern Evolutionary Thought* in 1991. These important contributions are examples of combining the scholarship of historical research with the experience and point of view that come from practicing the scientific discipline under study. Among his many honors were a National Medal of Science in 1970 and the Balzan Prize in 1984.

See also Dobzhansky, Theodosius; Evolution; History of Science; Zoology

References

Greene, John, and Michael Ruse, guest editors. "Ernst Mayr at Ninety." *Biology and Philosophy* 9 (1994): 263–427.

Mayr, Ernst. "How I Became a Darwinian." Pp. 413–423 in *The Evolutionary Synthesis: Perspectives on the Unification of Biology.* Edited by Ernst Mayr and William B. Provine. Cambridge: Harvard University Press, 1980.

McClintock, Barbara (1902–1992)

Barbara McClintock discovered transposable genes, or "jumping genes," in the late 1940s, though her discovery was not appreciated for another two decades. Barbara McClintock was born in Hartford, Connecticut, where her father was a physician. Her mother opposed any of her daughters' gaining higher education, fearing that it would hurt their marriage prospects. Her father supported her entry into Cornell University in 1919. Two years later McClintock took the only course in genetics available to undergraduates and so impressed her instructor with her enthusiasm

that he invited her to take the class in genetics offered to graduate students. Her focus on cytology, genetics, and zoology led to a B.S. in 1923, an M.A. in 1925, and a Ph.D. in 1927, all from Cornell.

McClintock specialized in cytogenetics, the chromosomes of plants and their genetic makeup. She used maize for her genetics studies because the plant could be self-fertilized. McClintock grew her own maize and by applying new staining techniques, found that she could stain the chromosomes of the maize and identify them individually under a microscope. Her dedication to her research, her long hours and complete concentration, became the stuff of legend. After she graduated with her doctorate, McClintock remained at Cornell for four years as an instructor in the botany department. She formed the core of a productive group of graduate students and faculty engaged in genetics research, including Marcus M. Rhoades (1903–1991), Harriet Creighton, and George W. Beadle (1903–1989). Beadle later won a Nobel prize in 1958 for his work in genetics. McClintock and Creighton were the first to show that chromosomes carried hereditary information.

From 1931 to 1936, McClintock conducted research in California, Missouri, Germany, and at Cornell, funded by fellowships from the National Research Council, the Rockefeller Foundation, and the Guggenheim Foundation. Bias against women in science made it difficult for her to obtain a faculty appointment, even though she was by then respected as one of the top geneticists in the world and the foremost expert in maize genetics. She finally obtained a faculty position as an assistant professor at the University of Missouri in 1936. After five years, she realized that the administrators of the university would not allow her to advance there, so she quit, and friends found her a position at the Cold Spring Harbor Laboratory, a research laboratory sponsored by the Carnegie Institution. She remained there for the rest of her life, happy to engage in full-time research. In 1944 she became the president of the Genetics Society of America and was also elected to the National Academy of Sciences, only the third woman so honored.

During the 1940s at Cold Spring Harbor, McClintock discovered that the genes in chromosomes could move from chromosome to chromosome. These transposable genes, or "jumping genes," were regulated by "controlling genes," which suppressed or allowed the expression of a gene. She presented her theory in 1951, but then-current scientific opinion held that chromosomes were immutable. McClintock failed to persuade her colleagues that her discovery of transposable genes was significant until, in the 1960s and early 1970s, molecular biologists discovered a similar phenomenon in bacteria.

After Francis Crick (1916–2004) and James D. Watson (1928–) discovered the structure of DNA in 1953, research in genetics increasingly focused on the molecular level. Traditional biologists like McClintock, who looked at the whole organism with a holistic sense of intuition, were increasingly pushed to the margins. Later in life, when her discovery of transposable genes was recognized as significant, McClintock received the 1983 Nobel Prize in Medicine or Physiology, one of many awards and honorary doctorates recognizing her influence.

See also Crick, Francis; Foundations; Genetics; Nobel Prizes; Watson, James D.

References

Keller, Evelyn Fox. *A Feeling for the Organism: The Life and Work of Barbara McClintock.* San Francisco: W. H. Freeman, 1983.

McGrayne, Sharon Bertsch. *Nobel Prize Women in Science.* Second edition. Secaucus, NJ: Carol, 1998.

Nobel e-Museum. "Barbara McClintock—Autobiography." http://nobelprize.org/medicine/laureates/index.html (accessed February 13, 2004).

Medawar, Peter (1915–1987)

The immunologist Sir Peter Medawar demonstrated acquired immunological tolerance, an important step in understanding the immune system in order to engage in organ transplants. Peter Brian Medawar was born in Rio de Janeiro, where his father was a businessman. Educated in English schools, Medawar attended Magdalen College at Oxford University, where he received his first-class degree in zoology in 1936. He became a fellow of Magdalen College, teaching and conducting research in the growth of tissues and pathology. Medawar married in 1937 and fathered two sons and two daughters.

During World War II, Medawar studied the problem of skin graft rejection in burn victims and developed techniques that made possible the use of a patient's own skin for grafts. After the war, Medawar earned his D.Sc. from Oxford in 1947 and that same year became a professor of zoology at the University of Birmingham. He continued to research transplantation issues, especially why grafts were rejected. He developed a concentrated form of fibrinogen, a compound found in blood, which could be used to glue the nerve endings in skin grafts. Medawar drew attention for his work and was elected a Fellow of the Royal Society in 1949 and knighted in 1952. He moved to University College, London, in 1951.

When Medawar learned of the proposed theory of acquired immunological tolerance developed by the Australian medical physician Frank Macfarlan Burnet (1899–1985), he chose to investigate further. This theory argued that the immune system learned while in the womb and shortly after birth how to distinguish which substances are foreign to the body and thus what to reject. Most scientists thought that this ability was innate. Medawar proved that Burnet's theory was correct. This led other immunologists to realize that they could prevent the immune system from rejecting foreign substances and led to the development of techniques widely used to transplant organs.

Medawar and Burnet shared the 1960 Nobel Prize in Physiology or Medicine. From 1962 to 1971, Medawar served as director of the National Institute for Medical Research at

Mill Hill. Though a stroke forced him into retirement, he remained active in research. With the publication of *The Uniqueness of the Individual* in 1957, Medawar began a publishing career of ruminations on the nature of science that made him a well-known philosophical and popular writer. His *Advice to a Young Scientist* (1979) and *The Limits of Science* (1984) are particularly thoughtful and lucid.

See also Medicine; Nobel Prizes; Organ
  Transplants
References
Medawar, Peter. *Memoir of a Thinking Radish: An Autobiography.* Oxford: Oxford University Press, 1986.
Nobel *e*-Museum. "Peter Medawar—Biography." http://nobelprize.org/medicine/laureates/index.html (accessed February 13, 2004).

Media and Popular Culture

The media (newspapers, magazines, and television) are often the primary means by which members of the lay public learn about science and scientists learn about developments in fields other than their own. Magazines like *National Geographic, Scientific American, Nature, Discover, New Scientist,* and *Natural History* are targeted toward different market segments, expecting different levels of education from their readers. Popularizers of science, who translate complex terminology and concepts for the lay audience, include television personalities, with one of the latest being the popular "Bill Nye the Science Guy" in the United States.

The science fiction writers Isaac Asimov (1920–1992) and Arthur C. Clarke (1917–) gained as much prominence for their popular science writings as for their fictional tales of scientific wonder. Active scientists, often among the top of their field, have also written popular works to communicate with a larger audience. Carl Sagan (1934–1996), Stephen Hawking (1942–), and Fred Hoyle (1915–2001) wrote effective popular works on astronomy and cosmology; Stephen Jay Gould (1941–2002) and Edward O. Wilson (1929–) wrote on evolutionary theory, zoology, and paleontology; George Gamow (1904–1968) covered many subjects in his *Mr. Tomkins* books; and Jacques Cousteau (1910–1997) brought the wonders of the ocean to television viewers as well as readers. Science fiction short stories, novels, movies, and television shows often incorporate the scientific process and scientific theories into their stories, though with mixed success at creating deeper scientific understanding. Often the writers of science fiction movies and television shows seem to have only a vague understanding of the scientific ideas they are trying to use.

In *The Two Cultures and the Scientific Revolution* (1959), the English molecular physicist and novelist C. P. Snow (1905–1980) described the gulf of understanding between scientists and literary intellectuals: the people within these two cultures understand their own cultures, but scientists often do not appreciate the humanities, and intellectuals in the humanities do not understand science. Snow advocated education to overcome the ignorance on both sides.

Critics in the 1980s and 1990s often accused the media of scientism, which is the uncritical acceptance of the scientific point of view, which assumes that the practice of science is a rational endeavor uninfluenced by social or ideological factors. This criticism contains echoes of Snow's point about the two cultures. The writings of social constructionists and others who relentlessly pointed out the human foibles that they saw as intrinsic to scientific endeavors led to the so-called science wars of the 1980s and 1990s.

The portrayal of science is also distorted by the considerable public attention and media coverage that big science projects receive, giving the impression that most science occurs in these big projects, when most science is actually done by individuals or small teams. Pharmaceutical companies engage in advertising, vigorously promoting medicinal drug use and trying to define the public discussion over new drugs and new diseases. Portrayals of scientists at times are pure hagiography, especially when the scientists are medical researchers who have solved a serious medical problem; at other times, scientists are portrayed as engaged in amoral tampering with the natural order, especially scientists engaged in nuclear or weapons research.

See also Asimov, Isaac; Big Science; Clarke, Arthur C.; Cousteau, Jacques; Gamow, George; Gould, Stephen Jay; Hawking, Stephen; Hoyle, Fred; National Geographic Society; Sagan, Carl; Science Fiction; Social Constructionism; Sociology of Science; Wilson, Edward O.

References
Labinger, Jay A., and Harry Collins, editors. The One Culture? A Conversation about Science. Chicago: University of Chicago Press, 2001.
McRae, Murdo William, editor. The Literature of Science: Perspectives on Popular Science Writing. Athens: University of Georgia Press, 1993.
Nelkin, Dorothy. Selling Science: How the Press Covers Science and Technology. Revised edition. San Francisco: W. H. Freeman, 1995.
Tuomey, Christopher. Conjuring Science: Scientific Symbols and Cultural Meanings in American Life. New Brunswick, NJ: Rutgers University Press, 1996.

Medicine

Modern medicine emerged in the nineteenth century, when the development of the germ theory of disease, the discovery of anesthesia to control the pain of surgery, and improved sanitation began to curb the horrors of infectious diseases. The first half of the twentieth century brought more advances, including the discovery of penicillin in 1928 and its widespread use during World War II, after scientists developed a way to mass-produce penicillin in beer vats. The further development of specialized antibiotics allowed physicians to target particular types of bacteria and has banished many diseases from developed countries.

The second half of the century saw an increasing number of discoveries, and the practice of science and medicine grew ever closer. In developed countries, the expansion of private insurance programs and government-sponsored coverage have delivered effective health care to most citizens of those countries. Children in the United States now typically receive twenty immunizations against eleven diseases before they are two years old. Whereas at the start of the twentieth century, no more than half of all children reached adulthood, basic sanitation and vaccination campaigns make it possible for most children in industrialized nations to grow up.

Chronic diseases of affluence (obesity, cancer, heart disease, and geriatric conditions), as opposed to infectious diseases, now preoccupy health care providers in developed countries. Plastic surgery enables the affluent to pursue culturally conditioned images of beauty. The birth control pill, other forms of birth control, the availability of abortions, and antibiotics for the more common sexually transmitted diseases have enabled a change in sexual mores throughout the developed world.

In developing countries, improvements in basic sanitation and basic vaccination against childhood diseases have dramatically dropped mortality rates and helped these countries surge in population. Infectious diseases have continued to be a major problem. Many developing countries lie along the equator, and malaria, the greatest remaining scourge, still kills millions annually and incapacitates many millions more. Development of an effective vaccination against malaria has struggled in the face of neglect by the citizens of developed countries, who are rarely personally affected by the disease, and the intrinsic difficulty of the problem. Mosquito control by synthetic pesticides such as DDT (dichloro-diphenyl-trichloro-ethane) proved effective, but too expensive, and the side effects of the pesticides are now better appreciated.

The Austrian-born Canadian endocrinologist Hans Selye (1907–1982) demonstrated how physiological stress affects the human body, fundamentally changing the perspectives of scientists on how hormones and other chemicals in the body react to stress. Physi-

A United Nations doctor examines a child in Mzimzina, Uganda, in 1967. (Corel Corporation)

cians came to accept that physiological stress could contribute to many problems, including heart attacks, arthritis, allergies, kidney disease, and inflammatory tissue diseases.

Improvements in technology provided new tools for diagnosis and treatment. The heart-lung machine, invented by the physician John H. Gibbon (1903–1956), was first successfully used in 1953, enabling the heart and lungs to be stopped, their functions performed by the machine, while doctors operated. The development of computerized tomography (CAT) scans, magnetic resonance imaging (MRI), and ultrasound technologies has allowed doctors to noninvasively examine the inside of the body. Improved radiation therapy is used to fight cancer. Lasers have been applied to eye surgery and other types of surgery that require extremely fine control.

The first successful organ transplant occurred in 1954 in Boston when a patient received the kidney of his twin brother and lived for another eight years. Improved surgical techniques and the development of antirejection drugs have created a situation in which the transplantation of hearts, livers, kidneys, blood vessels, bone marrow, corneas, and

other organs or tissues is now commonplace in developed countries.

International cooperation in health care, fostered by private foundations and organizations like the World Health Organization (WHO), has led to considerable successes in delivering basic health and sanitation care to developing countries. The worldwide smallpox vaccination campaign, led by WHO, eradicated that dreaded disease in 1979, except for a few samples retained in government laboratories. Jonas Salk (1914–1995) and Albert Sabin (1906–1993) both created polio vaccines in the 1950s, ending a scourge that crippled children. WHO has targeted polio as the next disease to be eradicated through worldwide vaccination.

The evolutionary struggle between humanity and pathogens has led to exotic new threats, such as acquired immunodeficiency syndrome (AIDS), Ebola, and Lassa fever. Creutzfeldt-Jakob disease (related to mad cow disease and scrapie in animals) is in a class of its own, in that it is apparently caused by prions, pathogens the size of a molecule. The overuse of antibiotics in developed countries eventually led to the evolution of new antibiotic-resistant strains of bacteria. In the United States, nonprescription antibiotics could be purchased up until the 1950s, and many countries still have no effective control over the purchase of antibiotics. The rise of multiple-drug resistant bacteria has brought back diseases, such as tuberculosis, that doctors had thought remained only among the poor and negligent. The new antibiotic-resistant strains of diseases are an excellent example of Darwinist natural selection in action, and have led to efforts by physicians to reduce the use of antibiotics when not warranted.

The physiologist Robert Edwards (1925–) and gynecologist Patrick Steptoe (1913–1988) developed in vitro fertilization, a medical technique for removing a mature egg from a woman, fertilizing it with donor sperm in the laboratory, allowing it to divide several times until it becomes a blastocyst, and implanting the blastocyst in a uterus. The patient Lesley Brown was impregnated with her own egg fertilized by her husband's sperm. On July 25, 1978, a daughter named Louise Joy was delivered by caesarian section, becoming the world's first "test-tube baby." In vitro fertilization and other assisted reproductive technologies have raised new ethical issues and laid the essential groundwork for the possible future genetic engineering of children.

Advances in genetics and the Human Genome Project increased our understanding in all areas of medicine. In 1989, the team led by Lap-Chee Tsui (1950–) discovered the gene that caused the genetic disorder cystic fibrosis, a necessary first step to eventually overcoming that disease. The study of the immune system has led to new therapies and ways to combat autoimmune diseases. The issue of cloning, long a staple of science fiction, became a very real problem in the 1990s when animals were successfully cloned. The new field of bioethics has been created to deal with the difficult ethical and moral issues that medical advances have created.

See also Acquired Immunodeficiency Syndrome; Bioethics; Birth Control Pill; Carson, Rachel; Centers for Disease Control and Prevention; Cloning; Edelman, Gerald M.; Edwards, Robert; Elion, Gertrude B.; Genetics; Human Genome Project; In Vitro Fertilization; Kübler-Ross, Elisabeth; Lasers; Levi-Montalcini, Rita; Medawar, Peter; Microbiology; National Institutes of Health; Neuroscience; Organ Transplants; Pharmacology; Population Studies; Prions; Psychology; Sabin, Albert; Salk, Jonas; Selye, Hans; Smallpox Vaccination Campaign; Steptoe, Patrick; Tsui, Lap-Chee; World Health Organization; Yalow, Rosalyn

References

Bloom, Samuel William. *The Word as Scalpel: A History of Medical Sociology.* New York: Oxford University Press, 2002.

Cooter, Roger, and others, editors. *Medicine in the Twentieth Century.* New York: Routledge, 2002.

Garrett, Laurie. *The Coming Plague: Newly Emerging Diseases in a World out of Balance.* New York: Farrar, Straus and Giroux, 1994.

Lerner, Barron H. *The Breast Cancer Wars: Hope, Fear, and the Pursuit of a Cure in Twentieth-century America.* New York: Oxford University Press, 2001.

McKeown, Thomas. *The Role of Medicine: Dream, Mirage, or Nemesis?* Oxford: Basil Blackwell, 1979.

Porter, Roy. *The Greatest Benefit to Mankind: A Medical History of Humanity.* New York: Norton, 1997.

Meteorology

The Norwegian meteorologist Vilhelm Bjerknes (1862–1951) strongly urged that the central purpose of meteorology was accurately predicting future weather. Based in Bergen, Norway, Bjerknes founded an influential scientific movement called the Bergen School of meteorology, emphasizing measures of atmospheric pressure and vigorously promoting Norwegian methods of air-mass and frontal analysis. These methods understood the weather in terms of distinct air masses and the fronts formed on the boundaries between the masses.

The United States established a national weather service in 1870, initially located within the War Department. The army Signal Office established a network of around 500 observers who sent in reports via telegraph. In 1891, the functions of the weather service moved to a new National Weather Bureau, part of the Department of Agriculture. Other nations established similar national organizations. The Americans eventually adopted the practices of the Bergen School and began to publish weather charts with the now familiar isobar lines showing pressure differences and weather fronts. Military operations during World War II demanded accurate weather information, and the U.S. Army and Navy trained 8,000 weather officers. In 1943, the U.S. Navy flew the first airplane into the eye of a hurricane to collect scientific data. Sverre Petterssen (1898–1974), the chief weather forecaster for the British bombers flying from Britain over Germany, discovered the jet stream. After the war, surplus radar equipment and air-planes, sounding rockets, and advances in electronics all gave meteorologists more tools.

In the 1950s, meteorologists in the United States became concerned about the anemic condition of their profession. Only about ten doctorates in meteorology were granted each year from 1953 to 1957, and many practical meteorologists relied on experience rather than education, leading to meteorology having the lowest education level of any scientific discipline. The National Academy of Sciences and the National Science Foundation (NSF) both studied the problem of meteorology's underdevelopment, and beginning in 1958, the NSF more than doubled its funding for what became known as the atmospheric sciences. In 1962, the American Meteorological Society changed the name of its *Journal of Meteorology* to the *Journal of Atmospheric Sciences,* and started a new journal, the *Journal of Applied Meteorology,* not only indicating the growth in the field and the recognition that meteorology was more than just weather forecasting, but also reflecting an expanded scientific effort to understand the complete nature of the atmosphere. The NSF established the National Center for Atmospheric Research (NCAR) in 1960 at the University of Colorado in Boulder. The center hired a permanent staff, purchased a powerful computer, and became a strong central laboratory for the expanded professional vision of meteorologists.

Open-air nuclear testing from the late 1940s to the early 1960s pushed radioactive particles high into the atmosphere, and monitoring of these particles allowed meteorologists to trace upper-air wind patterns. These studies led to concern about the effects of open-air testing on public health and directly to the 1963 Nuclear Test Ban Treaty between the Soviet Union and United States. The International Geophysical Year of 1957–1958 included significant atmospheric science efforts. Other international weather monitoring efforts included the Global Atmospheric Research Programme (GARP) in the 1960s and 1970s.

In the 1950s, some scientists became excited by the idea of weather modification. Attempts to promote rainfall by cloud seeding with silver iodide and other chemicals spread from aircraft showed poor results. The American military continued this effort during the Vietnam War, trying to encourage rainfall in the Vietnam region to inhibit their enemy's ability to move. As a result of these efforts, the United Nations General Assembly passed a "Convention on the Prohibition of Military or Any Other Hostile Use of Environmental Modification Techniques" in 1976, and many nations, including the two superpowers, signed and ratified the treaty.

Computers and satellites completely transformed the practice of meteorology. The mathematician John von Neumann (1903–1957), the "father of computers," used early electronic computers in the early 1950s to try to model the weather. Since then, computers have come to be used to manage the mass of data involved in meteorology, including measurements of temperature, pressure, and humidity from collection stations, and they are also used in sophisticated atmospheric modeling efforts. In 1960, the National Aeronautics and Space Administration (NASA) launched *Tiros 1,* the first satellite of many designed for meteorology. Transmitting back television and infrared images, the Tiros series proved its worth when *Tiros 3* tracked Hurricane Carla in 1963, allowing hundreds of thousands to evacuate before Carla reached land, proving the value of weather satellites. In addition to the United States, the former Soviet Union, the European Space Agency, Japan, China, and India have all launched weather satellites, which have revealed unknown cloud formations and allowed scientists to follow the development of storm systems. The World Meteorological Organization (WMO), part of the United Nations, coordinates the exchange of information between different countries and their weather satellites. Since 1966 no tropical storm has escaped detection and tracking, and meteorologists now know exactly what is happening anywhere in the world. New instruments are regularly developed for weather satellites, allowing them to measure temperatures at different altitudes, wind speed, and other types of ever more granular data.

The meteorologist Edward N. Lorenz (1917–) applied computers to modeling the atmosphere and mathematically showed in 1963 that a small change in initial conditions can have a dramatic effect on later events. This is why weather forecasting is so difficult and becomes more unreliable the further out the meteorologist tries to forecast. Lorenz's work became a foundation for the later development of chaos theory, which is a new mathematical understanding of physical phenomena represented by systems of nonlinear equations. The dynamics of the atmosphere are often used as a classic example of chaos theory in action.

Advances in oceanographic studies since World War II have also helped meteorologists understand the role of the oceans and seas in the atmosphere. The geologist and oceanographer Wallace S. Broecker (1931–) showed that a thermohaline circulation system in the northern Atlantic delivers heat from the tropics up to the water off northwest Europe. This effect is now called Broecker's Conveyor Belt; it has a major impact on the weather of Europe, making that part of the world much warmer and much more habitable than would be the case without the thermohaline circulation system.

In the 1920s, Sir Gilbert Walker, who headed the Indian Meteorological Service, documented that when the air pressure over the Pacific Ocean was high, the air pressure over the Indian Ocean was low, and vice versa. In 1969, another meteorologist expanded on this observation with his description of the El Niño Southern Oscillation (ENSO). The El Niño episodes, coming every four or five years, are caused by a rise in the water temperature off the western coast of South America. This temperature phenomenon has predictable effects on the weather of both North and South America.

The study of atmospheric chemistry led to a better understanding of the ozone layer high in the stratosphere. In the early 1970s, scientists discovered that man-made chlorofluorocarbons (CFCs) reacted strongly with ozone and could deplete the ozone layer, which protects life on the ground from harmful ultraviolet radiation. Further research found that the ozone layer was already deteriorating, including the formation of a hole in the ozone layer over Antarctica. The Montreal Protocol international agreements of 1987, 1990, and 1992 agreed to phase out the use of CFC gases.

In the 1980s, scientists determined that the average temperature on Earth was rising, on the order of approximately 0.6 degree Celsius (1 degree Fahrenheit) during the twentieth century. This led to concerns about global warming, which became a major issue in the scientific community and popular political discourse in the 1980s and 1990s and on into the twenty-first century. Meteorologists have joined scientists from other disciplines in an effort to build better mathematical models of Earth's atmosphere, many so complex that only supercomputers can run them. Deep-core drilling, especially in Greenland, has been used to recover data on past climate fluctuations.

See also Associations; Broecker, Wallace S.; Chaos Theory; Cold War; Computers; Deep-Core Drilling; El Niño; Environmental Movement; Global Warming; International Geophysical Year; Lorenz, Edward N.; National Science Foundation; Neumann, John von; Oceanography; Ozone Layer and Chlorofluorocarbons; Satellites

References

Fleming, James Rodger, editor. *Historical Essays on Meteorology, 1919–1995.* Boston: American Meteorological Society, 1996.

Friedman, Robert Mare. *Appropriating the Weather: Vilhelm Bjerknes and the Construction of Modern Meteorology.* Ithaca, NY: Cornell University Press, 1989.

Mazuzan, George T. "Up, Up, and Away: The Reinvigoration of Meteorology in the United States: 1958–1962." *Bulletin of the American Meteorological Society* 69, no. 10 (October 1988): 1152–1163.

Nebeker, Frederik. *Calculating the Weather: Meteorology in the 20th Century.* San Diego: Academic, 1995.

Smith, W. L., and others. "The Meteorological Satellite—Overview of 25 Years of Operation." *Science* 231, no. 4737 (January 31, 1986): 455–462.

Williams, James Thaxter. *The History of the Weather.* Commack, NY: Nova Science, 1999.

Microbiology

Microbiology became a field of study because of the invention of the microscope in the early seventeenth century. Advances in microscopy in the twentieth century—the transmission electron microscope (1931), scanning electron microscope (1942, commercialized 1965), and scanning tunneling microscope (1979)—helped scientists see microbes, such as protozoans, algae, molds, bacteria, and viruses, in more detail. The earliest life-forms were microbes, and the Precambrian Era, prior to 550 million years ago, lacked any known fossils until the paleobotanist Elso Barghoorn (1915–1984) and others discovered microfossils in the 1950s. Evolutionists think that only single-celled life existed until the evolution of multicellular life at the beginning of the Cambrian Era.

In the 1940s, a collection of scientists, including the German-born physicist Max Delbrück (1906–1981), the Italian-born physician Salvador E. Luria (1912–1991), and the American microbiological chemist Alfred D. Lewis (1908–1997), became interested in bacteriophages, which are viruses that infect bacteria. Working at different institutions in the United States, but freely sharing their work, the so-called phage group often met at the Cold Spring Harbor Laboratory on Long Island, funded by the private Carnegie Foundation. By 1952 the trio and their teams had proven that bacteriophages contain deoxyribonucleic acid (DNA) surrounded by a protein shell and that DNA is what contains the elusive genes of heredity. Delbrück, Luria, and Lewis shared the 1969 Nobel Prize in Physiology or Medicine.

The zoologist James D. Watson (1928–) and the biologist Francis Crick (1916–2004) drew on the work of the phage group when they discovered the structure of the DNA molecule in 1953. Microbiology continued to serve an important role in developing the science of genetics. Microbes, especially the bacterium *Escherichia coli (E. coli),* found in the human colon, serve as living laboratories for molecular biologists; they became the basis for the discovery of genetic engineering and remain the continuing foundation of the multibillion-dollar biotechnology business.

Because some microbes act as pathogens in the human body, microbiologists have always been engaged in medical research. In the 1940s, fungal infections were a rare problem for humans, outside of minor problems like athlete's foot. Fungal infections soon became a problem following surgery, especially organ transplants, from the increased use of indwelling catheters and the increased use of broad-spectrum antibacterial drugs. In the 1950s and 1960s, a number of effective antifungal agents were discovered and synthesized. Beginning in the late 1960s, microbiologists found to their surprise that some fungi secreted chemicals that acted as antimicrobials. Later research found antimicrobials being produced in bacteria for use against other bacteria and to defend themselves from being consumed by protozoa. Fungi also create chemicals that kill insects that consume them.

Since the late nineteenth century, microbiologists have been aware of symbiosis, a situation in which two different organisms live together and serve each other's purposes. An example of symbiosis occurs in lichens, where a fungus and algae mutually exist together, with the algae providing food through

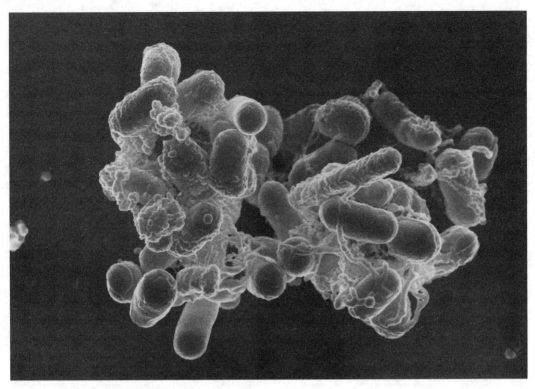

Group of Escherichia coli *bacteria under an electron microscope (Charles O'Rear / Corbis)*

photosynthesis and the fungi providing shelter. A later researcher proposed in 1927 the principle of symbiogenesis, the creation of new species when older species combine on the microbial level. The microbiologist Lynn Margulis (1938–) revived this theory forty years later as her serial endosymbiosis theory (SET). Initially the theory was ignored, but by the end of the 1970s, microbiologists had come to accept that bacteria exchanged genes with each other.

This ability for bacteria to accelerate the process of natural selection through gene transfer is thought to have important implications for the theory of evolution. Even without symbiogenesis, the process of natural selection is seen most readily in bacteria and viruses, which undergo quicker evolution because of the short time spans between generations. Natural selection is apparent in the flu, which is a different virus every year as the virus rapidly mutates.

Biological weapons have only rarely been used in wars, but their possible impact as weapons of mass destruction has encouraged governments to fund scientific research into the offensive use of bacteria and viruses as weapons and how to defend against such weapons. Anthrax, botulism, smallpox, glanders, and *Yersinia pestis* (cause of the Black Death) are among the many microbes that have been developed into weapons by military scientists. The 1972 Biological and Toxin Weapons Convention banned the continued manufacture and development of biological weapons and required the destruction of existing stocks, though the Soviet Union and its successor state, Russia, did not adhere to the treaty and continued a large biological weapons program well into the 1990s.

See also Barghoorn, Elso; Biology and the Life
   Sciences; Biotechnology; Crick, Francis;
   Enders, John; Evolution; Genetics; Margulis,
   Lynn; Organ Transplants; Pharmacology;
   Symbiosis; War; Watson, James D.

References

Chadarevian, Soraya de. *Designs for Life: Molecular
   Biology after World War II*. Cambridge:
   Cambridge University Press, 2002.
Dixon, Bernard. *Power Unseen: How Microbes Rule
   the World*. Cambridge: Oxford University
   Press, 1996.
Koprowsky, Hilary, and Michael B. A. Oldstone,
   editors. *Microbe Hunters—Then and Now*.
   Bloomington, IL: Medi-Ed, 1996.
Society for General Microbiology. *Fifty Years of
   Antimicrobials*. Cambridge: Cambridge
   University Press, 1995.
Wakeford, Tom. *Liaisons of Life: From Hornworts to
   Hippos, How the Unassuming Microbe Has Driven
   Evolution*. New York: Wiley, 2001.

Miller, Stanley L. (1930–)

Stanley L. Miller conducted a famous experiment that showed that the prebiotic chemicals present on early Earth could form amino
acids, the building blocks of life. Stanley
Lloyd Miller was born in Oakland, California. After earning a B.S. in chemistry from
the University of California at Berkeley in
1951, Miller entered graduate school at the
University of Chicago. He heard a lecture by
Harold C. Urey (1893–1981), a 1934 Nobel
laureate in chemistry for his discovery of
heavy hydrogen, in which Urey proposed that
the atmosphere of early Earth, when life was
still absent, was chemically reducing, composed of water vapor, methane, hydrogen,
and ammonia. Urey suggested that someone
should perform an experiment to see if this
primeval soup could lead to organic compounds. Unknown to Urey, the Russian biochemist Aleksandr Oparin (1894–1980) and
the British scientist J. B. S. Haldane
(1892–1964) had also arrived independently
at similar conclusions in the 1920s. Given
their priority, their view of the early atmosphere of Earth became known as the
Oparin-Haldane theory. The German chemist

Walther Löb (1872–1916) also conducted
chemistry experiments similar to what Urey
proposed in the early part of the twentieth
century, though his motivation was unrelated
to theories of the early Earth's atmosphere.

A year later, Miller asked Urey if he could
perform the experiment. Urey thought that
the experiment carried a high risk of failure
and discouraged the graduate student, but
Miller persisted, and Urey finally agreed that
the young man could devote no more than a
year to the effort. Miller designed a simulation of the early Earth's atmosphere using
two glass flasks connected via glass tubes.
One flask contained water that Miller heated
to create water vapor, representing the ocean,
and the other flask contained methane, hydrogen, and ammonia, representing the atmosphere. He reasoned that lightning would have
been common and introduced a continuous
electrical discharge to simulate lightning and
encourage chemical reactions.

Within a couple of days, the experiment
changed color, and further investigation
showed that amino acids, the essential building blocks of life, had been formed as a result
of the experiment. After some difficulty convincing others of the validity of their experiment, Urey and Miller published their results
in the journal *Science* in May 1953. Though
many scientists now think that the early
Earth's atmosphere was not chemically reducing and did not contain the chemicals used by
Miller, the influential Miller-Urey experiment remains an iconic classic. The astronomer Carl Sagan (1934–1996) believed
that the Miller-Urey experiment was the key
event that persuaded scientists that perhaps
life on Earth was not a unique event in a vast
universe. The discovery of amino acids in a
carbonaceous meteorite recovered in Murchi-
son, Australia, in 1969 lent further credence
to the idea that forming amino acids in the
universe was easier than originally thought.

Earning his doctorate in 1954, Miller
spent a year as a fellow at the California Institute of Technology, and five years on the

faculty of the department of biochemistry at Columbia University, before settling in the chemistry department at the University of California, San Diego. He has remained active in studying early conditions in Earth's atmosphere and in the emerging field of astrobiology, also known as exobiology. Miller received the 1983 Oparin Medal from the International Society for the Study of the Origin of Life.

See also Chemistry; Geology; Sagan, Carl; Search for Extraterrestrial Intelligence
References
Henahan, Sean. "From Primordial Soup to the Prebiotic Beach." http://www. accessexcellence.org/WN/NM/miller.html (accessed February 13, 2004).
"Stanley Miller's 70th Birthday." http:// exobio.ucsd.edu/birthday_70.htm (accessed February 13, 2004).

Minsky, Marvin (1927–)

The mathematician Marvin Minsky is one of the pioneers of research in artificial intelligence. Marvin Minsky was born in New York; his father was an eye surgeon, and his mother was a Jewish activist. Minsky attended private schools where his gifts were recognized, and after brief service in the U.S. Navy, Minsky entered Harvard University in 1946. Initially majoring in physics, he expanded his interests into psychology after becoming fascinated with how the mind worked, and finally graduated in 1950 with a B.A. in mathematics. Minsky moved to Princeton University, where a colleague and he built in 1951 the Snarc (Stochastic Neural-Analog Reinforcement Computer). Made out of 400 vacuum tubes, the machine was an early attempt at creating a learning system based on neural nets modeled on those the human brain uses. He earned a Ph.D. in mathematics from Princeton in 1954, then returned to Harvard as a junior fellow, where he worked on microscopes and patented a scanning microscope. Minsky married in 1952 and fathered two daughters and a son.

The term *artificial intelligence* (AI) was first coined in a two-month summer workshop at Dartmouth College in 1956 organized by John McCarthy (1927–) and Minsky. In 1958, Minsky joined the faculty of the Massachusetts Institute of Technology (MIT). McCarthy joined him at MIT and a year later they founded the MIT Artificial Intelligence Project. The project became the Artificial Intelligence Laboratory in 1964. For a time Minsky served as director of the laboratory, but found he did not like administrative tasks, so he became Donner Professor of Science in the department of electrical engineering and computer science in 1974 and Toshiba Professor of Media Arts and Sciences in 1990.

The first two decades of AI research were dominated by researchers at MIT. Minsky made important contributions in robotics, visual perception, computational geometry, and basic theories of AI and computation. Minsky also promoted the computer language Logo, used to teach programming to children, and later helped found the Thinking Machines Corporation. Minsky devoted and continues to devote much of his intellectual energy to developing and expounding a mechanistic theory of the mind, arguing that consciousness is actually an illusion, and thus there is no difference between machines and humans. Many researchers in the fields of AI and cognitive science do not support this more extreme view. Among Minsky's many awards are the 1970 Turing Award from the Association of Computing Machinery.

See also Artificial Intelligence; Cognitive Science; Computers
References
"Marvin Minsky." http://web.media.mit. edu/~minsky/ (accessed February 13, 2004).
Minsky, Marvin. *The Society of Mind*. New York: Simon and Schuster, 1988.

Molina, Mario (1943–)

The chemist Mario Molina discovered, with F. Sherwood Rowland (1927–), that chlorofluorocarbon (CFC) gases damaged the ozone layer of the atmosphere. Mario José Molina was born in Mexico City, where his father was a lawyer and also involved in politics. His aunt, a chemist, encouraged his early interest in chemistry, and his family sent him to a boarding school in Switzerland at the age of eleven so that he might learn German, reasoning that a chemist could use a knowledge of that language. Molina entered the National University of Mexico (UNAM) in 1960 and majored in chemical engineering. After earning his B.S. in 1965, he went to the University of Freiberg in Germany for two years of graduate research. After a year as an assistant professor at UNAM, Molina moved to the University of California at Berkeley in 1968 to work with George C. Pimentel (1922–1989) on a doctorate in physical chemistry. Pimentel's group had invented the chemical laser several years earlier, and Molina became an expert on chemical lasers. After earning his Ph.D. in 1972, Molina remained at Berkeley for a year to continue his research on chemical dynamics. Luisa Tan, a fellow graduate student in Pimentel's group, became Molina's wife in 1973. They later had a son.

In 1973, Molina moved to the University of California at Irvine as a postdoctoral fellow on Rowland's research team. Rowland offered Molina several possible projects, and Molina decided on the topic most outside his area of expertise so that he might learn more. At a workshop, Rowland had listened to a presentation by the biophysicist and inventor James Lovelock (1919–) on recent measurements of the level of CFCs in the atmosphere. All CFCs are made by human industry, and they were used at that time mainly in aerosol cans and air conditioners. Molina and Rowland began to investigate what happened to CFC molecules in the atmosphere. They found that when the CFCs reach the ozone layer, high in the atmosphere, ultraviolet radiation from the Sun breaks down the CFC molecules. Each CFC molecule consists of one atom of carbon, one of fluorine, and three of chlorine ($CFCl_3$). The released chlorine atoms rapidly react with ozone molecules, converting them to normal oxygen.

Small amounts of CFCs have the potential to rapidly reduce the amount of ozone. The ozone layer protects life on Earth from the harmful effects of solar ultraviolet radiation. Molina and Rowland published their findings in the journal *Nature* in 1974, and their laboratory discovery was later confirmed by measurements of the ozone layer. The realization that continued use of CFCs would deplete the ozone layer, with catastrophic results for human civilization, led to the Montreal Protocol international agreements of 1987, 1990, and 1992 to phase out the use of these gases.

In 1975, Molina became a faculty member at Irvine and continued to work on environmental chemistry. His wife often collaborated with his research team. Tiring of the duties of academia, Molina moved to the Molecular Physics and Chemistry Section at the Jet Propulsion Laboratory in 1982 in order to devote himself to research. In 1989 he returned to academia as a professor at the Massachusetts Institute of Technology. Molina was elected a member of the National Academy of Sciences in 1989 and a member of the Academia Mexicana de Ingenieria in 1990. In 1995, Paul Crutzen (1933–), Rowland, and Molina received the Nobel Prize in Chemistry, the only Nobel prize in the sciences won by a Mexican.

See also Chemistry; Crutzen, Paul; Environmental Movement; Jet Propulsion Laboratory; Lovelock, James; Meteorology; Nobel Prizes; Ozone Layer and Chlorofluorocarbons; Rowland, F. Sherwood

References
Newton, David E. *The Ozone Dilemma: A Reference Handbook.* Santa Barbara, CA: ABC-CLIO, 1995.
Nobel e-Museum. "Mario Molina—Autobiography." http://nobelprize.org/medicine/laureates/index.html (accessed February 13, 2004).

Monod, Jacques (1910–1976)

The microbiologist Jacques Monod and his colleagues proposed the existence of messenger ribonucleic acid (mRNA) and operons to

Jacques Monod, microbiologist, 1975 (Sophie Bassouls/
Corbis Sygma)

explain how enzymes and proteins are created and used within cells. Jacques Lucien
Monod was born in Paris, where his father, a
painter, had married an American. At the age
of seven, the family moved to Cannes, and
Monod came to think of himself as a southern
Frenchman instead of a Parisian, even though
he spent most of his professional life in Paris.
His father's interest in science and Darwinism
led Monod to a career in biology, though he
considered a career as a musician and remained an avid amateur musician throughout
his life. Monod moved to Paris in 1928 to attend the Faculty of Science at the University
of Paris for a degree in the natural sciences.
He graduated in 1931, then went to work at
the Roscoff marine biology station. His mentors there led him to realize that his university education was sadly out of date, and he
developed a keen interest in microbiology
and genetics, along with a belief in the importance of the chemical and molecular understanding of biological processes. Time
spent at the California Institute of Technology

also brought him up to date with the rapidly
changing fields of microbiology and genetics.
Monod returned to Paris to complete his
doctorate in 1941 at the Sorbonne. With the
German occupation of France during World
War II, Monod left science and became a
leader in the anti-Nazi Resistance. For a time
he was a member of the French Communist
Party, but resigned because of the insistence
by Josef Stalin (1879–1953) that the non-
Darwinian ideas of agricultural geneticist
Trofim Lysenko (1898–1976) be supported
by all party members.

After the war, Monod joined the Institut
Pasteur as a laboratory director in a department run by André Lwoff (1902–1994).
Monod continued his studies of the actions of
enzymes and proteins in bacteria. With the
discovery of the structure of DNA in 1953 by
Francis Crick (1916–2004) and James D.
Watson (1928–), the field was revolutionized. In 1957, François Jacob (1920–) joined
Monod in his research, and they developed
the theories of mRNA, operons, and allosteric transitions to explain how information was transferred from DNA to form enzymes and proteins inside cells. A key result
of these theories is an understanding of how
repressor genes turn on and turn off the creation of enzymes and proteins.

Monod married an archaeologist in 1938
and fathered twin sons, Olivier and Philippe,
both of whom became scientists, one a geologist and the other a physicist. Monod received many honors for his scientific work
and his wartime activities, including the 1965
Nobel Prize in Physiology or Medicine,
which he shared with Lwoff and Jacob. In
1959, Monod became a professor of the
chemistry of metabolism at the Sorbonne; in
1967, he became a professor at the College of
France; and in 1971, he became director of
the Institut Pasteur, where he vigorously and
dogmatically pursued a program of reform,
emphasizing the pure pursuit of science. He
published a book, Chance and Necessity, in
1971 that promoted his existentialist humanist beliefs, arguing that all biological evolu-

tion was the result of pure chance. Diagnosed with a terminal disease, Monod returned to Cannes to die at the age of sixty-six.

See also Cold War; Crick, Francis; Genetics; Lysenko, Trofim; Microbiology; Nobel Prizes; Watson, James D.

References
Lwoff, André. "Jacques Lucien Monod, 9 February 1910–31 May 1976." *Biographical Memoirs of the Fellows of the Royal Society* 23 (1977): 385–412.
Nobel e-Museum. "Jacques Monod—Biography." http://nobelprize.org/medicine/laureates/index.html (accessed February 13, 2004).

Müller, K. Alex (1927–)

K. Alex Müller and his colleague discovered that certain materials can achieve superconductivity at higher temperatures than previously suspected. Karl Alexander Müller was born in Basel, Switzerland. When he was eleven years old, his mother died, and he then attended the Evangelical College in Schiers for seven years. After his obligatory military service, Müller entered the Swiss Federal Institute of Technology (ETH), receiving a doctorate in physics in 1958. He joined the Battelle Memorial Institute in Geneva and became manager of a group using magnetic resonance to investigate layered compounds. In 1962 he accepted an appointment on the faculty of the University of Zürich and was invited to become a researcher at the IBM Zürich Research Laboratory a year later. Müller became a professor at the University of Zürich in 1970 and head of the physics group at the laboratory in 1972. Müller married in 1956 and fathered a son and a daughter.

A young student, Georg Bednorz (1950–), came to the laboratory in 1972 as a summer researcher. Bednorz worked on the same strontium-titanium oxide that interested Müller, and the older man became Bednorz's mentor, supervising his doctoral work and bringing Bednorz to the laboratory permanently in 1977. In 1982, Müller became an IBM Fellow, and in 1985, Müller gave up his management position to devote himself completely to their joint research. Their goal was to find new materials that would superconduct electricity at higher temperatures.

Superconductivity had been discovered in 1911. By the time Müller became interested in the problem, the best superconducting metal available had to be cooled to 23 degrees kelvin before the electricity flowed with no loss of energy to resistance from the metal. This was too cold for commercial applications. Müller and Bednorz chose to start working with oxides instead of metals. The work required great precision and attention to detail. Reading of the French development of a new barium-lanthanum-copper oxide, Müller and Bednorz tried it and obtained superconductivity at 35 degrees kelvin in January 1986. They published their results in a minor journal, hoping to obtain priority of discovery yet keep quiet enough so that they could continue to pursue their studies without competition. They failed to escape notice, and their discovery set off a rush among physicists to find other superconducting materials. Because of the economic potential of superconducting materials, allowing the transmission of electricity without loss, the Nobel committee quickly awarded the prize—one of the few times in their history. Müller and Bednorz shared the 1987 Nobel Prize in Physics.

See also Nobel Prizes; Physics
References
Matricon, Jean, and Georges Waysand. *The Cold Wars: A History of Superconductivity*. New Brunswick, NJ: Rutgers University Press, 2003.
Nobel e-Museum. "Alex Muller—Autobiography." http://nobelprize.org/ physics/laureates/index.html (accessed February 13, 2004).

N

Nambu, Yoichiro (1921–)

The Japanese-born physicist Yoichiro Nambu made important contributions to understanding quarks and was the first to propose that quarks act as if they are connected by strings. Yoichiro Nambu was born in Tokyo, Japan, but the great earthquake of 1923 compelled the family to move back to his father's hometown of Fukui, near Kyoto. His father, a schoolteacher with a nonconformist attitude, owned an eclectic personal library. Nambu suffered through his regimented schooling, and he only found a more congenial intellectual atmosphere when he went to college in Tokyo in 1937.

Drafted into the army, Nambu was eventually rescued from digging trenches to work on the army's radar project. The Japanese navy and army pursued separate radar projects with no interservice cooperation, and Nambu was even assigned the task of stealing scientific documents from the navy project, which he did by deceiving an unsuspecting professor. Nambu's grandparents died in an American bombing raid, though his parents survived. After the war, he married his assistant, and then returned alone to Tokyo as a research assistant at the University of Tokyo, where he slept on his desk for three years. Nambu later fathered two children.

The vibrant prewar Japanese physics community, led by future Nobel laureate Shinichiro Tomonaga (1906–1979), gradually revived. In 1952, after receiving his D.Sc. from the University of Tokyo, Nambu came to the United States at the invitation of the Institute for Advanced Study at Princeton University. He chose to remain in America and became a faculty member at the Enrico Fermi Institute at the University of Chicago. Nambu proposed the existence of gluons to hold together quarks, worked on superconductivity, and successfully predicted the existence of new particles, including the boson.

In 1970, the same year that he became a U.S. citizen, Nambu realized that quarks act as if they are connected by strings. This insight led to further work by Nambu and others and the development of string theory in the 1980s. String theory provided a promising avenue to the development of a grand unified theory to unify the four known forces of quantum physics, a goal long sought after by physicists. Nambu retired from the University of Chicago in 1991; received the National Medal of Science, the Dirac Medal, and the Order of Culture (the latter from the Japanese government); and shared the 1994–1995 Wolf Physics Prize.

See also Bardeen, John; Gell-Mann, Murray; Grand Unified Theory; Particle Physics; Physics

References

Brown, Laurie M., and Yoichiro Nambu. "Physicists in Wartime Japan." *Scientific American* 279, no. 6 (December 1998): 96–103.

Etori, Akio. "Elemental Genius." *Look Japan* 42 (May 1996): 24–25.

Makerjee, Madhusree. "Profile: Yoichiro Nambu." *Scientific American* 272, no. 2 (February 1995): 37–39.

Nanotechnology

Research on manufacturing objects on the scale of 100 nanometers or less is called nanoscience or nanotechnology. The 1986 book by the molecular nanotechnologist K. Eric Drexler (1955–), *Engines of Creation: The Coming Era of Nanotechnology,* popularized the term *nanotechnology* and introduced his vision of a world dominated by nanoengineering. Drexler first published a paper on molecular manufacturing systems in 1981, drawing on the ideas of the Nobel laureate physicist Richard P. Feynman (1918–1988). Drexler later earned the first doctorate in molecular nanotechnology in 1991 from the Massachusetts Institute of Technology (MIT), where his thesis supervisor was the artificial intelligence pioneer Marvin Minsky (1927–).

By building microscopic objects atom by atom, nanoengineers hope for revolutionary advances in the material sciences, leading to advances in computers. The ultimate dream is to create "nanobots," microscopic robots with the ability to manipulate matter on a molecular level. The next step is replicating assemblers, generic nanobots that can create more copies of themselves. Medicine might benefit from advanced sensors for disease detection, new therapies, and new implants, as well as by using nanobots to operate on internal organs or repair the interior of human cells. Vigorous advocates of the potential for nanotechnology often paint scenarios of future possibilities that are derided by critics as mere science fiction, and indeed, science fiction authors in the 1990s used nanotechnology extensively in their stories, often as a kind of magic.

In 1979 the German physicist Gerd Binnig (1947–) and the Swiss physicist Heinrich Rohrer (1933–), working at the International Business Machines (IBM) Zürich Research Laboratory, invented the scanning tunneling microscope (STM), an important tool for the nascent field, capable of imaging down to atomic level. The STM can manipulate conductive material, and in 1989, the IBM physicist Don Eigler used an STM to arrange thirty-five xenon atoms to spell the microscopic word *IBM*. IBM later developed the Atomic Force Microscope (AFM) in 1986, which can manipulate nonconductive material on the atomic level. More practical examples of nanotechnology emerged in the 1990s, with molecular-engineered nanoproducts eventually making up a worldwide market of over $40 billion.

Arguments similar to those opposing genetically modified organisms are also offered against nanotechnology. Critics of nanotechnology fear that nanobots that could self-replicate might run out of control, like bacteria run amok. What mechanisms in nature might stop nanobots? Nanotubes, microscopic tubes made of carbon atoms, are a major advance in nanotechnology with many applications. Researchers at DuPont injected nanotubes into the lungs of rats, and 15 percent of the rats died within a day from blockage of their airways. Studies such as the DuPont study lend credibility to the arguments of critics.

In the 1990s, the National Science Foundation (NSF) coordinated nanotechnology research and funding under a National Nanotechnology Initiative. The National Institutes of Health (NIH) also funded nanoscience and nanotechnology research project grants in biology and medicine through the NIH Bioengineering Consortium. Research in nanotechnology gained a boast, becoming a big science project, when the U.S. Congress allocated $3.7 billion in 2003, to be spent over a four-year period, on nanotechnology research. The funds aimed to establish a net-

Scanning tunneling microscope photograph of the word IBM *spelled in xenon atoms (Courtesy of IBM)*

work of university-based technological research centers, and included an effort to fund research into the ethical issues that might arise from nanotechnology. Observers expect many of the ethical issues to be similar to issues already familiar from the field of bioethics.

See also Artificial Intelligence; Big Science; Bioethics; Biotechnology; Feynman, Richard; Medicine; Minsky, Marvin; National Institutes of Health; National Science Foundation; Particle Physics; Science Fiction

References
Drexler, K. Eric. *Engines of Creation: The Coming Era of Nanotechnology.* New York: Anchor, 1986.
Editors of Scientific American. *Understanding Nanotechnology.* New York: Warner, 2002.
"The Foresight Institute: Preparing for Nanotechnology." http://www.foresight.org/ (accessed February 13, 2004).
McCarthy, Wil. *Hacking Matter: Levitating Chairs, Quantum Mirages, and the Infinite Weirdness of Programmable Atoms.* New York: Basic, 2003.
Mulhall, Douglas. *Our Molecular Future: How Nanotechnology, Robotics, Genetics, and Artificial Intelligence Will Transform Our World.* Amherst, NY: Prometheus, 2002.
Rieth, Michael. *Nano-Engineering in Science and Technology: An Introduction to the World of Nano-Design.* River Edge, NJ: World Scientific, 2003.

Nash, John F. (1928–)

The mathematician John F. Nash made important contributions to game theory, which became an important part of later economic theories. John Forbes Nash Jr. was born in Bluefield, a prosperous town in West Virginia, where his father was an electrical engineer and his mother was a former schoolteacher. His mathematics skills were recognized at a young age, and his parents arranged for him to take extra classes at the local Bluefield College while still in high school. In 1945, he entered the Carnegie Institute of Technology (now Carnegie-Mellon University) with a full scholarship, the George Westinghouse Scholarship, majoring in chemical engineering. He soon switched to chemistry, then to mathematics. In two years he earned his B.S. and the faculty awarded him an M.S. in addition because of his recognized genius and ability. Harvard University and Princeton University both offered him fellowships; he chose Princeton because it was closer to home and they seemed more eager to have him come to the school.

The economist Oskar Morgenstern (1902–1976) and mathematician John von Neumann (1903–1957) had developed the concept of game theory, and published in 1944 their influential *Theory of Games and Economic Behavior*. Nash took their ideas, expanded them, gave them a firmer mathematical basis, and wrote his dissertation, receiving his Ph.D. in 1950. His work was published in that same year as "Noncooperative Games" in the *Annals of Mathematics*. At the age of twenty-two, Nash had already done the work that would make him famous. He published three more papers on game theory in the next three years, expanding his work to cooperative games, two-person bargaining games, and mutually optimal threat strategies in those kinds of games, as well as creating what became known as Nash equilibria. The Hungarian exile John C. Harsanyi (1920–2000) and the German Reinhard Selten (1930–) both expanded on Nash's work, and in the 1960s, game theory became an integral part of economic theory.

Nash accepted a position as a faculty member at the Massachusetts Institute of Technology (MIT), married a physics student in 1957, gained tenure in 1958, and began to experience hallucinations. In 1959, while his wife was pregnant, Nash resigned from MIT and admitted himself to the psychiatric ward of a hospital. He was diagnosed as a paranoid schizophrenic, and psychoanalysis failed to help him, since his problem was biological in origin. After he was released from the hospital, Nash spent some time wandering Europe. His wife divorced him in 1963, and he lived with his mother and sister when not in mental hospitals. Eventually his wife allowed him to live with her and their son, but they did not remarry.

Nash eventually came to control his mental illness by force of rationality, without drugs or psychotherapy. An important part of this process was rejecting any thoughts about political matters. He began to work on mathematical problems again and to visit old friends and haunts at Princeton. Nash shared the 1994 Nobel Prize in Economics with Harsanyi and Selten. Nash had been considered for the prize ever since the mid-1980s, but his mental illness caused the Nobel committee to exercise caution. Nash's lack of a formal university affiliation also concerned them, so Princeton created the new title of visiting research specialist to give Nash an academic home. His life and illness were portrayed in the Oscar-winning 2001 movie, *A Beautiful Mind.*

See also Economics; Mathematics; Neumann, John von; Nobel Prizes; Psychology

References
"John F. Nash, Jr.—Autobiography." http://nobelprize.org/medicine/laureates/index.html (accessed February 13, 2004).

Nasar, Sylvia. *A Beautiful Mind: A Biography of John Forbes Nash, Jr., Winner of the Nobel Prize in Economics, 1994.* New York: Simon and Schuster, 1998.

National Aeronautics and Space Administration (NASA)

The National Aeronautics and Space Administration (NASA) is an example of Big Science; under its aegis, the U.S. government spends billions of dollars a year on the scientific exploration of space, the development of space technology, and sustaining a manned presence in space. The launch of the first artificial satellite, *Sputnik 1,* by the Soviet Union on October 4, 1957, shocked the United States. The American answer was *Explorer 1,* launched by the United States Army on January 31, 1958. Both efforts were part of the International Geophysical Year of 1957–1958, an international effort to learn about the planet, but the underlying sense of competition came from the cold war. Putting satellites into orbit, then people into orbit, and finally men on the Moon, as part of the Apollo project, was driven by the competition for international prestige between the two superpowers, attempting to establish the ideological superiority of either communism or democracy through demonstrations of technological superiority.

The United States Congress passed the National Aeronautics and Space Act of 1958, and NASA was born on October 1, 1958. The earlier National Advisory Committee on Aeronautics (NACA), with 8,000 employees and three research centers, became part of NASA. Elements of the rocket research programs of the navy and army joined NASA, and the contract with the California Institute of Technology's Jet Propulsion Laboratory, builder of *Explorer 1,* was transferred from the U.S. Army to NASA.

The Soviets again beat the Americans by putting the first man into space when Yuri A. Gagarin performed a single orbit of Earth on April 12, 1961. NASA's Project Mercury, already under way, put Alan B. Shepard into space in a suborbital flight on May 5, 1961, and the first American into orbit on February 20, 1962. On May 25, 1961, in his first state of the union address, President John F. Kennedy

Aerial view of NASA's Johnson Space Center near Houston, Texas (Courtesy of NASA)

urged Congress to commit the United States to landing a man on the Moon and returning him safely to Earth before the end of the decade. The Apollo project consumed NASA's energies for a decade. On July 21, 1969, men walked on the Moon. From 1969 to 1972, a total of twelve astronauts walked on the Moon, the only men to ever do so.

After $25 billion was spent on the Apollo project, congressional appropriations for NASA declined in the 1970s. NASA ran three Skylab missions in 1973 and 1974, during which American astronauts lived in space aboard a small space station. During the Apollo-Soyuz mission of 1975, American astronauts and Soviet cosmonauts rendezvoused in space and shook hands inside a temporary tunnel connecting the two spacecraft. Most of the 1970s was devoted to designing and building a reusable space shuttle. On April 12, 1981, the first space shuttle lifted off. Getting the most complex technical invention of all time to work on the first try was an impressive engineering achievement. The reusable shuttle proved to be so expensive that expendable rockets remained cheaper for putting most satellites into orbit. Losses of the space shuttles *Challenger* in 1986 and *Columbia* in 2003 marred the program.

In 1982, NASA began planning for a permanent space station. Over the years, international partners joined the effort, and in 1998 the International Space Station began construction in orbit. Permanent crews began habitation in 2000. NASA was joined by Russia, the European Space Agency, Japan, and Canada in this effort. Proposals for a manned lunar base and a manned Mars mission have been stymied by projected high costs and the lack of a cold war to provide motivation.

Besides manned spaceflight, NASA devoted considerable attention to scientific exploration. The Jet Propulsion Laboratory administered all of the interplanetary spacecraft that NASA built, eventually visiting all of the planets except for Pluto. Orbiting telescopes, like the Hubble Space Telescope, gave astronomers a view of the cosmos unobscured by the atmosphere. Various efforts searching for extraterrestrial life (SETI) were also fully funded or partially funded by NASA.

The efforts of NASA-funded contractors and NASA employees led to major advances in microelectronics, computers, rocket technologies, sensors, and spin-off products. Considerable medical research gave NASA doctors an understanding of the long-term effects of zero gravity on the human body. NASA also continued the efforts of the earlier NACA in creating new aircraft technologies and conducting research in aviation. A notable aviation achievement came in the 1960s with the X-15 program, which involved a rocket-powered airplane that actually left the atmosphere and then glided back to Earth.

See also Apollo Project; Big Science; Cold War; Computers; European Space Agency; International Geophysical Year; Jet Propulsion Laboratory; Satellites; Search for Extraterrestrial Life; Shoemaker-Levy Comet Impact; Soviet Union; Space Exploration and Space Science; Telescopes; Voyager Spacecraft

References

Bilstein, Roger E. *Orders of Magnitude: A History of the NACA and NASA, 1915–1990* (NASA SP-4406). Washington, DC: Government Printing Office, 1989.

Burrows, William E. *This New Ocean: The Story of the First Space Age.* New York: Random House, 1998.

McDougall, Walter A. *The Heavens and the Earth: A Political History of the Space Age.* New York: Basic, 1985.

National Aeronautics and Space Administration. http://www.nasa.gov/ (accessed February 13, 2004).

National Geographic Society

Founded in 1888 in Washington, D.C., by a group of renowned explorers and scientists, the National Geographic Society is the largest popular science society in the world. The Society originally focused on elite members of society, promoting understanding of geographical issues. Beginning in 1900, Gilbert H. Grosvenor (1875–1966) transformed the *National Geographic* magazine from a professional journal into a mass-market magazine. Circulation quickly grew, reaching 10 million members of the Society in 1980. Grosvenor served as editor in chief of the magazine from 1903 to 1954 and as president of the Society after 1920. His son and grandson continued in his footsteps. The magazine grew famous for its photographs of exotic things, peoples, places, and animals, becoming the primary window that many Americans used to view their world. National Geographic maps became standard fare in most American schoolrooms and homes. Appalled at geographic illiteracy among the American population, the Society began creating educational materials and training teachers in geography in the 1970s.

The Society sponsored exploration and scientific expeditions around the world. The magazine brought reports of these expeditions back to Society members, and television specials reached an even larger audience. The Society became a key sponsor of the scientific investigations of ancient hominids and contemporary primates by the Leakey family, beginning with Englishman Louis S. B. Leakey (1903–1972) and his protégés: the Englishwoman Jane Goodall (1934–), the Canadian Biruté Galdikas (1946–), and the American Dian Fossey (1932–1985). The Society also sponsored the oceanographic expeditions of the French explorer Jacques Cousteau (1910–1997). Long active in the conservation movement, the Society has embraced the environmental movement, though some critics believe that the Society is not aggressive enough in the cause of environmentalism.

See also Anthropology; Archaeology; Cousteau, Jacques; Environmental Movement; Fossey, Dian; Goodall, Jane; Leakey Family; Media and Popular Culture

References

Bryan, C. D. B. *The National Geographic Society: 100 Years of Adventures and Discovery.* New York: Abrams, 1987.

Grosvenor, Gilbert M. "A Hundred Years of the National Geographic Society." *The Geographical Journal* 154, no. 1 (March 1988): 87–92.

National Geographic Online. http://www.nationalgeographic.com/ (accessed February 13, 2004).

National Institutes of Health

In 1798 the federal government established the Marine Hospital Service for sick and disabled seamen. The Service grew, and in 1887 a laboratory for the study of bacteria, the Laboratory of Hygiene, was created at the Marine Hospital on Staten Island, New York. The laboratory later moved to Washington, D.C., and took on ever greater responsibilities. In 1930, the Hygienic Laboratory became the National Institute of Health. A 1935 private grant of land led to the laboratory moving to Bethesda, Maryland, in 1938. Congress kept adding responsibilities and funding to the organization, and in 1948 it became the National Institutes of Health (NIH). One of the darker days in NIH history occurred in 1954 when a batch of NIH-approved Salk polio vaccine included viruses that had not been deactivated. Eighty children received the vaccine, who then infected about 120 more people, leaving 11 dead and three-fourths of the victims paralyzed with polio.

The NIH eventually grew to include twenty-seven institutes and centers. Their names and founding dates are the National Cancer Institute (1937); Center for Scientific Review (1946); National Heart, Lung, and Blood Institute (1948); National Institute of Dental and Craniofacial Research (1948); National Institute of Allergy and Infectious Diseases (1948); National Institute of Diabetes and Digestive and Kidney Diseases (1948); National Institute of Mental Health (1949); National Institute of Neurological Disorders and Stroke (1950); Warren Grant Magnuson Clinical Center (1953); National Library of Medicine (1956); National Institute of Child Health and Human Development (1962); National Institute of General Medical Sciences (1962); National Center for Research Resources (1962); Center for Information Technology (1964); National Eye Institute (1968); John E. Fogarty International Center (1968); National Institute of Environmental Health Sciences (1969); National Institute on Alcohol Abuse and Alcoholism (1970); National Institute on Drug Abuse (1973); National Institute on Aging (1974); National Institute of Arthritis and Musculoskeletal and Skin Diseases (1986); National Institute of Nursing Research (1986); National Institute on Deafness and Other Communication Disorders (1988); National Human Genome Research Institute (1989); National Center for Complementary and Alternative Medicine (1992); National Center on Minority Health and Health Disparities (1993); and National Institute of Biomedical Imaging and Bioengineering (2000).

The federal Centers for Disease Control and Prevention (CDC), founded much later than the NIH, overlaps at times with the NIH in its responsibilities, causing duplication of efforts and bad feelings between the organizations. In general, though, the NIH concentrates on basic research, and the CDC works on recognizing and controlling communicable diseases.

Federal funding of the NIH exceeded $13 billion in 2000. In terms of funding, the NIH is an example of Big Science, though about 80 percent of their funds are allocated in grants to medical researchers at universities, hospitals, and medical centers around the nation, and the balance is used to run their own extensive laboratories and administer the NIH. The development of NIH represents one of the first cases in which the United States Congress ceded control of research money to the scientific community, allowing scientific review committees to decide where money should be spent. More than eighty Nobel laureates have received funding from the NIH, and five Nobel laureates have come from NIH laboratories. In 1984, a team headed by Robert Gallo (1937–) at the NIH and a French team both discovered the retrovirus human immunodeficiency virus (HIV), which causes AIDS. The Human Genome Project, started in 1990 as a joint project between the United States Department of Energy (DOE) and the NIH, led to the successful mapping of the genes in the human body.

See also Big Science; Biotechnology; Centers for Disease Control and Prevention; Human Genome Project; Medicine; Nobel Prizes; Pharmacology; Psychology; Salk, Jonas

References
National Institutes of Health. http://www. nih.gov/ (accessed February 13, 2004).
The NIH Almanac. Bethesda, MD: National Institutes of Health, 2000.
Strickland, Stephen P. *The Story of the NIH Grants Program.* Lanham, MD: University Press of America, 1988.

National Science Foundation

The American electrical engineer Vannevar Bush (1890–1974) directed the Office of Scientific Research and Development during World War II. As the war drew to an end, Bush wrote an influential report, *Science— The Endless Report,* for the president of the United States that encouraged continued federal involvement in developing new science and technology. The cold war helped motivate the American government in this area, and in 1950 the Congress created the National Science Foundation (NSF). Bush wanted an independent board to control the NSF, but the federal government retained control. The president appointed, and the Senate confirmed, the NSF director and members of the policymaking National Science Board, following a practice common for other senior federal positions.

The physicist Alan T. Waterman (1892–1967) served as the first director (1951–1963), choosing to downplay the role of the NSF as a coordinator of federal science research. He did not want to antagonize the other established agencies in the federal government, such as the National Institutes of Health (NIH) and the Atomic Energy Commission (AEC), by creating turf battles. Waterman concentrated on funding basic research efforts and developing new scientific talent through graduate fellowships and postdoctoral positions. The NSF accepted proposals from scientists and used peer review boards to judge their merits and allocate funding. In 1962, officials in the Office of the President of the United States assumed the little-used coordination function of the NSF.

At first the NSF tended to favor the elite universities, but that gradually changed, especially since Congress wanted to spread around the largesse. The NSF started with an annual budget of only $225,000, much less than the $33.5 million that Bush had proposed in 1945. The NSF did not control research laboratories of its own, as the NIH and AEC did, but contracted with universities to establish and run laboratories funded by the NSF.

When the Soviet Union launched the first artificial satellite, *Sputnik I,* in 1957, the United States reacted with alarm. Funding for scientific and technological research increased, as well as funding to educate scientists and engineers. The NSF benefited from this trend, and an increasingly larger portion of the NSF budget went to big science projects, such as the failed Mohole Project of the early 1960s, which aimed to drill through the mantle of the Earth, specialized ships for oceanography, and the Very Large Array (a system of 27 radio telescopes in New Mexico) for radio astronomy. The NSF participated in the International Geophysical Year (IGY) of 1957–1958, especially concentrating on research in Antarctica. After the IGY ended, the NSF remained the lead federal agency funding a continuing presence and research on the frozen continent. The NSF budget in 1983 passed $1 billion, and reached $3.7 billion by the end of the century.

In general the NSF avoided controversy. In the early 1950s, however, when Senator Joseph McCarthy questioned the loyalty of academic scientists who received NIH funding, the NSF took note and declared that the NSF would not require security checks for their scientists, though they would not fund the research of a known communist. McCarthyism faded before any possible negative political consequences of this decision emerged. Funding of social science projects occasionally ran afoul of criticism from members of Congress, encouraging the NSF to minimize projects in the social sciences. In the 1970s the NSF moved beyond basic research to fund Research Applied to National Needs (RANN), such as efforts to alleviate pollution, solve urban problems, and handle energy issues. Critics feared that RANN could dilute the traditional emphasis on basic research, and the NSF canceled the program in 1978, returning to its primary mission, though engineering and other applied disciplines did continue to get funding on a reduced scale.

Part of the charter of the NSF required the organization to help increase communications among scientists. Besides sponsoring publications, workshops, and educational programs, the NSF created the Computer Science Network (CSNET) in 1981 to provide universities that did not have access to ARPAnet with their own network. ARPAnet, the network created by the Advanced Research Projects Agency (ARPA), later became the Internet. In 1986 the NSF sponsored the NSFNET "backbone" to connect five supercomputing centers together. The backbone also connected ARPAnet and CSNET together. The funding from the NSF at this time played a crucial role in the expansion of the Internet. When ARPAnet was dismantled in 1990, the Internet was thriving at universities and technology-oriented companies. The NSF backbone was later dismantled, when the NSF realized that commercial and nonprofit entities could keep the Internet running and growing on their own.

See also Big Science; Cold War; International Geophysical Year; Internet; National Institutes of Health; Universities

References
Appel, Toby A. Shaping Biology: The National Science Foundation and American Biological Research, 1945–1975. Baltimore: Johns Hopkins University Press, 2000.
England, J. Merton. A Patron for Pure Science: The National Science Foundation's Formative Years, 1945–1957. Washington, DC: National Science Foundation, 1982.
National Research Council. 50 Years of Ocean Discovery: National Science Foundation, 1950–2000. Washington, DC: National Academy, 2000.
National Science Foundation. http://www.nsf.gov/ (accessed February 13, 2004).

Neumann, John von (1903–1957)

The mathematician John von Neumann made important contributions to many areas of mathematics and quantum physics before turning to computers and developing the theoretical foundations of digital electronic computers. Johann (later anglicized to John) von Neumann was born to a Jewish family in Budapest, Hungary, where his father was a banker. His family recognized his extraordinary intelligence as a child and hired a private tutor to supplement his education. During 1919, when a brief communist government gained control of Budapest in the chaos after World War I, the family found it prudent to stay away, leaving von Neumann with a lifelong abhorrence of communism.

Von Neumann entered the University of Budapest in 1921 to study mathematics, but his father feared that employment opportunities in that field were limited, so von Neumann agreed to also study chemistry. While pursuing his Budapest studies, he also studied at the University of Berlin and the Swiss Federal Institute of Technology in Zurich. He earned a degree in chemical engineering from the latter institution in 1925. When he received his doctorate in mathematics from the University of Budapest in 1926, von Neumann was only twenty-two years old and already publishing mathematical articles. His dissertation described the axiomatization of set theory. In 1927, he studied at the University of Göttingen as a Rockefeller Fellow and met the American physicist J. Robert Oppenheimer (1904–1967). Von Neumann then worked as a lecturer in mathematics at the University of Berlin for three years, while conducting research into the relationship between quantum physics and operator theory. In 1930, von Neumann moved to Princeton University in America, where he began as a visiting lecturer, becoming a full professor and one of the original members of Princeton's Institute for Advanced Study only three years later. His prowess in mathematics and numerous original contributions made him a leading theorist. With the economist Oskar Morgenstern (1902–1976), von Neumann developed the concept of "game theory," and the two men published in 1944 their influential *Theory of Games and Economic Behavior.*

Von Neumann became a U.S. citizen in 1937, allowing him to obtain the necessary security clearances during World War II. He worked on the Manhattan Project with Oppenheimer, where he championed the implosion method of detonating an atomic bomb, a method initially ignored by most other scientists, but later used. He also lent his wide expertise as a consultant to other defense projects. A chance encounter with an army liaison officer at a railroad station led to his involvement with the U.S. Army's ENIAC (Electronic Numerical Integrator and Computer)

project in 1944 at the University of Pennsylvania Moore School of Electrical Engineering. The ENIAC was a high-speed electronic calculator designed to create artillery ballistic tables, which was being built by J. Presper Eckert (1919–1995) and John Mauchly (1907–1980). While building ENIAC, Eckert and Mauchly developed for their next computer the concept of a stored program, where data and program code resided together in memory. This concept allowed computers to be programmed dynamically so that the actual electronics and connections did not have to be changed with every program.

Von Neumann expanded on the concept of stored programs and laid the theoretical foundations of all modern computers in a 1945 report and later work. His ideas were embodied in what came to be known as the von Neumann Architecture. The center of the architecture is the fetch-decode-execute repeating cycle, in which instructions are fetched from memory and then decoded and executed in a processor. The results of the instruction change data that are also in memory. Eckert and Mauchly deserve equal credit with von Neumann for their innovations, though von Neumann's elaboration of their initial ideas and his considerable prestige lent credibility to the budding movement to build digital computers. Eckert and Mauchly went on to build the first commercial computer, the UNIVAC, in 1951. Von Neumann went back to Princeton and persuaded the Institute for Advanced Study to build their own pioneering computer, the IAS (derived from the initials of the institute), which he designed. He also developed a theory of automata, including the idea of self-replicating automata.

In 1954, von Neumann was appointed as a member of the Atomic Energy Commission (AEC) and moved to the Washington, D.C., area. His anticommunist convictions informed his aggressive approach to the continued development of nuclear weapons and civilian uses of nuclear power. He successfully advocated for the creation of a ballistic missile program to deliver nuclear weapons

Neuroscience

The term *neuroscience* emerged after World War II in an effort to expand the traditional boundaries of neurology to include anatomical, physiological, and psychological perspectives. In the first half of the century, neurologists had determined that the brain is made of independent nerve cells called neurons and that these neurons are divided by synapses. Electrical signals across the synapses make up brain activity. In the 1920s scientists began to find chemicals called neurotransmitters that can either help or hinder the ability of neurons to make electrical connections. Later research determined that the brain is born with about 100 billion neurons and that no new neurons are created during an individual's life. Invented in the 1920s by the German neuropsychiatrist Hans Berger (1873–1941), the electroencephalograph created tracings on a paper called an electroencephalogram (EEG), which enabled neurologists to make crude evaluations of brain activity.

With new tools and insights, neurologists began to move away from just the physiology of reflex action, in which a given stimulus to the brain or nervous system is seen to result in a consistent outcome. Neurologists also used brain surgery on primates to understand which parts of the brain controlled which function. Applying their knowledge to humans, they learned that the frontal lobe controlled the higher functions and that more primitive functions were located deeper inside, a finding consistent with the idea of biological evolution. In the 1940s and 1950s, lobotomy operations that severed the neural connections to the frontal lobe and even lobectomy operations that removed part of the frontal lobe were used on difficult-to-control mental patients. The zoologist Roger Sperry (1913–1994) shared the 1981 Nobel Prize in Physiology or Medicine for showing through surgical procedures on animals and humans during the 1960s that the left and right hemispheres of the brain perform different functions.

The ability of computers to translate large

John von Neumann, mathematician (Library of Congress)

over long distances. Along with honorary doctoral degrees, von Neumann was elected to the National Academy of Sciences in 1937 and received several national honors for his defense work. In 1957 he died of cancer, perhaps contracted during his work on the Manhattan Project. Von Neumann was a gregarious man with sophisticated tastes, a command of four languages, a prodigious memory, and an amazing ability to perform calculations in his head. He married in 1930, fathered a daughter, and divorced in 1937, before remarrying in 1938. Both of his wives came from Budapest.

See also Cold War; Computers; Mathematics; Nuclear Physics; Particle Physics

References

Aspray, William. *John von Neumann and the Origins of Modern Computing.* Cambridge: MIT Press, 1990.

Goldstine, Herman H. *The Computer: From Pascal to von Neumann.* Princeton: Princeton University Press, 1972.

Macrae, Norman. *John von Neumann.* New York: Pantheon, 1992.

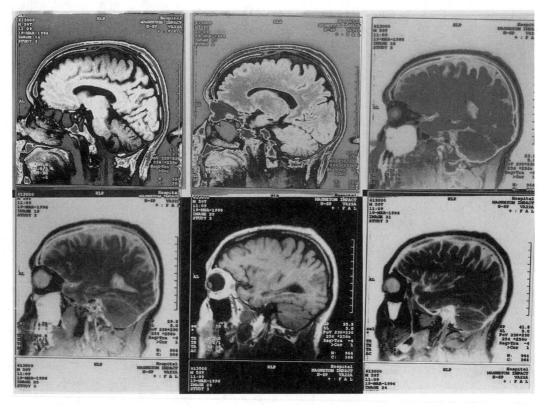

Multiple MRI brain scans (Tom and Dee Ann McCarthy / Corbis)

amounts of data into images, along with advances in nuclear physics, led to the development of much more sophisticated tools for analyzing the brain as it functioned. The nuclear physicist Allan M. Cormack (1924–1998), a South African who later moved to America, developed the basic theory of computer tomography (CT) in the early 1960s, where x-rays are interpreted by mathematical algorithms to create a two-dimensional image of a slice of the human body. The electrical engineer Godfrey N. Hounsfield (1919–), a researcher at Electric and Musical Industries in England, patented a CT device in the late 1960s, and by the mid-1970s, computed axial tomography (CAT) scanners were being commercially produced. Cormack and Hounsfield shared the 1979 Nobel Prize in Physiology or Medicine for their invention.

The nuclear chemist Michael E. Phelps (1939–) invented positron emission tomog-raphy (PET) in 1973, which uses radioactive tracers and can show neural activity in the brain. Because of the cost of a cyclotron necessary to create the radioactive pharmaceuticals and the cost of the machine, there were only about 150 PET installations worldwide at the end of the century. Phelps was among the first to use his invention to show in intricate detail that different parts of the brain are activated when performing the different mental tasks of talking, listening, reading, and thinking. This kind of evidence pointed to the localization of function within the brain. Newer scanners, such as functional magnetic resonance imaging (fMRI), do not use a radioactive tracer.

Neurologists once thought that the brain inevitably atrophied with age, with a decreasing ability to create new neural connections, but neurologists now recognize that neurons retain the potential to establish new connections throughout the entire life span. This is a return to the idea of the earlier plasticity, where the brain can redevelop damaged functions despite localized damage. Neuroscientists also moved away from accepting the localized, connectionist model of the brain that had been dominant in the 1970s, a model according to which only specific parts of the brain were thought to do specific functions.

Brain scans and a greater understanding of brain chemistry have led to a better understanding of mental disorders with a biological basis, such as schizophrenia, some forms of depression, Alzheimer's, and Parkinson's disease. In the 1950s, neurologists suspected that excessive amounts of the neurotransmitter dopamine contributed to schizophrenia. Drugs designed to address the dopamine problem had mixed success. In the 1960s, neurologists noticed that abusers of phencyclidine (PCP) and people who received low doses of ketamine, an anesthetic, exhibited symptoms similar to schizophrenia. Further research found that these drugs block the NMDA (N-methyl-D-asparate) receptors in the brain. The discovery of endorphins in the 1970s, the natural opiates of the brain, helped neurologists understand how some people can so readily achieve a natural high and how addiction works. Some drugs for biologically based mental disorders have been developed, but they often have substantial side effects.

Studies from scans have led some neurologists to the conviction that men's and women's brains function differently in many respects. Cognitive psychologists have also come to the same point of view. These conclusions, flying in the face of the traditional view within feminism that biological differences are unimportant, remain controversial.

The biochemist Gerald M. Edelman (1929–), who earned his 1972 Nobel prize for his work on antibodies, later turned to neuroscience. He developed a theory that groups of neural connections in the brain use a process akin to natural selection to select from among other groups because they are more effective at producing consciousness, memory, and other higher functions of the mind. This idea is not widely accepted. The Nobel laureate Francis Crick (1916–2004) was among other prominent scientists who also turned to neuroscience in their later years, an indication that the field is striving to answer fundamental questions and promises exciting developments in the future.

See also Cognitive Science; Computers; Crick, Francis; Edelman, Gerald M.; Feminism; Medicine; Nobel Prizes; Nuclear Physics; Pharmacology; Psychology

References
Corsi, Pietro, editor. *The Enchanted Loom: Chapters in the History of Neuroscience.* New York: Oxford University Press, 1997.
Gillick, Muriel R. *Tangled Minds: Understanding Alzheimer's Disease and Other Dementias.* New York: Dutton, 1998.
LeDoux, Joseph. *Synaptic Self: How Our Brains Become Who We Are.* New York: Viking, 2002.
Moir, Anne, and David Jessel. *Brain Sex: The Real Difference between Men and Women.* New York: Penguin, 1989.
Restak, Richard. *The New Brain: How the Modern Age Is Rewiring Your Mind.* Emmaus, PA: Rodale, 2003.

Neutrinos

In 1931, Wolfgang Pauli (1900–1958), an Austrian and professor of theoretical physics at the Federal Technical Institute in Zürich, Switzerland, suggested that an undiscovered subatomic particle was emitted during beta decay, a form of radioactivity. His proposed particle would be almost impossible to detect, and he quickly regretted his hasty suggestion, fearing possible embarrassment at such a notion. The Italian physicist Enrico Fermi (1901–1954) took this idea and developed it mathematically, calling the unknown particle a neutrino. In 1934, Fermi proposed that in the process of beta decay, a neutron decayed into three particles: a proton, an electron, and a neutrino. The proposed neutrino carried a small charge and reacted so weakly with other matter that it would be incredibly difficult to detect any evidence of it.

The theory also predicted that neutrinos would have no mass or an incredibly minuscule amount of mass.

With the invention of nuclear reactors during World War II, scientists had a strong source of beta decay, and some realized that nuclear reactors provided a wonderful source of possible neutrinos. In 1956, the physicists Frederick Reines (1918–1998) and Clyde Cowan (1919–1974) used a solution of water and cadmium near a nuclear reactor at Savannah River, South Carolina, to detect the photons emitted from the reactions that formed neutrinos. Reines and Cowan sent a telegram to Pauli about their discovery, confirming the suggestion that Pauli had regretted, and Reines received a share of the 1995 Nobel Prize in Physics for his work. (Cowan was dead by that time and Nobel prizes are not awarded posthumously.)

The neutrino detected by Reines and Cowan occurred during the formation of an electron. As scientists used particle accelerators to discover an ever greater number of subatomic particles, they realized that there was more than one type of neutrino. Julian Schwinger (1918–1994) proposed in 1957 that two forms of neutrinos existed, one associated with electrons and the other with muons. A neutrino associated with the formation of a muon particle was discovered in 1962. In the 1970s, a neutrino associated with the formation of a tau particle was established in theory. Though the tau neutrino has not been found experimentally, it is commonly accepted as the third type of neutrino.

The Sun, powered by nuclear reactions, also emits neutrinos. A series of expensive experiments have tried to detect these solar neutrinos as well as neutrinos emitted by distant supernovas. The first experiment was organized by the radiochemist Ray Davis (1914–) in the Homestake gold mine in South Dakota. Solar neutrinos that passed through a huge tank filled with chlorine reacted with chlorine-37 isotopes to create argon 37. Davis was puzzled to find only

about one fourth of the neutrinos expected according to the prevailing theories. Possible explanations included the possibility that our solar models are slightly flawed or that neutrinos do possess a slight mass. The Japanese physicist Masatoshi Koshiba (1926–) of the University of Tokyo organized the Kamiokande experiment using similar principles and found similar results. Davis and Koshiba shared the 2002 Nobel Prize in Physics with x-ray astronomer Riccardo Giaconni (1931–) for their work.

During the 1990s, other experiments around the world tried to find the missing solar neutrinos, including the Russian-American Gallium Solar Neutrino Experiment in the Caucasus Mountains and an effort at the Los Alamos National Lab in New Mexico. These efforts occasionally announced success at finding neutrinos with mass, but were not able to confirm their results. A new theory, based on results from the Sudbury Neutrino Observatory in Ontario, Canada, has suggested that solar neutrinos change type after they are emitted from the Sun but before being detected by the Earth-based experiments, leading to the undercount, since earlier experiments had looked for only a certain range of neutrinos.

See also Astronomy; Nuclear Physics; Particle Physics

References

Franklin, Allan. *Are There Really Neutrinos? An Evidential History.* New York: Perseus, 2001.

McDonald, Arthur B., Joshua R. Klein, and David L. Wark. "Solving the Solar Neutrino Problem." *Scientific American* 288, no. 4 (April 2003): 40–49.

Solomey, Nickolas. *The Elusive Neutrino: A Subatomic Detective Story.* New York: Scientific American Library, 1997.

Nobel Prizes

Alfred Nobel (1833–1896) patented his invention of dynamite in 1867, one of 355 patents made during a vigorous life of research and entrepreneurship that earned

The Nobel prize functions as the highest honor that scientists can receive in their field of expertise. (Corel Corporation)

him a fortune. His secret will stipulated that on his death, his wealth was to be used to fund annual prizes in physics, chemistry, physiology or medicine, literature, and peace. A Nobel Foundation was set up in Stockholm, Sweden, to invest Nobel's fortune and coordinate the work of the institutions that actually choose the recipients of the prizes. The Nobel Peace Prize laureate is selected by the Norwegian Shorting (parliament); the Nobel Prize in Physiology or Medicine is selected by the Karolinska Institutet; the Nobel Literature Prize is selected by the Swedish Academy; and the Nobel Prize in Physics and the Nobel Prize in Chemistry are both selected by the Royal Swedish Academy of Sciences.

The king of Sweden presents the prizes. The first prizes were awarded in 1901, and for the first twenty-five years laureates were not publicly announced until the award ceremony itself. Though the rules have occasionally changed, in practice Nobel prizes are not awarded posthumously and may be divided among up to three laureates. The co-laureates do not have to have worked on the same area

in their discipline. Occasionally a laureate is not selected, and the prize is not awarded for that year. Records of committee deliberations on prize awards are kept sealed for at least fifty years. Oddly enough, the Nobel Foundation was not exempted from Swedish taxes until 1946 and for years was the top taxpayer in Stockholm, sapping its economic vitality. Less constrained investments and freedom from taxes allowed the foundation to regain a firm footing in the 1950s.

In 1968, the Bank of Sweden made a large donation to the foundation to fund The Bank of Sweden Prize in Economic Sciences in Memory of Alfred Nobel, selected by the Royal Swedish Academy of Sciences. The first economics prize in 1969 was shared by Ragnar Frisch (1895–1973) of the University of Oslo and Jan Tinbergen (1903–1994) of the Netherlands School of Economics in Rotterdam for their work, beginning in the 1920s and 1930s, in helping develop economics into a mathematically rigorous and quantitatively oriented discipline.

The Nobel prizes function as the highest honor that scientists can receive in their field of expertise. Scientists have occasionally won the Peace Prize, notably the chemist Linus Pauling (1901–1994) in 1962, for his efforts to end open-air nuclear weapons testing, and Norman E. Borlaug (1914–) in 1970, for his efforts with the Green Revolution. Other science prizes, such as the Japan Prize and Kyoto Prize, both established in 1985, are modeled on the Nobel prizes. The Swedish businessman Holger Crafoord (1908–1982) and his wife, Anna-Greta Crafoord (1914–1994), funded the annual Crafoord Prize to be awarded in those areas not covered by the Nobel Prizes, such as mathematics, astronomy, the biosciences, ecology, and the geosciences. The Royal Swedish Academy of Sciences selects who will receive the Crafoord Prizes.

See also Chemistry; Economics; Green Revolution; Medicine; Pauling, Linus; Physics

References

Hargittai, István. *The Road to Stockholm: Nobel Prizes, Science, and Scientists.* New York: Oxford University Press, 2002.

Les Prix Nobel annual series. Stockholm: Almqvist and Wiksell, International.

Levinovitz, Agneta Wallin, and Nils Ringertz. *The Nobel Prize: The First 100 Years.* River Edge, NJ: World Scientific, 2001.

Nobel *e*-Museum. http://nobelprize.org/medicine/laureates/index.html (accessed February 13, 2004).

Nuclear Physics

Fifteen Nobel laureates worked on the Manhattan Project, directed by the physicist J. Robert Oppenheimer (1904–1967), making it perhaps the single project in human history with the greatest concentration of brainpower. Only in 1938 had the idea of nuclear fission in a chain reaction, making a bomb possible, even been broached. The success of the project in producing atomic weapons in 1945 and their use on the Japanese cities of Hiroshima and Nagasaki in August 1945, hastening the end of the war, had multiple consequences: it set the stage for an arms race within the context of the cold war; it firmly established physics as the queen of the sciences for the next several decades; it created the potential for other uses of nuclear power; and it created a climate of global fear.

After the Soviets detonated their own atomic bomb in 1949, the United States embarked on building a fusion-based device, the hydrogen bomb. The Hungarian-born physicist Edward Teller (1908–2003) led the American effort, which exploded the first hydrogen bomb on Eniwetok Atoll in 1952. The Soviet physicist Andrei Sakharov (1921–1989) designed the first Soviet hydrogen bomb, exploded only a year later in 1953. The nuclear arms race began in earnest, with the two superpowers pouring billions of dollars and enormous resources into making tens of thousands of nuclear weapons. Other countries developed their own nuclear arsenals, though these were small compared with the superpower stockpiles. The United States and Soviet Union experimented with putting nuclear weapons not only in bombers and atop ballistic missiles, but into depth charges, torpedoes, anti-aircraft missiles, and artillery shells. A considerable amount of effort went into developing small enough nuclear weapons to place atop missiles or in suitcases. Some weapons maximized their electromagnetic pulse (EMP) to destroy electronics and electrical systems; other weapons were designed to minimize their radioactive fallout, and the neutron bomb, developed in the 1970s, was designed to maximize the radiation effect and minimize the blast effect, killing people and not destroying things. The doctrine of mutual assured destruction (MAD) made both superpowers reluctant to use their nuclear weapons for fear that a retaliatory strike would destroy their own nation. Some scholars have argued, considering the ideological intensity of the cold war between the democracies and communism, that the fear of nuclear weapons actually prevented a third world war.

Studies of radioactive fallout from nuclear tests in the 1950s led to efforts to ban aboveground testing, which succeeded with the Nuclear Test Ban Treaty of 1963. In 1962, the Nobel laureate and chemist Linus Pauling (1901–1994) earned a second Nobel Prize, the Peace Prize, for his leadership in the antitesting movement. In the 1980s a vigorous antinuclear movement emerged in Western Europe and the United States, which led to further scientific research, including the finding by the planetary astronomer Carl Sagan (1934–1996) and other scientists that a full-scale nuclear war would throw up so much dust into the atmosphere that the planet would be plunged into a nuclear winter. After the fall of the Soviet Union, the superpowers signed a series of treaties during the 1990s to sharply reduce their stockpiles. With a diminished threat of superpower conflict, concerns among the informed turned to the proliferation of nuclear technology and the possibility that an unstable country or terrorist group might obtain nuclear weapons.

The demand for nuclear physicists increased sharply after World War II, especially within the superpowers. Graduate programs in nuclear physics at universities expanded to accommodate the demand. The physicists used their newfound connections to government to successfully lobby for additional funding for other big science projects, such as particle accelerators and other large devices to detect and study ever smaller pieces of matter and energy, including solar neutrinos and cosmic rays. After the end of the cold war, the cancellation of the U.S. Superconducting Supercollider (SSC) in 1993 signaled that the government largesse for physicists had limits. A downside of the increased prominence of physicists was an emerging sense that scientists were irresponsible in developing science and technology that could destroy all life. This ambivalence about scientists is easily recognized in postwar movies, books, and popular culture.

A team led by Enrico Fermi (1901–1954), the Italian-born Nobel laureate and physicist, built the first fission-based nuclear reactor in 1942 as part of the Manhattan Project. More nuclear reactors followed to create the fuel for the atomic bombs. During the 1950s, nuclear scientists and nuclear engineers eagerly turned to applying nuclear reactors to every possible situation. In the military arena, nuclear power plants replaced oil engines on submarines, aircraft carriers, and cruisers, especially on U.S. ships. Experimental vessels included a nuclear-powered airplane and a nuclear-powered merchant freighter. Scientists also recognized the potential for nuclear reactors to supply electricity, and a small experimental breeder reactor in Idaho generated the first electricity in 1951, though just enough power to light the reactor building.

The Soviet Union started building the first civilian nuclear power plant in 1954, with the United Kingdom following a year later, and the United States started building its first civilian reactor in 1956. Nuclear engineers

The Three Mile Island nuclear facility in Pennsylvania (Greenpeace)

expended considerable effort to make the reactors safe, believing that their designs for light-water reactors would automatically shut down if controls failed. By 1998, 437 nuclear power plants operated in twenty-nine countries, providing perhaps 5 percent of the world's electrical supply. France and Japan, lacking significant domestic sources of energy, invested heavily in nuclear plants. Expensive to build because of safety measures, nuclear plants proved to supply cheap electricity with no air pollution, but at the cost of generating large amounts of radioactive waste. Occasional accidents, such as those at Windscale in the United Kingdom (1957), Mayak in Siberia (1957), Three Mile Island in Pennsylvania (1979), and Chernobyl in the Soviet Ukraine (1986), plus opposition from the environmental movement beginning in the 1970s, led many to question the wisdom of nuclear reactors. Curiously enough, French scientists in the 1970s discovered the remains of a natural fission reactor in Oklo, Gabon, in Africa, where an underground bed of uranium ore supported intermittent chain reactions over a billion years ago.

The process of nuclear fusion combines atoms rather than splitting them apart as nuclear fission does. The German-born physicist Hans Bethe (1906–), who fled the Nazis to America, proposed in 1938 that the process of fusion powered the Sun. A claim in 1951 by the Argentine president Juan Perón (1895–1974) that his country had built a fusion reactor astonished the world and turned out to be false. In the early 1950s, American scientists began to develop a way to use magnetic fields to confine the high-temperature plasma necessary for fusion to occur, literally putting the power of the Sun into a magnetic bottle. The United States Atomic Energy Commission sponsored the secret work. The United States dropped the shroud of secrecy around the fusion effort after the Soviet success with *Sputnik I* in 1957, as part of a publicity effort to show that the United States still engaged in groundbreaking scientific research, though the effort eventually foundered on technical difficulties.

American scientists revived their fusion research effort in the early 1970s as the nuclear industry came under concerted assault by the new environmental movement. Fusion enthusiasts expected a commercial fusion reactor to be safer and generate less radioactive waste than fission reactors. The radioactive waste generated by fusion reactors would also have a much shorter half-life than the radioactive waste of fission reactors. The 1970s effort continued into the 1980s before funding fizzled and the United States Department of Energy elected to begin to work with other countries on the problem, spreading the cost of research. The Soviets, Japanese, French, and other European countries had already worked on fusion, with the Soviet tokamak method of confinement be-

coming a favored approach even in the United States. Despite billions of dollars in research, the promise of fusion remains only a hope for cheaper, cleaner electricity. At the end of the century, fusion researchers hoped to build a large research fusion reactor in either France or Japan, and negotiations on the ITER ("the way" in Latin) project are continuing in 2004.

The announcement in 1989 that two respected chemists at the University of Utah had discovered a way to generate fusion at room temperatures electrified nuclear physicists. Within months this "cold fusion" experiment had been dismissed, after other laboratories failed to replicate the claimed results. Though there are indications that something of interest occurred, and some research still continues, nuclear physicists returned to working on the problem of hot fusion.

Nuclear reactors and cyclotrons are used to manufacture radioactive isotopes for medical treatment. The German physicist Wilhelm Conrad Roentgen (1845–1923) discovered x-rays in 1895, and physicians soon learned to use x-rays in their medical practices, including using radium to treat different forms of cancer. After World War II, nuclear medicine developed ever more sophisticated methods of radiation treatment for cancer. Computed axial tomography (CAT) scanners and positron emission tomography (PET) scanners, both developed in the 1970s, offered even more powerful tools for physicians to peer inside the human body.

See also Big Science; Cold Fusion; Cold War; Dyson, Freeman; Environmental Movement; Ethics; Hydrogen Bomb; Nobel Prizes; Oppenheimer, J. Robert; Particle Accelerators; Pauling, Linus; Physics; Sagan, Carl; Sakharov, Andrei; Soviet Union; Teller, Edward; War

References
Balogh, Brian. *Chain Reaction: Expert Debate and Public Participation in American Commercial Nuclear Power 1945–1975*. Cambridge: Cambridge University Press, 1991.
Herman, Robin. *Fusion: The Search for Endless Energy*. Cambridge: Cambridge University Press, 1990.

Kenneth, Fowler, T. *The Fusion Quest.* Baltimore: Johns Hopkins University Press, 1997.

Mould, Richard F. *Chernobyl Record: The Definitive History of the Chernobyl Catastrophe.* Bristol, UK: Institute of Physics Publishing, 2000.

Rhoades, Richard. *Dark Sun: The Making of the Hydrogen Bomb.* New York: Simon and Schuster, 1995.

Wellock, Thomas Raymond. *Critical Masses: Opposition to Nuclear Power in California, 1958–1978.* Madison: University of Wisconsin Press, 1998.

Nüsslein-Volhard, Christiane (1942–)

The geneticist Christiane Nüsslein-Volhard made important contributions to understanding developmental biology by explaining the genetics of the embryonic development of fruit flies. She was born Christiane Volhard during World War II, in Magdeburg, Germany, where her father, an architect, served as an air force pilot during the war. By the age of twelve, she determined that she wanted to be a biologist, and her parents supported her in her interests. Though she was obviously intelligent, poor grades dogged her academic career, and she struggled with boredom. She attended Frankfurt University before moving to the University of Tübingen in 1964, which had just launched a biochemistry major. She studied at the Max Planck Institute for Virus Research in Tübingen, learning chemistry, genetics, and molecular biology. She earned a diploma in biochemistry in 1969 and her doctorate in genetics in 1973. A marriage to a physics student, which ended in divorce, resulted in her hyphenated last name, Nüsslein-Volhard.

She asked the biologist Walter Gehring (1939–) if she could join his research laboratory in Basel, Switzerland, to conduct postdoctoral research. She met the American biologist Eric Wieschaus (1947–) at the laboratory, and they worked on the embryonic development of the fruit fly, *Drosophila melanogaster.* A fellowship at Freiburg in 1977

Christiane Nüsslein-Volhard, geneticist, 1995 (Bossu Regis / Corbis Sygma)

followed, before she and Wieschaus were offered a joint research position at the newly founded European Molecular Biology Laboratory in Heidelberg. For three years they worked on fruit flies, developing new techniques to screen genetic mutations and publishing important papers. Nüsslein-Volhard and Wieschaus bred 40,000 fruit fly families, each with a single chemically induced genetic defect, then used a microscope to examine thousands of embryos and larvae. The fruit fly contains 20,000 genes, and Nüsslein-Volhard and Wieschaus found that only 139 are essential to embryonic development. Other researchers demonstrated that these same genes are also used for the same purposes in other species, including the vertebrates.

Wieschaus moved back to the United States to a position at Princeton University, and Nüsslein-Volhard returned to Tübingen in 1981. Their research was recognized as significant because it showed the way to understanding more complex genetic disorders, many of which have their source in embryonic development. Nüsslein-Volhard became a division director at the Max Planck Institute for Molecular Biology and extended her work from insects to vertebrates, raising over 100,000 zebra fish in order to understand their embryonic development. Her research using the zebra fish continues today. Nüsslein-Volhard, Wieschaus, and the American Edward B. Lewis (1918–), an early pioneer in fruit fly genetics, shared the 1995 Nobel Prize in Physiology or Medicine.

See also Biology and the Life Sciences; Biotechnology; Genetics; Nobel Prizes

References
"Christiane Nüsslein-Volhard—Autobiography." http://nobelprize.org/medicine/laureates/index.html (accessed February 13, 2004.)
McGrayne, Sharon Bertsch. *Nobel Prize Women in Science*. Second edition. Secaucus, NJ: Carol, 1998.

O

Oceanography

Oceanography includes the study of the physical attributes of the oceans and the biological attributes of the ocean. Studies sponsored by the Allied military during World War II greatly enhanced understanding of waves in order to help in troop landings, ocean acoustics to help in the struggle with submarine warfare, and undersea geography. After the war, the National Science Foundation, founded in 1950, became the main source of funding for civilian oceanographic research in the United States, while the U.S. Navy continued a vigorous research program that included classified and unclassified activities.

Oceanographic institutes, many founded in the 1920s, grew after World War II into major centers of research. Some of the more prominent institutes include the Scripps Institution in California, Woods Hole in Massachusetts, Bedford Institute in Nova Scotia, Alfred-Wegener Institute in Bremerhaven, Germany, and the P. P. Shirshov Institute in Moscow. The International Geophysical Year of 1957–1958 included intensive studies of the oceans by many nations. The Soviet research ship *Vityaz* recorded the deepest spot in any ocean in 1957 in the Challenger Deep of the Mariana Trench in the Pacific Ocean. The British research ship *Challenger II* had found the Challenger Deep in 1951, and the

American bathyscaphe *Trieste* took two crew down to the ocean bottom at that site in 1960.

The International Decade of Ocean Exploration in the 1970s also promoted international research. Other international efforts, part of Big Science, have included the Intergovernmental Oceanographic Commission (IOC), founded in 1960 as part of the United Nations. The International Council for the Exploration of the Sea (ICES), founded in 1902, expanded its initial focus on fisheries to a broader environmental research agenda. The 1983–1993 Tropical Ocean Global Atmosphere study measured the interaction between the tropical oceans and the atmosphere. The long-running World Ocean Circulation Experiment measured the flow of water around the globe.

Bruce Heezen (1924–1977) and Marie Tharp (1920–) of Columbia University created the first maps of ocean floor topography in 1952. Though eventually the whole world was mapped, Heezen and Tharp started with the Mid-Atlantic Ridge. Tharp discovered a rift valley in the middle of the ridge, strong evidence of continental drift, but because that theory was vigorously rejected by geologists of the time, her own interpretation of the theory did not become accepted until the theory of plate tectonics was accepted in the

Beluga whales (SeaWorld of California / Corbis)

1960s. Studies of the geology of the ocean bottom became the key evidence that proved plate tectonics.

New technologies allowed scientists to sample and observe previously inaccessible depths and enabled major new scientific discoveries. Jacques Cousteau (1910–1997) invented the Aqua-Lung in 1943, leading to scuba (*self-*contained *underwater breathing apparatus*) gear. Undersea habitats, manned and unmanned submersibles, new sonar technologies, deep-core drilling, and satellites have all extended the human senses. Satellites have increased understanding of meteorology, ocean temperatures, circulation patterns, and numerous other areas.

The Columbia University geologist Wallace S. Broecker (1931–) has pioneered important studies of ocean circulation, leading to the realization of how important the oceans are to transferring heat and maintaining global temperatures. Our emerging understanding of global warming is intertwined with the oceans, since the oceans are the main mechanism regulating the temperature of the planet. For millennia, the oceans have been used to wash away the refuse of civilization, and the post–World War II era saw a flourishing of scientific efforts and political activism to cope with ocean pollution.

The second half of the twentieth century saw the expansion of national fishing fleets funded by government subsidies. By the 1990s over two-thirds of all fisheries in the world's oceans were either fully exploited or fish populations had collapsed from overfishing. By the application of science and technology—using sonar, factory ships, and enormous nets—more fish have been harvested from the oceans in the twentieth century than in all previous centuries combined. The decline of desirable fish populations has led to the growth of aquaculture, much of it in inland ponds, but some in enclosed pens on ocean coasts, growing carp, shrimp, and salmon. Aquaculture grew from a worldwide production of 5 million tons in 1980 to 25 million tons in 1996. Aquaculture uses fish meal derived from so-called trash fish as food and thus continues to strain the ability of the oceans to produce enough fish.

Although much of the effort of oceanographers has been devoted to the physical attributes of the oceans, biologists have also made their own contributions. Only in the mid-1980s were the microscopic *Prochlorococcus* cyanobacteria found, which are responsible for more than half of all photosynthesis in the oceans. Numerous new species have been discovered, many of them existing only at great depths. In 1977, the deep-sea submersible *Alvin* found deep-sea hydrothermal vents at about 2,700 meters in the Galapagos Rift near the Galapagos Islands. The unique ecology surrounding the vents includes large tubeworms, giant clams, and other new species. Bacteria around the vents use chemosynthesis to create energy, converting the sulfides into organic carbon. The more complex life-forms form symbiotic relationships with the chemosynthesis bacteria and survive off them.

Cetacean research illustrates how biological research in the ocean has changed. Initial research into cetaceans came from whalers, who were concerned about cetacean behavior only insofar as it helped them in catching more whales for their meat and blubber. Scientists first studied cetaceans by examining the carcasses of killed whales and dolphins. As aquariums acquired cetaceans for public shows, scientists began to study the live behavior of captive cetaceans. Intrigued by the evidence of strong social behaviors and sophisticated communications, researchers moved to observing cetaceans in the wild, and since the 1970s, studies of captive cetaceans have virtually ended. Just as with studies of land animals, cetacean studies have increasingly turned to identifying individuals, tagging them, and understanding their behavior. Beginning in the 1970s, intensive efforts by environmental groups to limit and even ban whaling, including the well-known "Save the Whale" campaigns, have met with substantial success.

See also Big Science; Broecker, Wallace S.; Cousteau, Jacques; Deep-Sea Hydrothermal Vents; Environmental Movement; Geology; Global Warming; International Geophysical Year; Meteorology; National Science Foundation; Plate Tectonics; Undersea Exploration

References
Bascom, Willard. *The Crest of the Wave: Adventures in Oceanography.* New York: Harper and Row, 1988.
Borgese, Elisabeth Mann. *Ocean Frontiers: Explorations by Oceanographers on Five Continents.* New York: Abrams, 1992.
Broad, William J. *The Universe Below: Discovering the Secrets of the Deep Sea.* New York: Simon and Schuster, 1997.
Lawrence, David M. *Upheaval from the Abyss: Ocean Floor Mapping and the Earth Science Revolution.* New Brunswick, NJ: Rutgers University Press, 2002.
Mann, Janet, and others, editors. *Cetacean Societies: Field Studies of Dolphins and Whales.* Chicago: University of Chicago Press, 2000.
Rozwadowksi, Helen M. *The Sea Knows No Boundaries: A Century of Marine Science under ICES.* Seattle: University of Washington Press, 2002.

Oppenheimer, J. Robert (1904–1967)

The influential physicist J. Robert Oppenheimer founded and directed the Los Alamos Laboratory during World War II, where the first atomic bombs were designed, and so Oppenheimer is known as the "father of the atom bomb." Oppenheimer was born to a Jewish family in New York City, where his father was a wealthy merchant and his mother was a painter. His younger brother, Frank, also became a noted physicist. Oppenheimer earned a B.A. in physics from Harvard University in 1926, graduating summa cum laude. Fascinated by atomic physics, he spent a year at the Cavendish Laboratory at Cambridge University, then a year at the University of Göttingen, where he earned his doctorate in 1927. After further postgraduate study, he accepted faculty appointments at both the University of California at Berkeley and the California Institute of Technology (Caltech). Married in 1940, Oppenheimer fathered a son and a daughter.

Oppenheimer's scientific interests ranged widely, and he contributed to many areas, including work on quantum mechanics, relativity, cosmic rays, and positrons, and made some of the initial mathematical studies that later led to the theory of black holes. He served as a mentor to many students and built Berkeley into the largest center for theoretical physics in the United States. Independently wealthy through his inheritance, Oppenheimer contributed to the leftist political causes that he favored. He was keenly antifascist. During World War II, the American and British collaborated on the Manhattan Project to build the atomic bomb. Oppenheimer initially led the theoretical study group, then became responsible for organizing and building the main research laboratory of the project. He selected Los Alamos, near Santa Fe, New Mexico, for the laboratory and proved to be a brilliant scientific administrator, organizing a successful effort that led to the first atomic bombs in 1945.

In 1947, Oppenheimer moved to Princeton University, where he became director of the Institute for Advanced Study. Oppenheimer also served as the chair of the General Advisory Committee within the Atomic Energy Commission (AEC), which formulated military and civilian nuclear policy for the United States. With the cold war becoming more intense, the physicist Edward Teller proposed developing the hydrogen bomb. Oppenheimer opposed the project on both moral and practical grounds, and given the anticommunist tenor of the times, his loyalties were called into question. His enemies recalled past associations with the American Communist Party and in a famous 1954 hearing, where fellow physicist Edward Teller (1908–) testified against him, Oppenheimer's security clearance was revoked. Although physicists rallied to his support, the United States government effectively exiled him

from any of his former influence on nuclear policy. In 1963, partially as a gesture of reconciliation, Oppenheimer was awarded the AEC Fermi Medal, which Teller nominated him for.

> See also Big Science; Black Holes; Cold War; Ethics; Hydrogen Bomb; Neumann, John von; Nuclear Physics; Teller, Edward

References
Goodchild, Peter J. *Robert Oppenheimer: Shatterer of Worlds*. Boston: Houghton Mifflin, 1981.
Herken, Gregg. *Brotherhood of the Bomb: The Tangled Lives and Loyalties of Robert Oppenheimer, Ernest Lawrence, and Edward Teller*. New York: Henry Holt, 2002.
Smith, Alice Kimball, and Charles Weiner, editors. *Robert Oppenheimer: Letters and Recollections*. Cambridge: Harvard University Press, 1980.

Organ Transplants

Hindu surgeons in the sixth century B.C. performed the first known organ transplants, peeling off patches of skin from the arms of a patient to apply to his nose during reconstructive surgery. It was not until the twentieth century, however, that anesthesia, blood transfusions, and advanced medical techniques allowed more complex organ transplants. The first successful organ transplant occurred on December 23, 1954, in Boston, when a patient received the kidney of his identical twin brother and lived for another eight years. The donor also survived the experience. Joseph Edward Murray (1919–) performed the procedure and later shared the 1990 Nobel Prize in Physiology or Medicine for his achievement.

In a Denver hospital, the surgeon Thomas E. Starzl (1926–) performed the first successful human liver transplant on May 5, 1963, though the patient died three weeks later. Liver transplants are considered especially challenging because up to half of the body's blood passes through the liver every minute. Not until the introduction of cyclosporine, a new antirejection drug, in 1980 was Starzl able to push the survival rate for liver transplant patients to over 70 percent.

Starzl's team performed the first combined heart and liver transplant on a six-year-old Texas girl, Stormie Jones, in 1984. Starzl cited her death at age thirteen as one of the reasons for his retirement from active surgery.

Using heart-lung machines, surgeons developed coronary bypass methods and then moved on to heart transplants. Christiaan Barnard (1922–2001) in South Africa performed the first successful cardiac transplant in 1967, replacing the heart of a fifty-four-year-old man with that of a twenty-four-year-old woman who had died in an automobile accident. The patient died of pneumonia eighteen days later. In January 1968, Barnard tried again, and the patient lasted twenty months, eventually dying of chronic rejection. More than 100 other attempts were made in 1968 by more than sixty centers worldwide, with dismal success rates. After understanding how to control rejection, success rates improved, until over 90 percent of heart transplant patients now survive for more than a year. Barney C. Clark received an artificial heart in March 1983 at the University of Utah and survived 112 days. Although research has continued on artificial organs, xenotransplantation (that is, using organs from animals) currently offers more promise.

In the 1950s, E. Donnall Thomas (1920–) developed methods to store and transfuse bone marrow from animal to animal. In March 1969, Thomas and his colleagues performed the first successful bone-marrow transplant after correctly matching the patient to a donor and using drugs to suppress rejection by her immune system. Thomas shared the 1990 Nobel Prize in Physiology or Medicine with Murray.

On October 26, 1984, a twelve-day-old girl known only as Baby Fae received a baboon heart transplant, and though she died shortly thereafter, the issue of using animal organs was raised. Some commentators expressed horror at the very thought, others worried that such a mixing of tissue between species might spawn the creation of deadly

Richard Herrick (left) and his identical twin brother, Ronald. Richard received a kidney from Ronald in 1954 in the first successful organ transplant. (Bettmann / Corbis)

new viruses, and a few objected to using animals as organ factories. In 1992, Starzl coordinated the transplantation of baboon livers into two men. The first lived seventy-one days and the second only twenty-six days. In both cases, complications from surgery and not liver rejection were the cause of death. Also in 1992, scientists at Cambridge University created a breed of transgenic pigs by adding a human gene to the pig's genome. The goal of this and other efforts is to create genetically modified pigs that will be better suited for organ donations because their organs will not react to human proteins.

Improved surgical techniques and the development of antirejection drugs have led to the point that the transplantation of hearts, livers, kidneys, blood vessels, bone marrow, corneas, and other organs or tissues is now commonplace in developed countries. In 1998, a New Zealander named Clint Hallam received the first hand transplant, which his body eventually rejected.

Organ transplants raise a host of ethical issues, especially considering the shortage of donated organs. Bioethicists and patients struggle with the questions: where do we get the organs, who gets to receive them, and is it appropriate to increase the organ supply by using animal organs?

See also Bioethics; Medawar, Peter; Medicine; Nobel Prizes

References

Caplan, Arthur L., and Daniel H. Coelho, editors. *The Ethics of Organ Transplants: The Current Debate.* Amherst, NY: Prometheus, 1998.

Hakim, Nadey S., and Vassilios E. Papalois. *History of Organ and Cell Transplantation.* River Edge, NJ: World Scientific, 2003.

Munson, Ronald. *Raising the Dead: Organ Transplants, Ethics, and Society.* New York: Oxford University Press, 2002.

Starzl, Thomas E. *The Puzzle People: Memoirs of a Transplant Surgeon.* Pittsburgh: University of Pittsburgh Press, 1992.

Veatch, Robert M. *Transplantation Ethics.* Washington, DC: Georgetown University Press, 2000.

Ozone Layer and Chlorofluorocarbons

The ozone layer is created by the presence of an isotope of oxygen (O_3) in the stratosphere, which reduces the amount of ultraviolet radiation emitted by the Sun that reaches the Earth's surface. Ozone molecules are created naturally in the stratosphere, and without their protection, too many ultraviolet rays could lead to DNA mutations, cancer, and blindness. In 1970, the Dutch-born chemist Paul Crutzen (1933–) demonstrated in an article for the *Quarterly Journal of the Royal Meteorological Society* that nitrogen oxides formed naturally by soil bacteria can rise up to the stratosphere. Once in the stratosphere, the nitrogen oxide molecules are broken apart by sunlight and react with ozone, accelerating the chemical processes that lead to the natural decay of ozone molecules.

Crutzen's research was not generally accepted until the Mexican chemist Mario Molina (1943–) and the American chemist F. Sherwood Rowland (1927–), working at the University of California at Irvine, discovered in 1974 that chlorofluorocarbon (CFC) gases accelerate the decay of the ozone layer. When CFCs reach the ozone layer, high in the atmosphere, ultraviolet radiation from the Sun breaks down the CFC molecules. Each CFC molecule consists of one atom of carbon, one of fluorine, and three of chlorine ($CFCl_3$). The released chlorine atoms rapidly react with ozone molecules, converting them to normal oxygen. Small amounts of CFCs have the potential to rapidly reduce the amount of ozone. First invented in 1928, CFCs were manufactured for use in air conditioners and aerosol cans, and became a multibillion-dollar industry by the 1970s.

This laboratory discovery led to hearings by the United States Congress in that same year. A movement began to ban CFC manufacture and use. In 1976, the National Academy of Sciences published a report supporting Molina and Rowland's hypothesis. In 1977, James G. Anderson (1944–) found that the concentration of chlorine oxide in the upper atmosphere above Texas was much higher than predicted. On October 21, 1978, the United States banned the manufacture of CFC propellants. Other countries followed suit and inaugurated international efforts to create a far-reaching international accord. Scientists are by no means united on the threat from CFCs. Even the biophysicist James Lovelock (1919–) published an article in 1980 that argued that the threat from CFCs and possible ozone depletion was exaggerated, though he later changed his mind.

As part of the International Geophysical Year, in 1957 the British Antarctic Survey Team began measuring the ozone layer over that frozen continent. In the 1970s, the levels of ozone began to plummet, down by a hundredfold by 1985. In 1987, Anderson and his team from Harvard University used a high-flying U-2 aircraft to measure the ozone layer over Antarctica and found a hole in the ozone layer. Later a similar hole was found over the Arctic. Scientists still disagree on exactly why these holes exist, but the majority think that depletion of the ozone layer, probably due to CFCs, is the cause. The National Aeronautics and Space Administration uses satellites and aircraft to monitor ozone around the world, increasing the amount of data for researchers to draw upon.

The growing realization that continued use of CFCs might deplete the ozone layer with catastrophic results for human civilization led to the Montreal Protocol international agreements of 1987, 1990, and 1992 to phase out the use of these gases. Most industrialized countries signed the Protocol, though China and India refused. Alternate chemicals were created to substitute for CFCs in air conditioners and industrial processes. These accords are held up as monuments to environmental action. Chlorofluorocarbons are also greenhouse gases, leading to concerns that they might contribute to global warming. The effect of ozone depletion on global warming is also uncertain. Crutzen, Molina, and Rowland were awarded the 1995 Nobel Prize in Chemistry for their work on this problem.

See also Chemistry; Crutzen, Paul; Environmental Movement; Global Warming; International Geophysical Year; Lovelock, James; Meteorology; Molina, Mario; National Aeronautics and Space Administration; Rowland, F. Sherwood

References

Cagin, Seth, and Phillip Dray. *Between Earth and Sky: How CFCs Changed Our World and Endangered the Ozone Layer.* New York: Pantheon, 1993.

Christie, Maureen. *The Ozone Layer: A Philosophy of Science Perspective.* Cambridge: Cambridge University Press, 2000.

Molina, Mario J., and F. Sherwood Rowland. "Stratospheric Sink for Chlorofluoromethanes: Chlorine Atom–Catalyzed Destruction of Ozone." *Nature* 249 (June 28, 1974): 810–812.

Newton, David E. *The Ozone Dilemma: A Reference Handbook.* Santa Barbara, CA: ABC-CLIO, 1995.

Reid, Stephen J. *Ozone and Climate Change: A Beginner's Guide.* Amsterdam, the Netherlands: Gordon and Breach Science, 2000.

P

Particle Accelerators

The various forms of particle accelerators all use electromagnets to accelerate atomic or subatomic particles into a beam. In 1994, an estimated 10,000 particle accelerators existed in the world; by 2000, an estimated 15,000 particle accelerators existed. Most of the machines are small, used in industrial or medical applications. Only 110 of these machines are used in nuclear and particle physics research, in which subatomic particles are accelerated to near the speed of light and smashed against other particles or objects, creating a shower of smaller particles. The information derived from studies of atomic and subatomic particles in particle accelerators has been the main source of experimental evidence for the development of particle physics since 1950.

The American physicist Ernest O. Lawrence (1901–1958) and a graduate student, David Sloan, built the first linear accelerator, a cyclotron, in 1931 at the University of California at Berkeley. Lawrence drew on earlier work in Germany and Britain, and received the 1939 Nobel Prize in Physics for this work. Lawrence created the Radiation Laboratory at Berkeley in 1936, which the university renamed the Lawrence Berkeley Laboratory after his death. One of his cyclotrons artificially produced technetium, the first element discovered that does not occur naturally in nature.

Research after World War II led to the development of numerous types of particle accelerators, some designed to accelerate protons, others electrons or positrons. Some particle accelerators are linear, though many are circular in construction. The energy produced is usually measured in MeV (millions of electron-volts) or GeV (billions of electron-volts). The largest linear accelerator in the world is the Stanford linear accelerator (SLAC), completed in 1966, 2 miles (3.2 kilometers) long, and after several upgrades, it now produces 50 GeV. Other major centers of research in America include the Fermi National Accelerator Laboratory in Batavia, Illinois; the Brookhaven National Laboratory in Upton, New York; and the national laboratory in Los Alamos, New Mexico.

Before World War II, scientists in Europe led the world in nuclear physics research. The flight of physicists from the Nazis and the chaos of World War II pushed the United States into the lead in this field. In 1954, twelve European nations created CERN (Conseil Européen pour la Recherche Nucléaire), located near Geneva on the French-Swiss border, in an effort to revive European research in nuclear and particle physics. CERN's first particle accelerator began

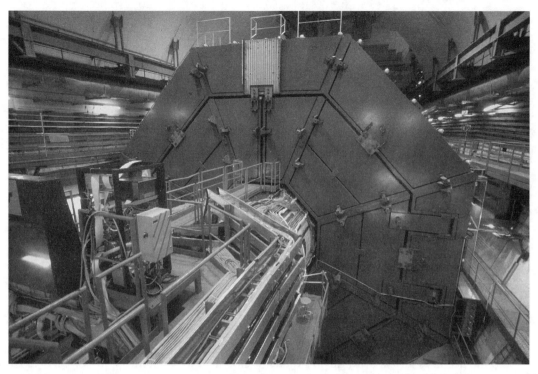

The Large Electron-Positron Collider (LEP) at CERN in Switzerland (Pierre Vauthey / Corbis Sygma)

operation in 1957. As more accelerators have been built over the years, CERN has lived up to its promise, creating scientific discoveries and numerous Nobel Prizes. The Large Electron-Positron Collider (LEP), opened in 1989, is the largest scientific instrument ever constructed, a 16.2-mile (27-kilometer) ring producing 50 GeV. As a serendipitous spin-off of CERN's efforts, the World Wide Web was invented by a software developer, Tim Berners-Lee (1955–), at CERN in 1991.

The Soviet Union and its successor state, Russia, as well as Japan, China, India, and several other countries, have also built particle accelerators for studying nuclear physics and particle physics. As an example of Big Science, large particle accelerators require substantial government funding, teamwork among specialists in diverse fields, and government support, partially motivated by the desire to enhance national prestige. An ambitious attempt to build the Superconducting Supercollider (SSC) in Texas, with a circumference of 52 miles (87 kilometers) and plans to produce collisions of protons and antiprotons at 40 TeV (trillions of electronvolts), began in 1989. Disputes within the physics community about where limited government funding should go, escalating costs, and federal politics caused the SSC to founder on federal budget cuts in 1993 after being partially built. A major rationale for the SSC, national prestige, no longer mattered as much after the cold war ended in 1991.

See also Berners-Lee, Timothy; Big Science; Cold War; Particle Physics; Physics

References

Adams, Steve. *Frontiers: Twentieth-Century Physics.* London: Taylor and Francis, 2000.

Herman, Armin, and others. *History of CERN.* 3 volumes. Amsterdam: North-Holland, 1987, 1990, 1996.

Panofsky, Wolfgang. "The Evolution of SLAC and Its Program." *Physics Today* 36, no. 10 (1983): 34–41.

Pickering, Andrew. *Constructing Quarks: A Sociological History of Particle Physics.* Chicago: University of Chicago Press, 1984.

Particle Physics

Particle physics is the branch of physics that deals with subatomic particles. As the twentieth century progressed, particle physicists deconstructed the atom into ever smaller particles. The German Werner Heisenberg (1901–1976), the Englishman P. A. M. Dirac (1902–1984), and the Austrian Wolfgang Pauli (1900–1058) developed quantum electrodynamics (QED) in the 1920s to explain the interactions between electrons and photons. In the late 1940s and early 1950s, the German-born physicist Maria Goeppert Mayer (1906–1972) developed the shell model of the nucleus, explaining how protons and neutrons occupy orbital shells around the atomic nucleus.

After World War II, a second generation of theorists refined quantum electrodynamics by mathematically renormalizing it and fixing flawed equations that led to infinite results and problems with how to handle the self-energy of particles. The physicists Richard P. Feynman (1918–1988) and Julian Schwinger (1918–1994), both from New York City, independently developed similar theories. The Japanese physicist Shinichiro Tomonaga (1906–1979), working in wartime isolation in Japan, earlier developed a similar system. Freeman Dyson (1923–) showed that the three theories were mathematically equivalent, which led to the 1965 Nobel Prize in Physics for Feynman, Schwinger, and Tomonaga. The physics community chose Feynman's mathematical notation system over Schwinger and Tomonaga's systems, and his method of graphically drawing particle interactions, dubbed Feynman diagrams, became widely used.

Postwar particle physicists heavily relied on two tools: the particle accelerator to smash subatomic particles into ever smaller particles and the electronic computer to organize the volumes of data from their research. As a matter of national prestige during the cold war, the U.S. federal government built ever larger particle accelerators in national laboratories, most of them attached to universities, in an example of what historians call Big Science. In 1954, twelve European nations created CERN (Conseil Européen pour la Recherche Nucléaire), located near Geneva on the French-Swiss border, which built multiple particle accelerators in a successful postwar effort to revive European research in nuclear and particle physics. The particle physics community has tended to divide into experimentalists who run the intricate experiments and theoreticians who develop mathematical models to understand the results of the experiments.

New subatomic particles had already been found in cosmic ray research, but the postwar particle accelerators revealed dozens of new subatomic particles during the early 1950s. There was little theoretical understanding of how to organize them until 1953, when the American theoretical physicist Murray Gell-Mann (1929–) proposed a new property for subatomic particles called "strangeness" to help categorize these subatomic particles. In the early 1960s Gell-Mann collaborated with the Israeli physicist Yuval Ne'eman (1925–), to devise the eightfold way to group the known particles into a table similar to the periodic table for chemicals. Just as the periodic table successfully predicted the existence of unknown chemicals to fill out the table, the eightfold way successfully predicted new particles with specific characteristics. By the early years of the twenty-first century, over two hundred subatomic particles have been found experimentally.

Theory also predicted the existence of antiparticles, with the same mass as the corresponding particle but an opposite value for the electric charge. When a particle and antiparticle meet, they annihilate each other in a complete conversion of mass to energy. In 1955, the American physicist Owen Chamberlain (1920–) and Italian-born physicist Emilio Sergè (1905–1989) used the Bevatron particle accelerator at Berkeley to confirm the theory by creating antiprotons. The two physicists shared the 1959 Nobel Prize in Physics, and numerous other antiparticles have been found since then. In 1995, researchers at CERN succeeded in creating nine atoms of antihydrogen, the first anti-atom, which existed for about 40 billionths of a second before being annihilated in collision with normal matter.

In 1964 Gell-Mann proposed a new type of fundamental matter that in turn formed the larger known subatomic particles. Gell-Mann whimsically labeled his proposed matter quarks and antiquarks, and suggested that three "flavors," or types, existed. The American theoretical physicist Sheldon L. Glashow (1932–), having published a theory in 1960 unifying the weak nuclear force and the electromagnetic force into an electroweak force, took Gell-Mann's work and proposed a new property for quarks called "charm" and the existence of a fourth quark. Since then the number of quarks has risen to six, and quark theory is now part of the standard model. The standard model attempts to combine the four basic forces (weak nuclear force, strong nuclear force, electromagnetism, and gravity) into a grand unified theory. Glashow's electroweak theory is also part of the standard model. Gell-Mann received an unshared Nobel Prize in Physics in 1969, and Glashow shared a 1979 Nobel Prize in Physics. In the 1970s, Gell-Mann expanded his theory of quarks into what he called quantum chromodynamics (QCD).

The Japanese-born physicist Yoichiro Nambu (1921–) also contributed to quark theory and realized in 1970 that quarks act as if they are connected by strings. Nambu and others developed this insight into string theory in the 1980s. String theory gets even smaller than quarks, dealing with hypothetical vibrating strings that are near the Planck length (about 10^{-33} centimeters or 10^{-34} inches). No instruments can measure objects that small. These strings in turn make up the larger known elementary particles, such as quarks, bosons, and gluons. String theory demands complex mathematics using up to twenty-six multiple dimensions, and provides a promising avenue to the development of a grand unified theory. String theory later combined with supersymmetry to create superstring theory.

In 1956, the physicists Frederick Reines (1918–1998) and Clyde Cowan (1919–1974) detected a theoretical particle called the neutrino at a nuclear reactor at Savannah River, South Carolina. Reines and Cowan had found the neutrino that is created occurred during the formation of an electron. In 1962 a neutrino associated with the formation of a muon particle was discovered. Neutrino theory in the 1970s predicted that a third type of neutrino, created during the formation of a tau particle, should also exist, though the tau neutrino has not been found experimentally.

The Sun, powered by nuclear reactions, emits neutrinos. In another example of Big Science, scientists built huge underground experiments to detect solar or stellar neutrinos. The radiochemist Ray Davis (1914–) organized the first experiment in the Homestake gold mine in South Dakota. Solar neutrinos that passed through a huge tank filled with chlorine reacted with chlorine-37 isotopes to create argon-37, though scientists were puzzled to detect only a fourth of the predicted number.

See also Big Science; Computers; Dyson, Freeman; Feynman, Richard; Gell-Mann, Murray; Glashow, Sheldon L.; Grand Unified Theory; Mathematics; Mayer, Maria Goeppert; Nambu, Yoichiro; Neutrinos; Nobel Prizes; Nuclear Physics; Particle Accelerators; Physics; Schwinger, Julian

References

Brown, Laurie M., Lillian Hoddeson, and Max Dresden. *Pions to Quarks: Particle Physics in the 1950s.* Cambridge: Cambridge University Press, 1989.

Crease, Robert P., and Charles C. Mann. *The Second Creation: Makers of the Revolution in Twentieth-century Physics.* New York: Macmillan, 1986.

Gribbin, John. *Q Is for Quantum: An Encyclopedia of Particle Physics.* New York: Free Press, 1998.

Pickering, Andrew. *Constructing Quarks: A Sociological History of Particle Physics.* Chicago: University of Chicago Press, 1984.

Schweber, Silvan S. *QED and the Men Who Made It: Dyson, Feynman, Schwinger, and Tomonaga.* Princeton: Princeton University Press, 1994.

Pauling, Linus (1901–1994)

The chemist Linus Pauling combined quantum mechanics with chemistry to explain chemical bonding and later turned his attention to the cause of world peace and nuclear disarmament. Linus Carl Pauling was born in Portland, Oregon, where his father was a pharmacist. He earned a B.Sc. in chemical engineering from Oregon State Agricultural College (now Oregon State University) in 1922. In 1925, he earned his Ph.D. in chemistry at the California Institute of Technology (Caltech). After studying for two years in Europe to learn the new science of quantum mechanics, he returned to Caltech as a faculty member. Pauling's unique interdisciplinary knowledge of quantum mechanics and chemistry enabled him to use x-ray diffraction photography to analyze the crystal structure of inorganic molecules and to revolutionize the understanding of chemical bonds. He became a professor of chemistry in 1931, the same year that he earned the Langmuir Prize from the American Chemical Society. His 1939 text, *The Nature of the Chemical Bond and the Structure of Molecules and Crystals,* became an influential classic in the field. After establishing his theories of chemical bonding, including the discovery of hybrid orbitals and resonance hybrids, he began to apply his techniques to organic molecules. His studies of hemoglobin led to the theory of native, denatured, and coagulated proteins. Pauling came close to beating Francis Crick (1916–2004) and James D. Watson (1928–) in the race to divine the structure of the DNA molecule. Pauling received the 1954 Nobel Prize in Chemistry for his work on chemical bonds, cementing his fame and influence.

During World War II, Pauling worked with the Office of Scientific Research and Development on developing explosives, rocket propellants, and an artificial substitute for human plasma. He declined to work on the Manhattan Project to develop the atomic bomb and after the war became a vocal opponent of further development of nuclear weapons. After becoming a Nobel

Linus Pauling, chemist and nuclear test ban activist (Library of Congress)

laureate, Pauling grew more aggressive in his antinuclear activities, earning considerable suspicion from the federal government and some other scientists that he was a communist sympathizer. His 1958 book, *No More War!,* emphasized his pacifist convictions. Becoming concerned about the accumulation of radioactive fallout in the atmosphere from open-air nuclear weapons testing, Pauling campaigned for the United States and Soviet Union to at least stop open-air testing in the interest of public health. His efforts helped lead to the 1963 Nuclear Test Ban Treaty, and he received the 1962 Nobel Peace Prize, the only person to earn two undivided Nobel prizes, and one of only three to win two Nobel prizes.

In 1964, Pauling left Caltech, spending three years at the Center for the Study of Democratic Institutions in Santa Barbara,

California, learning more about international relations while continuing his scientific studies. He then spent two years at the University of San Diego, and four years at Stanford University. In 1968, Pauling coined the term *orthomolecular medicine* to describe his conviction that vitamin supplements could boost the human immune system to protect against cancer and other ills. His best-known proposal asserted that megadoses of vitamin C could prevent the common cold and other illnesses. Pauling founded an institute to study orthomolecular medicine in 1973 and relentlessly supported these efforts for the rest of his life. Other scientists found mixed results in their own studies of Pauling's medical ideas, and most tend not to support them.

Always engaged in research and writing, Pauling published over one thousand papers in his lifetime. Pauling married in 1923, fathered three sons and a daughter, and credited the support of his wife, the former Ava Helen Miller (1903–1981), with his success, both as a helpmeet and in pushing him to be more aggressive about nuclear weapons disarmament.

See also Chemistry; Cold War; Crick, Francis; Nobel Prizes; Watson, James D.

References
Hager, Thomas. *Force of Nature: The Life of Linus Pauling.* New York: Simon and Schuster, 1995.
Marinacci, Barbara, editor. *Linus Pauling in His Own Words: Selected Writings, Speeches, and Interviews.* New York: Simon and Schuster, 1995.
Serafini, Anthony. *Linus Pauling: A Man and His Science.* New York: Paragon, 1989.

Penrose, Roger (1931–)

The British mathematician Roger Penrose contributed important insights to the theory of black holes, as well as to the theory of artificial intelligence. Roger Penrose was born in Colchester, Essex, England, where his father, a medical geneticist, and his mother, a medical doctor, temporarily lived while his father conducted a genetics survey. The family spent World War II across the ocean in London, Ontario, Canada. In 1952, Penrose graduated with a B.Sc. in mathematics from University College in London, where his father had become a professor of human genetics. Penrose and his father gained some popular recognition when they devised some geometrical figures later used by the Dutch surrealist artist M. C. Escher (1898–1972). Penrose earned his Ph.D. in mathematics at Cambridge University in 1957 and married in 1959. He spent time as a research fellow or visiting faculty at various universities in the United States and England before settling down as a professor of applied mathematics at Birbeck College, London, in 1967.

Penrose published his first article on singularities, later called black holes, in 1965. *Singularity* is the technical term used for what is believed to happen to certain stars at the end of their life cycle and what the primordial point at the beginning of the universe looked like before the Big Bang. Penrose's work intrigued the young Stephen Hawking (1942–), and the two mathematicians began to work on the problem together. The two mathematicians published a theory in 1970 that argued that though black holes could not be observed, relativistic radiation from near the event horizon of black holes should be detectable. (The event horizon is the boundary of black holes, at which incoming mass or light can no longer escape the intense gravity of the black hole.)

Penrose became the Rouse Ball Professor of Mathematics at Oxford University in 1973. His interests turned to computers and artificial intelligence, and he published the best-selling *The Emperor's New Mind: Concerning Computers, Minds, and the Laws of Physics* (1989) and a follow-up, *Shadows of the Mind: A Search for the Missing Science of Consciousness* (1994). He argued that the human brain can carry out processes that no computer can do, running counter to the general tendency among other researchers in the field of artificial intelligence. Penrose and Hawking shared the 1988 Wolf Prize in Physics for their work on black holes and relativity. Penrose became a fellow of the Royal Society in 1972, re-

Roger Penrose, mathematician (Bob Mahoney / Time Life Pictures / Getty Images)

ceived the Royal Society Royal Medal in 1985, and was knighted in 1994. Penrose retired from Oxford in 1998 and became the Gresham Professor of Geometry at Gresham College in London.

See also Artificial Intelligence; Astronomy; Black Holes; Cognitive Science; Hawking, Stephen

References

"Roger Penrose." http://turnbull.mcs.st-and.ac.uk/history/Mathematicians/Penrose.html (accessed February 13, 2004).

"Wolf Foundation Honors Hawking and Penrose for Work on Relativity." *Physics Today,* January 1989, 97–98.

Pharmacology

Pharmacology, the science of drugs, has a long history, which began with the study of traditional drugs derived from plants and chemicals. In the early nineteenth century, especially in France and Britain, scientists began to apply good scientific techniques to test the efficacy of old drugs and to develop new ones. Pharmaceutical companies arose in the mid-nineteenth century, some of them selling drugs with medicinal value, and others selling drugs based on quackery. A successful British pharmaceutical company, cofounded in 1880 by Henry Wellcome (1853–1936), led to one of the world's largest private foundations, The Wellcome Trust. In the early twentieth century, governments in industrialized nations stepped in to regulate the pharmaceutical industry, demanding purity in manufacture and trying to keep claims truthful. In the 1990s, the pharmacology industry came under criticism because its drug-testing programs focused on adult males and did not take into account that children, infants, women, and the elderly might react differently to drugs than did adult men in their prime. Some researchers have turned to focus on this issue, though pharmaceutical companies often see solving the problem as just increasing the costs of already expensive clinical trials.

Drug therapies became the foundation of modern medicine. The original scientific method to search for new drugs was based on a hit-or-miss process, in which researchers at universities or pharmaceutical companies tried many different compounds on many different diseases to see what would happen. A new method, called antimetabolite theory, developed before World War II, focused on learning how the disease worked, then trying to find a compound that interfered with the disease process. The pharmacologist Gertrude B. Elion (1918–1999) was among those who pioneered the use of this method. Elion developed purine compounds to extend the life of leukemia patients, and her 6-MP derivative azathioprine suppressed the immune system of the recipient of the first successful kidney transplant from a nonrelated donor. Advances in genetics that led to the development of genetic engineering created the next frontier for the pharmaceutical industry. Biotechnology companies, initially founded in the mid-1970s, are now in the forefront of pharmaceutical research.

The search for a vaccine for poliomyelitis (infantile paralysis) obsessed virologists after World War II. All the other major deadly childhood diseases had already been conquered, and polio represented the last great challenge. The leading polio researcher, Albert Sabin (1906–1993) at University of Cincinnati College of Medicine, effectively demonstrated that the polio virus entered the body through ingestion, rather than through the respiratory system. Work by American microbiologist John Enders (1897–1985) and others led to a way to more effectively culture mammalian viruses, a necessary advance in defeating polio. The virologist Jonas Salk (1914–1995) developed a killed-virus vaccine in 1952, and in 1955 the Salk vaccine passed its clinical trials and became widely used. Sabin's own live-virus vaccine later became more widely used because of its greater effectiveness.

Pharmacology is intimately connected with immunology, the study of the body's natural defenses against bacterial and viral invasions. Teams led by biochemists Gerald M. Edelman (1929–) and Rodney R. Porter (1917–1985) described the chemical structure of antibodies in the 1960s. Antibodies, or immunoglobulins, peptides made up of heavy and light chains, are formed by the immune system to destroy viruses and bacteria by binding to them. One of the key problems faced by pioneering surgeons when transplanting organs from one patient to another was that the new body sees the organ as foreign tissue and the immune system tries to reject it. Antirejection drugs partially solved this problem.

A German company introduced the drug thalidomide in the 1950s, prescribed as a sleeping pill and an aid to pregnant women with morning sickness. Unknown to the manufacturer and physicians, when used during the early phase of a pregnancy, thalidomide led to serious birth defects. Because the drug only affected the fetus during a limited phase and because of obfuscation by the manufacturer, the impact of the drug only became apparent after several years and the birth of thousands of affected babies, many born with stunted limbs. The drug was withdrawn, though it had never been offered for sale in the United States because the federal Food and Drug Administration (FDA), founded in 1906, has stricter requirements for the introduction of new drugs than any other nation. Critics argue that the strictness of the FDA, though effective with thalidomide, has also caused many to suffer or die while waiting for clinical trials to prove the efficacy of new drugs. Thalidomide was later reintroduced as an effective drug for other ailments, with stringent guidelines keeping it from pregnant women.

Founded in the early nineteenth century in Germany, homeopathy is a form of alternative medicine that is based on the theory that "like cures like"; drugs that produce a given set of symptoms in a healthy person are used to treat a sick person who has that same set of symptoms. Homeopathy also holds as a central theory that the potency of a drug increases as the drug is diluted into ever smaller amounts. Homeopathy remains popular in Germany, even though many of its tenets conflict with current medical understanding of physiology and drug interactions. Many homeopathic doses are in such small quantities that modern chemistry can barely detect them, and modern pharmacology cannot detect a measurable physiological effect.

In the United States, nonprescription antibiotics could be purchased up until the 1950s, and many countries still have no effective control over the distribution of antibiotics. The overuse of antibiotics in developed countries has eventually led to the evolution of new antibiotic-resistant strains of bacteria. The rise of multiple-drug-resistant bacteria has brought back diseases, such as tuberculosis, that doctors had thought remained only among the poor and negligent. The first incurable case of tuberculosis appeared in South Africa in 1977, and by 1985 the United States saw its first statistical rise in tuberculosis in over a century. In 1955, the World Health Organization launched their Global Malaria Eradication campaign to combat the most devastating disease in tropical countries. By the 1960s, health care workers began to encounter drug-resistant forms of malaria, requiring the development of new antimalarial drugs in what has become a biological arms race between humans and microbes.

Antibiotics have also become important tools in modern Western agriculture, especially on livestock and fowl farms where millions of animals live closely together and a disease can spread with startling rapidity. Critics of this industrialized farming have worried that the heavy use of antibiotics will lead to more drug-resistant diseases among these animals and also worry about the amount of antibiotics that may still remain in the animal or bird flesh after slaughter and sale.

Drug addiction has always existed, and the (then legal) trade in drugs was often an important element of international trade. As the pharmaceutical industry became regulated, the sale of drugs became restricted, and an illegal narcotics trade emerged. Some of this business is in drugs used as medicine, but many of the illegal narcotics are used as stimulants or depressants, spawning a black market worth tens of billions of dollars, and making the illegal trade in narcotics into a major global issue. Many of the drugs, such as opium, heroin, marijuana, and cocaine, come from plants. Other drugs have emerged from research laboratories, some of these drugs, such as methamphetamines, barbiturates, and phencyclidine (PCP, also called angel dust), with legitimate medical uses. To supply the burgeoning illegal drug market, chemists have begun to create designer drugs, which are similar to previous drugs but are just different enough not to be on any list of illegal drugs. These drugs are usually used for hedonistic purposes, and include 3,4 methylenedioxymethamphetamine (MDMA, also called ecstasy) and ketamine (GBH).

The fungus ergot grows on rye and similar grasses and is a traditional medicine. In the 1940s, researchers at a Swiss pharmaceutical company isolated lysergic acid diethylamide (LSD) while studying ergot. After accidentally ingesting a small amount of LSD, a Swiss researcher experienced vivid hallucinations. The drug interested psychologists and psychiatrists in the United States, Britain, and elsewhere, and during the 1950s and early 1960s, they experimented with the drug for psychotherapeutic purposes. Other people deliberately took the drug for entertainment or to experience altered states of consciousness. Many spoke of their hallucinations as a religious experience. In 1966, the United States banned the drug, with the last authorized psychotherapeutic study ending in 1975. Britain followed suit in banning LSD in 1971, as did other Western European countries in the 1970s. Despite its illegal status, the use of LSD to fuel "acid trips" played a significant role in the counterculture movement of the 1960s and 1970s in both the United States and Western Europe.

See also Agriculture; Biotechnology; Edelman, Gerald M.; Elion, Gertrude B.; Enders, John; Foundations; Genetics; Medicine; Organ Transplants; Sabin, Albert; Salk, Jonas

References

American Institute of the History of Pharmacy. http://www.pharmacy.wisc.edu/aihp/ (accessed February 13, 2004).

Davenport-Hines, Richard. *The Pursuit of Oblivion: A Global History of Narcotics.* New York: Norton, 2002.

Healy, David. *The Creation of Psychopharmacology.* Cambridge: Harvard University Press, 2002.

Landau, Ralph, Basil Achilladelis, and Alexander Scriabine, editors. *Pharmaceutical Innovation: Revolutionizing Human Health.* Philadelphia: Chemical Heritage Foundation with McGill-Queen's University Press, 1999.

Levy, Stuart. *The Antibiotic Paradox: How Miracle Drugs Are Destroying the Miracle.* New York: Plenum, 1992.

Weatherall, Miles. *In Search of a Cure: A History of Pharmaceutical Discovery.* New York: Oxford University Press, 1991.

Philosophy of Science

The philosophy of science strives to describe the ontological and empirical basis of science. Until recently, logical empiricism dominated the philosophy of science; the theory was that scientists empirically examined the natural world and developed natural laws by making logical deductions from the results of experiments and observations. Criticisms of this dominant positivist stance have come from historians, philosophers, and sociologists.

The physicist and historian Thomas S. Kuhn (1922–1996) became the most influential historian of science of his time with his 1962 book, *The Structure of Scientific Revolutions*. Kuhn adopted the term *paradigm* to describe the joint worldview, shared practices, and accepted theories of a group of scientists. Kuhn argued that scientists usually practiced "normal science," in that the current paradigm in their field was accepted and experiments either confirmed the paradigm or resulted in minor changes. As time passed, anomalies occurred in observations or experiments that could not be explained by the current paradigm. When enough of these anomalies accumulated, the old paradigm became ripe for revolution. A new paradigm emerged that explained the anomalies and incorporated the old paradigm inside it. Many scholars in various disciplines interpreted Kuhn's work as a demonstration that the practice of science was irrational, subjective rather than objective, and that perhaps science was really just a form of rhetoric. The later rise of social constructionism drew on Kuhn, though he disagreed with their epistemological relativism.

The philosopher Karl Popper (1902–1994) devised the idea of falsifiability to solve the problem of induction in the philosophy of science. Falsifiability is the idea that a scientific theory must include testable conclusions where if the experiment proves to be false, then the theory is false. The question of how to distinguish between valid science and pseudoscience formed the basis of his life of inquiry, and he considered himself to be an empiricist, but not a positivist. Popper's ideas are supported by many philosophers and scientists, even when they find creating falsifiable theories and observations difficult to implement in actual scientific work. Historical sciences, such as geology, paleontology, and evolutionary biology, have an almost impossible time meeting Popper's standards because they describe events that cannot be repeated or proven through experiments.

Paul K. Feyerabend (1924–1994), born in Vienna, Austria, became a protégé of Karl Popper's, but later rejected Popper's "critical rationalism" and came to argue in favor of "epistemological anarchism." In brief, Feyerabend believed that the scientific method was really a plurality of different approaches and methods, and that the best scientists picked whatever worked and did not follow any particular form of empiricist methodology. He later argued against the idea of progress in science, seeing only change. On this point, Feyerabend was in sympathy with the social constructionists.

Social constructionism emerged in the 1970s from the "strong programme" in the sociology of science. Social constructionists believe that all knowledge is subjective, a result of our social and cultural conditioning, and in the strongest form of social constructionism, thinkers question the ability of scientists to ever describe physical reality objectively. Social constructionism is about as far from logical empiricism as one can get. Feminist theories of science have also used the ideas of social constructionism to critique the gender bias in science, also arguing that a feminist approach to science would be more holistic and not as reductionist.

Besides having to defend the validity of their results in the face of deeper methodological musings of philosophers, scientists have also found themselves having to defend the merits of the scientific pursuits when arguing for the large taxpayer-funded projects of Big Science. Scientists have also had to defend the validity of scientific assumptions and conclusions against creationists, religious fundamentalists who interpret scriptures literally, radical feminist theorists, postmodernists, astrologists, and purveyors of pseudoscientific theories.

See also Big Science, Creationism, Feminism; Kuhn, Thomas S.; Popper, Karl; Religion; Social Constructionism

References

Boyd, Richard, Philip Gasper, and J. D. Trout, editors. *The Philosophy of Science.* Cambridge: MIT Press, 1991.

Feyerabend, Paul K. *Against Method: Outline of an Anarchistic Theory of Knowledge.* Third edition. New York: Verso, 1993.

Gower, Barry. *Scientific Method: A Historical and Philosophical Introduction.* New York: Routledge, 1997.

Gross, Paul R., Norman Levitt, and Martin W. Lewis, editors. *The Flight from Science and Reason.* Baltimore: Johns Hopkins University Press, 1996.

Kitcher, Philip. *The Advancement of Science: Science without Legend, Objectivity without Illusions.* New York: Oxford University Press, 1993.

Klee, Robert. *Introduction to the Philosophy of Science: Cutting Nature at Its Seams.* New York: Oxford University Press, 1997.

Kuhn, Thomas S. *The Structure of Scientific Revolutions.* Chicago: University of Chicago Press, 1962.

Physics

Physicists in the first half of the twentieth century completely revolutionized physics. The theory of relativity developed by Albert Einstein (1879–1955) led directly to the quantum theory of the atom, and ultimately, the development of atomic weapons, which pushed physics into the forefront as the premier science. After 1945, the cold war prompted massive government spending on nuclear weapons and the development of more sophisticated conventional weapons, resulting in more funding for scientific research and development. Nuclear physics became an important career area, as nuclear engineers and scientists developed civilian nuclear reactors and more sophisticated nuclear weapons, as well as applying nuclear physics to other areas, such as medicine and archaeology. By virtue of having invented the technology, physicists became important advisors on the direction of both military and civilian nuclear policy in the United States. A chemist, Linus Pauling (1901–1994), played such an important role in the campaign for the 1963 Nuclear Test Ban Treaty that he received the 1962 Nobel Peace Prize.

As the first science, physics of course became part of Big Science, a recipient of generous funding from the United States, the Soviet Union, and other industrialized nations. In the United States, the federal budget, both military and civilian, for physics grew until the late 1960s, then went into decline, surged under the presidency of Ronald Reagan (1911–2004) in the 1980s, and then went into decline again after the end of the cold war in 1989. The funding of research went not only into applied physics, but also into basic research, since the atomic bomb had taught everyone that seemingly innocent basic research could lead to powerful weapons.

In the second half of the twentieth century, physics expanded our knowledge of a wide range of things, from the smallest to the biggest. In the late 1940s, the German-born physicist Maria Goeppert Mayer (1906–1972) explained how protons and neutrons occupied orbital shells around the atomic nucleus. The revision of quantum electrodynamics (QED) theory in the 1950s by Richard P. Feynman (1918–1988), Julian Schwinger (1918–1994), Shinichiro Tomonaga (1906–1979), and Freeman Dyson (1923–) cleared up our understanding of how to explain the interactions between electrons and photons. New subatomic particles found in cosmic-ray research and particle accelerators showed that even smaller units of matter and energy existed. Eventually quark theory, developed in the 1960s and 1970s by Murray Gell-Mann (1929–) and others, showed that all these particles were composed of still smaller particles called quarks. String theory, begun in 1970 by Yoichiro Nambu (1921–), went beyond quarks, though the complex mathematics of the theory and the difficulty of finding physical proof of strings have kept the theory in the realm of informed speculation. By the end of the century, the standard model of elementary particle interactions had explained three forces (weak nuclear force, strong nuclear force, and electromagnetism) but could not integrate the force of gravity into these explanations. Theoretical physicists are still searching for a grand unified theory to explain all the forces simultaneously.

Astrophysicists saw the size of the known universe repeatedly expand after 1950, as newer optical telescopes, radio telescopes, and satellite observatories pushed back boundaries. Previously suspected physical phenomena, such as neutrinos and extrasolar planets, were discovered, and at the same time unsuspected phenomena revealed themselves to the new instruments: neutron stars, pulsars, and quasars. The details of quantum mechanics were found to be relevant to the largest structures in the universe, and theoretical physicists became cosmologists, speculating on the origin and structure of the universe. The Big Bang and the steady state theories of the origin of the universe vied for allegiance among cosmologists until the accidental discovery in 1964 by two Bell Labs researchers that the universe contained a low level of uniform background radiation, which matched the Big Bang mathematical model, but found no place in the steady state theory. The English physicist Stephen Hawking (1942–) became one of the best-known scientists of the century through his research into black holes and the Big Bang while he struggled with a debilitating medical condition.

The space age begin with the Soviet launch of the first artificial satellite in 1957, *Sputnik I*. The cold war inspired the superpowers to ideological rivalry and provided the funding to quickly develop ever more sophisticated space technology. The Apollo project landed twelve men on the Moon, but the real scientific value from space explo-

Two magnets based on superconducting technology demonstrate the lines of magnetic force with iron nails. (Bettmann/Corbis)

ration has come from satellites using their sensors on Earth, orbiting telescopes, and robotic spacecraft that explore other planets in the solar system, and from the profound technological inventions that the space race has encouraged. The impact of the Shoemaker-Levy comet on Jupiter in 1994 provided a powerful reminder that space exploration is about more than ideological propaganda. There are big rocks up there, and eventually Earth will get hit again, as has happened so many times before.

The Dutch physicist Heike Kamerlingh Onnes discovered superconductivity in 1911 when he found that some metals lost all resistance to the flow of electrons at temperatures near absolute zero. The American physicist John Bardeen (1908–1991) and his colleagues published in 1957 a theoretical foundation for superconductivity. The physicists K. Alex Müller (1927–) and Georg Bednorz (1950–) discovered a metal in 1985 that remained superconductive at a much higher temperature, setting off a successful race to find more such metals and promising the creation of more efficient methods of transmitting electricity.

Electronic computers became an important tool for physicists in all their endeavors and directly led to the development of chaos theory. At the end of the century the burgeoning field of nanotechnology also attracted physicists, given the enormous potential and challenges of manipulating matter on the molecular level.

In the late 1990s, when ruminations turned to grander thoughts, inspired by the impending end of the millennium, there were murmurs and even books that proposed that the end of physics was coming, because once a grand unified theory was able to explain all the known universe, physicists would not have anything new to do. Ironically enough, this very sentiment was common a century ago, just before Albert Einstein (1879–1955) upset everything that physicists thought they knew.

See also Apollo Project; Archaeology; Astronomy; Bardeen, John; Bell Burnell, Jocelyn; Big Bang Theory and Steady State Theory; Big Science; Black Holes; Chandrasekhar, Subrahmanyan; Chaos Theory; Cold Fusion; Cold War; Computers; Dyson, Freeman; Feynman, Richard; Gamow, George; Gell-Mann, Murray; Glashow, Sheldon; Grand Unified Theory; Hawking, Stephen; Hoyle, Fred; Hydrogen Bomb; Lasers; Libby, Willard F.; Mathematics; Mayer, Maria Goeppert; Medicine; Müller, K. Alex; Nambu, Yoichiro; Nanotechnology; National Science Foundation; Neutrinos; Nobel Prizes; Nuclear Physics; Oppenheimer, J. Robert; Particle Accelerators; Particle Physics; Penrose, Roger; Pulsars; Quasars; Sagan, Carl; Sakharov, Andrei; Schwinger, Julian; Space Exploration and Space Science; Teller, Edward; Tsui, Daniel C.; Wheeler, John A.; Wu, Chien-Shiung

References

Adams, Steve. *Frontiers: Twentieth-Century Physics.* London: Taylor and Francis, 2000.

Brown, Laurie, Abraham Pais, and Brian Pippard. *Twentieth Century Physics.* 3 volumes. New York: American Institute of Physics, 1995.

Crease, Robert P., and Charles C. Mann. *The Second Creation: Makers of the Revolution in Twentieth-Century Physics.* New York: Macmillan, 1986.

Kevles, Daniel J. *The Physicists: The History of a Scientific Community in Modern America.* Revised edition. Cambridge: Harvard University Press, 1995.

Kragh, Helge. *Quantum Generations: A History of Physics in the Twentieth Century.* Princeton: Princeton University Press, 1999.

Suplee, Kurt. *Physics in the 20th Century.* New York: Abrams, 1999.

Piccard, Auguste (1884–1962)
See Undersea Exploration

Plate Tectonics

The acceptance of the theory of plate tectonics, which explains continental drift, constitutes the most important scientific revolution in the earth sciences during the twentieth century. The Germany meteorologist Alfred Wegener (1880–1930) proposed in 1912 that the continents drift, as evidenced by the stratigraphical matches between the geology of continents across oceans, the location of fossils, and the obvious match on maps between landmasses separated by water. Specialists in geology and paleontology ridiculed his theory, especially because Wegener had no theory of a mechanism that would allow continents to move.

After World War II, oceanographers from many nations expanded their knowledge of the ocean floors, partly by taking samples of the magnetic characteristics of rocks in the Mid-Atlantic Ridge. Geologists and physicists from France, Japan, and America had occasionally proposed since early in the twentieth century that the polarity of Earth's magnetic fields periodically switched, resulting in telltale alignments within rocks formed from cooling magma. Many scientists did not believe that these flip-flops occurred, but new evidence soon changed that.

Oceanographers also came to realize that

the geologic patterns they found on the ocean floor were the result of sea floor spreading. In 1962, an influential article by the geologist Harry Hess (1906–1969) entitled "History of Ocean Basins" proposed that magma was the lubricant that allowed continental plates to move. Frederick Vine (1939–) and Drummond Matthews (1931–1997) in Cambridge, England, expanded the work of Hess by proposing the hypothesis that the magnetic strips on the ocean floor were a recording of the reversals in Earth's magnetic field. The geophysicist Lawrence Morley (1920–) of the Geological Survey of Canada independently proposed a similar hypothesis, and it became known as the Vine-Matthews-Morley (VMM) hypothesis. The VMM hypothesis was not immediately accepted. *Nature* and the *Journal of Geophysical Research* both rejected Morley's original article.

The theory of plate tectonics explains that as plates grid together, they create mountains from uplift, and where plates pull apart, they create rift valleys and opportunities for magma to well up from deeper in the Earth. Volcanoes and greater earthquake activity exist on the edges of continental plates. The Canadian geophysicist John Tuzo Wilson (1908–1993) developed the concept of hot spots, places in the center of plates where volcanic activity pushes upward. Hawaii and Yellowstone National Park are examples of such hot spots, leaving a trail of seamounts in the Pacific and lava fields in the American Northwest as continental plates drift across an upwelling of magma.

During the 1960s, geology experienced a revolution, as geologists became convinced that continents really do drift. A 1970 article first used the term *plate tectonics*. The theory proved to be a powerful explanatory mechanism that transformed geology from a science of particulars into a mature science with a body of integrated knowledge. Acceptance of the theory led to efforts to trace the continents as they have cycled between a single supercontinent and more scattered configurations. The last supercontinent, named Pangaea, existed 200 million years ago. With the exploration of other planets and moons by robotic spacecraft, scientists have looked for evidence that plate tectonics is active in mantles other than Earth's mantle. They have found evidence of past tectonic activity, but no convincing evidence of current activity. The story of plate tectonics is often used as an example of the theory of scientific revolutions developed by Thomas S. Kuhn (1922–1996); the ocean floor magnetic data represent just the kind of nagging anomaly that overthrows old paradigms, as it overthrew the old paradigm of static continents and replaced it with the new paradigm of continental drift.

See also Geology; Hess, Harry; Kuhn, Thomas S.; Oceanography; Space Exploration and Space Science; Volcanoes; Wilson, John Tuzo

References

Frankel, Henry. "The Development, Reception, and Acceptance of the Vine-Matthews-Morley Hypothesis." *Historical Studies in the Physical Sciences* 13 (1982): 1–39.

Glen, William. *The Road to Jaramillo: Critical Years of the Revolution in Earth Science.* Stanford, CA: Stanford University Press, 1982.

Hallam, Anthony. *A Revolution in the Earth Sciences: From Continental Drift to Plate Tectonics.* Oxford: Clarendon, 1973.

Menard, Henry W. *The Ocean of Truth: A Personal History of Global Tectonics.* Princeton University Press, 1986.

Oreskes, Naomi, editor. *Plate Tectonics: An Insider's History of the Modern Theory of the Earth.* Boulder, CO: Westview, 2002.

Wood, Robert Muir. *The Dark Side of the Earth.* London: Allen and Unwin, 1985.

Popper, Karl (1902–1994)

The philosopher Karl Popper devised the idea of falsifiability to solve the problem of induction in the philosophy of science. Karl Raimund Popper was born in Vienna, Austria, where his father was a successful lawyer with a library of 10,000 books. At the end of World War I in 1918, Austria entered an extended period of economic chaos, food shortages, and political upheaval, during which time Popper left the Vienna Gymnasium and enrolled in the University of Vienna, though he was too young to matriculate. He attended his classes sporadically, finding formal education boring, and learning mostly from reading books. In 1919, only seventeen years old, he formed the question that consumed his life: how does one distinguish between valid science and pseudoscience?

Believing that people should be able to work with their hands, Popper apprenticed as a cabinet maker. He completed his apprenticeship, though he found the work tedious and occupied his mind with ruminations on philosophy and speculations about the development of European classical music. Popper decided to become a schoolteacher and obtained a teacher's certificate in 1923, working for a year as a social worker with youth. In 1925, he enrolled full-time in the University of Vienna and the Institute of Education. He also published his first paper in that same year. In 1928, he wrote a thesis to get a certification from the Institute of Education, another thesis on axiomatic geometry to obtain a secondary school teacher's certificate, and a thesis to obtain a Ph.D. His work led him to an introduction to the Viennese Circle of Philosophers, who became famous for their positivist philosophy. Popper disagreed with their approach but profited from the association. In 1930, he obtained a job as a schoolteacher and married a fellow schoolteacher in that same year. They had no children.

Popper then wrote two books, leaving the first unpublished, but publishing the second in 1934; its English title is *The Logic of Discovery.* He set forth the answer to his earlier question, arguing that the problem of induction would be solved if scientists would frame their questions in terms of falsifiability. Repeated proofs of a scientific theory or scientific observation did not make it true; only experiments that attempted proof through falsifiability led to positive proof. His book attracted considerable attention, and he found himself accepting opportunities to lecture in London, Paris, Copenhagen, and other places in Europe. He accepted a uni-

versity lectureship in Christchurch, New Zealand, in 1937 and thus avoided the Nazi takeover of his homeland a year later. He remained in New Zealand until the end of World War II, then accepted a position as reader in logic and scientific method at the London School of Economics at the University of London. In that same year, 1945, he published two volumes on *The Open Society and Its Enemies,* attacking totalitarian political philosophies and defending the ideals of democracy. His 1957 *Poverty of Historicism,* a collection of essays, continued in the same vein, attacking political philosophies such as Marxism that derided individuals and assumed laws of inevitable political progress.

Popper became a full professor in 1949 and was knighted in 1965. His prolific publishing of articles and his two later books, *Conjectures and Refutations: The Growth of Scientific Knowledge* (1963) and the three-volume *Postscript to the Logic of Scientific Discovery* (1981–1982), continued a relentless campaign to promote his views in the philosophy of science. Many philosophers and scientists supported his ideas, though they found creating falsifiable theories and observations difficult to implement in actual scientific work. Historical sciences, such as geology, paleontology, and evolutionary biology, have an almost impossible task in meeting Popper's standards because they describe events that cannot be repeated.

See also Philosophy of Science
References
Magee, Bryan. *Karl Popper.* New York: Viking, 1973.
Popper, Karl. *Unended Quest: An Intellectual Autobiography.* La Salle, IL: Open Court, 1982.

Population Studies

Since 1950, scientists and other people have become increasingly worried about the growing human population of Earth. Estimates place the world population in 10,000 B.C. at 6 million humans, all living by hunting and gathering their food. With the invention

"One Child Only" billboard in Shanghai, China (Wolfgang Kaehler/Corbis)

of agriculture, irrigation systems, and other technologies, human population grew. The estimated world population stood at 252 million people in A.D. 1, with 170 million of those people living in Asia. By 1750, that number had grown to 771 million people. By 1950, population growth had accelerated, and the world total had reached 2.53 billion. Infant mortality and child deaths had declined, and life expectancy had gone up; people had a better chance of surviving to adulthood and living to old age.

Many commentators have pointed to modern medicine as the major cause of this upsurge in population. Thomas McKeown (1911–1988), professor of social medicine at the University of Birmingham Medical School, argued in 1976 that a careful study of the statistics shows that the growth curve started in Western Europe before modern Western medicine became effective in the late nineteenth century. McKeown argued that more abundant food from more efficient

agriculture, cleaner water, and better sanitation in the cities actually caused the population explosion. He demonstrated that birth rates did not change significantly, but the mortality rate dropped considerably in Western Europe in the eighteenth and nineteenth centuries. McKeown later expanded his argument in a 1979 book, *The Role of Medicine: Dream, Mirage or Nemesis?* Other scholars have argued that a natural cycle, part of the constant struggle between human immune systems and mutating diseases, led to a decrease in the virulence of diseases in the eighteenth century that also helped spark the population boom.

So many more people required more food. Increased amounts of land were placed under cultivation, and new strains of staple crops were created through the Green Revolution in the 1950s and 1960s. An American plant specialist, Norman E. Borlaug (1914–), and others developed new strains of wheat and rice beginning in the 1940s. Borlaug did not believe that the Green Revolution had solved the population crisis; rather, the increased food production had bought time to find a more permanent solution or set of solutions to the problem. Increased use of pesticides and artificial fertilizers also increased food production.

The philanthropist John D. Rockefeller III (1906–1978) founded The Population Council in 1952 to promote research into population control and aggressively promoted family-planning programs and contraceptive use throughout the world. The development of the birth control pill in the late 1950s provided another tool to control fertility and population in those countries where cultural attitudes had changed to favor smaller families. In the 1960s, many scientists grew ever more concerned about the burgeoning population. The entomologist Paul R. Ehrlich (1932–) became the best known of those scientists and writers sounding a warning call over human population growth when he published his best-selling book *The Population Bomb* in 1968. A zero population growth movement emerged with interesting consequences. A combination of social changes and economic forces encouraged smaller families in Western industrialized nations, and by the end of the century, most Western industrialized nations had achieved close to replacement-only fertility rates. Only immigration actually raised the population in the United States and some Western European countries. Populations continued to grow in the less industrialized European countries and in South America, Asia, and Africa.

World population passed 6 billion sometime during the month of October 1999. Total world population stood at 6.236 billion in the year 2000, with the bulk of the population, 3.736 billion, in Asia. Almost half of all these people lived in cities. If China had not implemented harsh population measures in 1978, often limiting each ethnic Chinese couple to a single child, China would have had an extra 250 million people in 1998. Despite the depredations of widespread fatalities

from acquired immunodeficiency syndrome (AIDS) in Africa, faster population growth is expected on that continent. Studies estimate that world population will reach 9 billion around the year 2050, with some 1.9 billion of those people on the continent of Africa. Many scientists believe that Earth's ecology cannot sustain the current levels of natural resource use by the planet's population, though there is no common agreement on what is a sustainable human population for the planet.

See also Acquired Immunodeficiency Syndrome; Agriculture; Birth Control Pill; Ecology; Ehrlich, Paul R.; Environmental Movement; Foundations; Medicine

References
Cohen, Joel. *How Many People Can the Earth Support?* New York: Norton, 1995.
Ehrlich, Paul R., and Anne H. Ehrlich. *The Population Explosion.* New York: Simon and Schuster, 1990.
Livi-Bacci, Massimo. *A Concise History of World Population.* Second edition. Oxford: Blackwell, 1997.
McKeown, Thomas. *The Role of Medicine: Dream, Mirage or Nemesis?* Oxford: Basil Blackwell, 1979.

Prions

Proteinaceous infectious particles (prions) are thought to be malformed proteins that cause a class of unique neurological diseases and behave like viruses in some ways. The Harvard-educated American physician Daniel Carleton Gajdusek (1923–) studied the odd neurological disease kuru among the Fore Highlanders people of Papua New Guinea in the 1950s. Kuru was called the laughing disease because victims descended into dementia before dying. Gajdusek believed that patients caught the disease when they practiced ritual cannibalism and ate the brains of the dead. Gajdusek showed in the 1960s that injecting brain cells from kuru victims into chimpanzees caused the chimpanzees to contract the disease. Brain tissues from autopsies of infected people showed characteristic holes, and similar symptoms were found in sheep that died from scrapie. In 1971, Gaj-dusek showed that victims of the rare Creutzfeldt-Jakob disease (CJD) showed symptoms similar to victims of kuru and scrapie. Scientists surmised that a virus caused these diseases and characterized it as a "slow virus" because it sometimes took decades after the initial infection to kill. Gaj-dusek shared the 1976 Nobel Prize in Physiology or Medicine for his work.

When a patient of the American neurologist Stanley B. Prusiner (1942–) died of CJD in 1972, Prusiner devoted his career to understanding the cause of the disease. In 1982, he published his theory that a protein caused CJD, as opposed to a virus. He named these proteins prions and argued that they were passed from infected people and destroyed brain tissue. Prusiner struggled to obtain and maintain his research funding from the National Institutes of Health (NIH) in the face of widespread skepticism about his theory. Prusiner later grouped CJD, scrapie, kuru, and possibly Alzheimer's together as prion diseases. In the 1980s, Prusiner's team and other scientists identified a gene for creating the protein, called the cellular prion protein, or PrPC. They determined that the PrPC was natural and harmless and that a variant called scrapie prion protein, or PrPSc, actually caused the prion diseases.

In 1996, a new variant of CJD appeared in Britain, apparently caused by eating meat from cows infected with bovine spongiform encephalopathy (BSE), or mad cow disease. The use of infected cattle brains and infected sheep brains in cattle feed to increase protein content was thought to be responsible for the spread of BSE. The death of a dozen people from CJD led to a massive campaign to destroy much of the cattle in England and discontinue the use of brains to supplement feed. Prusiner received the 1997 Nobel Prize in Physiology or Medicine for his work, though his theory of prions has remained controversial.

See also Genetics; Medicine; National Institutes of Health; Nobel Prizes

References
Rhodes, Richard. *Deadly Feasts: Tracking the Secrets of a Terrifying New Plague.* New York: Simon and Schuster, 1997.
Ridley, Rosalind M., and Harry F. Baker. *Fatal Protein: The Story of CJD, BSE, and Other Prion Diseases.* Oxford: Oxford University Press, 1998.

Psychology

The science of psychology emerged in the later part of the nineteenth century in Germany as a joint branch of physiology and philosophy. Psychiatry emerged from the practice in the nineteenth century of assigning physicians to be superintendents of insane asylums. The first scientific psychologists, such as the German Wilhelm Wundt (1832–1920), the Englishman Francis Galton (1822–1911), and the American E. B. Titchener (1867–1927), clearly focused on empirical studies of repeatable external phenomena that could be quantified, such as perception and sensory stimulus. They deliberately modeled their new science on the traditional physical sciences. With the exception of Galton, these pioneers frowned upon applied psychology, not wishing to challenge psychiatry in the art of healing, but rather concentrated on applying science to limited examinations of human behavior. But soon the American psychologist G. Stanley Hall (1884–1924) and others began to apply psychology to education and aptitude testing.

The theory of behaviorism, first established by the work of the famous Russian physiologist Ivan Pavlov (1849–1936) and greatly expanded in the 1920s by the American psychologist John B. Watson (1878–1958), became the first of many theories from the ranks of psychologists to compete with the all-encompassing nature of the theory of psychoanalysis promoted by the Austrian neuropsychiatrist Sigmund Freud (1856–1939). Following the lead of Hall, later psychologists moved beyond just education counseling and aptitude testing to create clinical psychology, in which the theories of psychology served as tools in mental health care. This innovation was bitterly contested within the profession, and it is no accident that members of the medical profession, under such leaders as Freud, rather than psychologists, were the first to extensively develop twentieth-century mental health care. Psychology, unsure of itself as a science, slowly made the transition into healing.

The psychologist B. F. Skinner (1904–1990) carried the banner of behaviorism that had been left by Watson. As a strict behaviorist, Skinner believed that nurture dictated all action and that nature played no role. He rejected free will or any notion of unconscious motivation. Mainstream psychologists did not accept his theory, but the extreme nature of his stance and the articulate manner in which he presented his ideas continue to make him a useful standard to measure later psychological theories against.

The American psychologist Carl R. Rogers (1902–1987) advocated client-centered therapy beginning in the 1940s, a form of therapy in which therapists urged clients to find their own paths to health. The therapist was a facilitator and was not to impose her or his own values on the client. This positive ideal of the individual as being able to self-heal became a characteristic of humanistic psychology, as did the ideal of self-realization. Rogers used the term *client,* rather than *patient,* to emphasize the nonmedical and noncoercive nature of his treatment. The American psychologist Abraham H. Maslow (1908–1970) also promoted humanistic psychology with his theory of self-actualization, developed in the 1950s and early 1960s. Maslow's theory postulated a pyramid of needs, with basic physiological requirements at the bottom, love and esteem in the center, and self-actualization as the capstone. Unless restrained by social norms or neuroses, people naturally strive to self-actualize once the other needs are fulfilled, though few ever achieve that pinnacle. Other humanistic psychological flavors of psychotherapy became prominent in the 1960s and 1970s, such as logotherapy, Gestalt therapy, existen-

tial analysis, self-directed encounter groups, and addiction recovery clinics.

Only after World War II did clinical psychology rise to become a significant scientific and social force, complementing and competing with the medically based discipline of psychiatry, especially in the United States. Universities opened counseling centers that offered educational counseling and psychotherapy. Mainline Protestant religions began to incorporate training in clinical psychology into their pastoral education programs, turning ministers and pastors into professionally trained psychotherapists. Religious leaders have always had a therapeutic role in society, though this embracing of modern psychology was ironic, in that surveys showed that of all the scientific and scholarly disciplines, psychologists and psychiatrists harbored the most negative attitudes toward organized religion, with the exception of philosophers. With the middle class growing more prosperous, psychotherapists inspired by Rogers, Freud, and other theorists began to offer psychotherapy to more than just institutionalized patients and the wealthy. Especially in the United States, psychotherapy became popular, and society at large became psychologized, using psychological terms and viewing relationships through pop psychology. Movies, books, magazine articles, seminars, and popular discourse spread the message.

Almost unnoticed amid the expansion of the field was a 1952 article by the English psychologist Hans J. Eysenck (1916–1997), who reported that after reviewing twenty-four studies he found no empirical evidence that neurotic patients who received psychotherapy recovered any more frequently than those not treated by a psychotherapist. Studies found that the form of psychological treatment often did not make a difference, and even indicated that at times, psychotherapy made patients, or clients, worse than if they had not been treated at all by a psychological professional. Appalled by the lack of scientific rigor in the selection of therapeutic modalities, the psychologists Sol Garfield (1918–) and Allen E. Bergin (1934–) edited a leading textbook, the *Handbook of Psychotherapy and Behavior Change: An Empirical Analysis* (1971), which went through multiple editions, in an effort to alert their colleagues to this problem.

So how are clients who seek psychotherapy healed? Multiple researchers have indicated that healing occurs when the psychotherapist creates an environment of trust and expertise, forming a therapeutic alliance. The psychotherapist's personal characteristics of understanding, empathy, and concern are the major factors in healing. The psychiatrist Jerome Frank (1909–) has pointed out that this alliance does not even have to be a face-to-face relationship. Ironically enough, a self-help book can also provide many of these same elements to a lesser degree and thus serve as a healing mechanism for the reader. This form of healing, whether based on verbal conversation or through a book, only works if the client accepts the causal explanations of the psychotherapist; the talking cure requires the patient's cooperation.

Without a doubt, psychopharmacological drugs can have a significant impact regardless of the cognitive cooperation of the patient. Doses of lithium carbonate have been successfully used to treat bipolar disorder, also called manic-depressive psychosis, in which the sufferer alternates between periods of manic euphoria and depression. The antidepressant Prozac, the first of the selective serotonin reuptake inhibitors (SSRIs), introduced in 1986, works by interfering with the reabsorption of the neurotransmitter serotonin in the brain. Prozac is also used to treat obsessive-compulsive disorder and bulimia nervosa, and is even used by some mentally healthy people as a psychological cosmetic to enhance their productivity. At the end of the century, very few psychological disorders were readily treated by drug therapy, despite a few outstanding successes. The social and cultural context of the disturbed person still determined most of a given problem's parameters.

The spirit of the 1960s looked to liberate and empower the victims of society wherever they were found. The awful conditions in mental hospitals drew the sympathy and indignation of social activists, and the activists also questioned the prerogative of psychiatrists to institutionalize people against their will. In the late 1960s, a vigorous antipsychiatry movement began, launching a well-meaning deinstitutionalization process. The asylums quickly emptied their patients onto the streets to be cared for by community-based treatment programs. Most of these programs have failed and homelessness has increased.

By the 1970s, the older rigid schools of psychoanalysis and behaviorism had begun to fade. Humanistic psychology and eclecticism have increasingly become the norm, though no unified psychological paradigm has emerged. Eclectic psychotherapists reject rigid adherence to any particular school of psychology or treatment modality. They choose from among a menu of options based on the needs of each individual client. The emphasis on self-realization and breaking social and psychological barriers within feminism also played a role in breaking the hold of the older schools. The notable book *Toward a New Psychology of Women* (1976), by the feminist psychiatrist Jean Baker Miller, helped promote the idea that psychologists had neglected to develop an accurate understanding of women.

Genuine advances in psychology are hard to achieve because human subjects are difficult to study and because repeatability is hard. There have been notable achievements. The experimental psychologist Harry F. Harlow (1905–1981) used studies of maternal deprivation with rhesus monkeys to show that love is a primary drive for primates and humans. He changed the way that modern psychologists look at the need for affection between adults and the need for affectionate mother-child relations, though his methods would not pass an ethics review board today. The Swiss-born psychiatrist Elisabeth Kübler-

Ross (1926–2004) found that terminally ill people and people in mourning pass through a series of five stages: denial and isolation, anger, bargaining, depression, and then acceptance. She also promoted the hospice movement and contributed to thanatology, the study of death.

The zoologist-turned-sexologist Alfred Kinsey (1894–1956) published two books, *Sexual Behavior in the Human Male* (1948) and *Sexual Behavior in the Human Female* (1953), that astonished Americans with their revelations that what was considered sexually deviant behavior, such as masturbation and homosexuality, was much more widespread than commonly assumed. Later critics found that Kinsey's scientific techniques inevitably led to biased results, overstating the incidence of different sexual practices in an effort to promote his social agenda. The detailed physiological and psychological work of the physician William H. Masters (1915–2001) and the psychologist Virginia Johnson (1925–) established a firm medical basis for sexology and the emergence of clinical sex therapy.

The study of primates by Jane Goodall (1934–), Dian Fossey (1932–1985), and others made important contributions to how we view human behavior. A form of behaviorism has been revived with the study of sociobiology and evolutionary psychology by the zoologists Edward O. Wilson (1929–) and Richard Dawkins (1941–). The role of dreams and their role in consciousness have always intrigued cognitively oriented psychologists. The discovery in 1953 that dreams correspond to rapid eye movement (REM) and that most other mammals also experience REM has raised many questions with no satisfactory answers.

Intelligence tests, often giving a single numeric intelligence quotient (IQ) score, were first created in the early part of the twentieth century. They fell into disfavor after the 1960s, when social activists objected to racist or social assumptions made in the tests, and when the realization began to grow that intelligence was a multifaceted concept, not easily reduced to a single number. In the 1990s, the idea of emotional intelligence (EQ) became common, though it was not quantified as a number; rather it was seen as a characteristic that indicated who would be more likely to be successful in life, regardless of their IQ score. A high EQ indicates the ability of an individual to control and channel emotions to serve long-term goals.

The 1980s saw the rise of the recovered-memory movement, based on the theory that the memory of traumatic events like sexual abuse is repressed. Some victims came forward with stories of abuse that they remembered years after the abuse, either as a result of psychotherapy or prompted by a flashback. These repressed memories were often vague or ambiguous, and the number of specific details usually increased with time. The issue of repressed memories was always controversial, but in the 1990s neurologists and psychologists who specialized in how memory actually works argued that there was not a mechanism for the kind of repression that the recovered-memory movement postulated. At times, sometimes using hypnosis, zealous psychotherapists had led their patients to believe that abuse had occurred when it really had not.

See also Bioethics; Cognitive Science; Dawkins, Richard; Feminism; Fossey, Dian; Goodall, Jane; Harlow, Harry F.; Kinsey, Alfred C.; Kübler-Ross, Elisabeth; Masters, William H., and Virginia Johnson; Medicine; National Institutes of Health; Religion; Rogers, Carl R.; Skinner, B. F.; Sociobiology and Evolutionary Psychology; Wilson, Edward O.

References

Danziger, Kurt. *Constructing the Subject: Historical Origins of Psychological Research.* New York: Cambridge University Press, 1990.

Fancher, Raymond. *The Intelligence Men: Makers of the I.Q. Controversy.* New York: Norton, 1985.

Goleman, Daniel. *Emotional Intelligence: Why It Can Matter More Than IQ.* New York: Bantam, 1995.

Hacking, Ian. *Rewriting the Soul: Multiple Personality and the Sciences of Memory.* Princeton: Princeton University Press, 1995.

Herman, Ellen. *The Romance of American Psychology: Political Culture in the Age of Experts.* Berkeley and Los Angeles: University of California Press, 1995.

Kramer, Peter D. *Listening to Prozac.* Revised edition. New York: Penguin, 1997.

Leahey, Thomas Hardy. *A History of Psychology: Main Currents in Psychological Thought.* Sixth edition. Upper Saddle River, NJ: Prentice Hall, 2003.

Wulff, David M. *Psychology of Religion: Classic and Contemporary Views.* New York: Wiley, 1991.

Pulsars

In October 1967, a Cambridge University graduate student, Jocelyn Bell Burnell, discovered the first pulsar as she analyzed the data from a radio telescope while looking for quasars. Bell Burnell noticed a half-inch of squiggles on a chart that intrigued her. She called it a bit of "scruff." Working bac kward, she found that the small signal corresponded to sidereal time, 23 hours and 56 minutes, indicating that it came from the same spot in the sky as Earth rotated beneath that spot. A more sensitive recorder installed in November allowed Bell Burnell and Antony Hewish, her academic advisor and the builder of the telescope, to look at the scruff in more detail. They found that the signal pulsed in intensity every 1.33 seconds. This was much too fast for a variable star, and its eerie regular intensity made some on their team wonder if they had actually stumbled across a transmission from an extraterrestrial intelligence. A second signal, discovered in December in a different part of the sky, pulsed at a rate of 1.2 seconds. With the signals so far apart, thoughts of extraterrestrial intelligence were laid to rest. Bell Burnell and the team had discovered a new phenomenon. When Hewish announced the discoveries in February, Bell Burnell had already found another two pulsars.

All stars radiate radio waves, but the intensity and period frequency of pulsars electrified astronomers. Three theories quickly emerged. The first two were variations of the idea that two white dwarfs or neutron stars were in such tight orbit about each other that they acted as a single lighthouse emitting radio waves. Although such a model produced radio pulses, it seemed unlikely that white dwarfs or neutron stars could orbit close enough to get such rapid pulses. The third theory proposed that the pulses were produced by a white dwarf that oscillated in intensity. Such stellar vibrations had already been detected in visible light from a white dwarf. (White dwarfs are dense stars smaller than Earth but as bright as the Sun.)

As for neutron stars, they were not yet actually known to exist; the theoretical possibility that such stars existed had been proposed in 1934 by the astronomers Walter Baade (1893–1960) and Fritz Zwicky (1898–1974). After a star reached the supernova stage of its stellar evolution, they theorized, the remnants collapsed into a neutron star, a star in which gravity became so strong that only closely packed neutrons remained. In November 1967, before the public announcement of the discovery of pulsars, the Italian astrophysicist Franco Pacini (1939–) speculated that a rotating neutron star, spinning very quickly, would broadcast radio waves because of its intense magnetic field. In June 1968, the American astronomer Thomas Gold (1920–) brought together Bell Burnell's discovery and Pacini's theory in an article for *Nature*.

In October 1968, another pulsar was discovered in the constellation of Vela, visible from the Southern Hemisphere of Earth. This pulsar pulsed ten times more rapidly than other pulsars and lay within the visible remnants of a supernova. A month later, a pulsar found in the midst of the colorful Crab Nebula burst thirty times a second. The Crab Nebula consists of the remnants of a star that went supernova in A.D. 1054, as recorded by Chinese astronomers. Fred Hoyle (1915–2001) and his colleagues had speculated in 1964 that a neutron star with an intense magnetic field might exist in the Crab Nebula, but they did not predict the existence of pulsars. These discoveries confirmed that neutron stars actually existed and that pulsars were rapidly rotating neutron stars, formed in the aftermath of supernovas. It is now thought that most pulsars do not have visible supernova remnants around them because the remnants have dissipated. A pulsar discovered in 1982 broadcast at a period of 1.6 milliseconds, indicating that the neutron star was spinning at a rate of more than 600 revolutions a second.

See also Astronomy; Bell Burnell, Jocelyn; Quasars; Telescopes

References

Greenstein, George. *Frozen Star: Of Pulsars, Black Holes, and the Fate of Stars.* New York: Freundlich, 1983.

Lyne, Andrew, and Sir Francis Graham-Smith. *Pulsar Astronomy.* Cambridge: Cambridge University Press, 1990.

Q

Quantum Mechanics
See Particle Physics; Physics

Quantum Physics
See Particle Physics; Physics

Quasars

Quasars, *quasi-stellar* radio sources, are the brightest and most distant objects known in the universe. In 1956, radio astronomers at Cambridge University undertook their third survey of the sky, seeking out radio sources and creating a catalogue. In the late 1950s and early 1960s, astronomers tried to identify visible objects that might be associated with known interstellar radio sources. The British astronomer Cyril Hazard, while at the University of Sydney, used the passage of the Moon in front of the location of the radio source 3C273 (the radio object numbered 273 in the third Cambridge University catalogue) to determine that a faint visible star (twelfth magnitude) was located in the same place as 3C273. The Dutch-born Maarten Schmidt (1929–), at the California Institute of Technology, obtained a spectral recording for 3C273 and found in 1963 that its hydrogen lines strongly shifted toward the red end

of the spectrum. Schmidt applied Hubble's Law and found that the object was very far away. Hubble's Law was named after Edwin Powell Hubble (1889–1953), who discovered that all galaxies are receding from each other as part of an expanding universe. Hubble also found that the faster a galaxy was receding, the farther away it was and the more its spectra shifted toward the longer wavelengths of the red end of the spectrum. Schmidt's analysis showed that 3C273 must be 3 billion light-years away, and to be visible, it must be much brighter than a normal galaxy.

The quasar 3C48 had been detected earlier, but misclassified as a star by the two astronomers who examined it. Within two years, ten such quasars had been located, and since then hundreds more possible quasars have been identified. The Einstein x-ray satellite observatory showed in 1980 that quasars are also powerful x-ray emitters. Photographs from the Hubble Space Telescope in the 1990s confirmed theories that predicted that quasars were located within galaxies and that they possessed jets of matter shooting out from them. In 1979, astronomers looked at two quasars that were close to each other and were so similar to each other that they proposed that an intermediate galaxy was

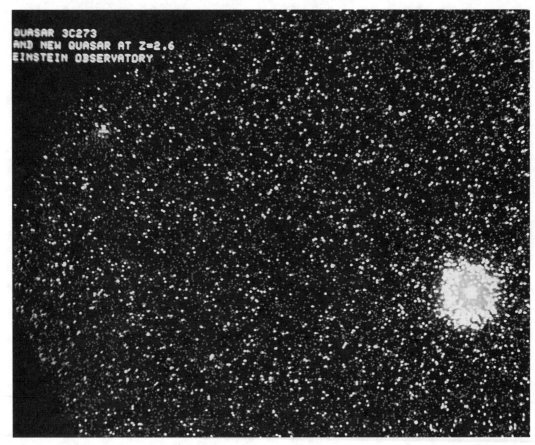

QUASAR 3C273
AND NEW QUASAR AT Z=2.6
EINSTEIN OBSERVATORY

X-ray picture of the quasar 3C273 (Corbis)

bending the light from a single quasar to create two images. This "gravitational lensing effect" has also been found with other quasars.

Such a great red shift invited suspicion that the quasars were actually much closer, but just receding from us. After this idea was discounted, theorists turned to the proposal that spinning black holes, a theoretical possibility with no positive proof at that time, were the source of the great energy of quasars. Quasars have gradually become identified as one of the proofs that black holes exist. The current thinking about quasars is that they are distant objects that give hints about the first couple of billion years after the Big Bang. The great energy output of quasars is the result of energy being emitted by large spinning black holes found at the center of galaxies, where stars are literally being consumed by the ravenous gravity pit. Matter shooting from the black hole forms jets thousands of light-years long, and the matter is moving so fast that relativistic effects are created.

See also Astronomy; Black Holes; Satellites; Telescopes

References

Kembhavi, Ajit K., and Jayant V. Narlkar. *Quasars and Active Galactic Nuclei*. Cambridge: Cambridge University Press, 1999.

Schmidt, Maarten. "3C 273: A Star-like Object with a Large Red-Shift." *Nature* 197 (1963): 1040.

R

Religion

People often view and experience religion and science as intellectual and cultural forces in conflict. This point of view emerged from the conflict between Western religious traditions and the theory of natural selection developed by Charles Darwin (1809–1882). Darwin and other geologically minded scientists argued that Earth had existed for far more than the six thousand years found by a literal interpretation of the Judeo-Christian Book of Genesis. Darwin and his successors also taught that humans were not unique creations of God, but were just an evolved form of primate, sharing a common ancestor with modern primates, with apes, chimpanzees, and monkeys. Darwinism, along with other aspects of modern thought, prompted a conservative backlash and the revitalization of fundamentalist and evangelical forms of Christianity at the beginning of the twentieth century. In the Islamic world, modern thought has inspired a similar phenomenon.

Although many religious people have accommodated themselves to the idea of Earth being billions of years old and the theory of Darwinist evolution, many fundamentalists, clinging to a literal interpretation of the Bible, have not. Creationist science is a strain of scientific endeavor devoted to proving that geological strata are the result of the flood described in Genesis, that biological evolution is wrong, and that Earth was divinely created within the last ten thousand years. Mainstream scientists have reacted strongly against creationism.

Contemporary scholars now recognize that the relationship between religion and science is a complex story and has varied throughout history. In the last century, science has become a dominant force in culture and society, though religion still maintains a significant hold. Some historians of science, such as the Englishman Herbert Butterfield (1900–1979), Stanley L. Jaki (1924–), and Reijer Hooykaas (1906–1994), have argued that Judeo-Christian culture actually laid the groundwork for the development of modern science. They have pointed out that the Jews abandoned the cyclic view of history, so common in other cultures, and that they viewed nature as a creation of one God, operating by laws and in accordance with justice, and not as created and governed by an animistic milieu of spirits and forces, guided by caprice. Thus, they have argued, thinkers came to the conclusion that God's laws and nature's laws could be discovered not only by revelation but also by human investigation. While science can trace its origins to ancient Greece, Greek philosophers believed in one god and much of Greek philosophy was adopted by early Christianity.

Many scientists have held or continue to hold strong religious views and have found minimal conflict between the two spheres, while others have abandoned the religious beliefs taught to them as children in favor of scientific materialism or some vague sense of an unknown creator. Most Judeo-Christian scientists who seek to reconcile their traditional religious beliefs with modern views on biological evolution have abandoned a strict literal interpretation of Genesis, arguing that the days in the biblical account are really geological ages, or even abandoning Genesis and arguing for a form of divinely guided evolution via natural selection. Other scientists, such as the zoologist Richard Dawkins (1941–), have strongly attacked religion and defended their own atheistic or agnostic points of view.

When the theory of the Big Bang began to gain prominence in the late 1940s, theologians were intrigued, since it seemed to be a variation on the idea of creation out of nothingness, as described in Genesis. Pope Pius XII (1876–1958) spoke positively in 1951 of the idea of a primeval atom. Fred Hoyle (1915–2001), creator of the steady state theory of the universe, complained bitterly that one reason people supported the Big Bang is that it resonated with their desires for divine creation. Ironically enough, other theologians found the steady state theory, which required the continuous creation of matter to fuel the expansion of the universe, to be more congenial to their theological point of view. The possibility of extraterrestrial life, a possibility seriously entertained by scientists, has obvious implications for religious theologies, since most theologies are very human-centered.

As our knowledge of physics advanced during the twentieth century, scientists noticed that their equations allowed many possible solutions. They also realized that certain physical constants were in such a range that they allowed life to exist. Any other values, and life would never have evolved on Earth. The astrophysicist Brandon Carter coined the term *anthropic principle* in 1970 to describe his idea that human beings observe the universe with those constants because of the fact that human beings exist in order to make such observations, otherwise human beings would not exist. Later theorists, such as John Barrow and James J. Tipler, who collaborated to write *The Anthropic Cosmological Principle* (1986), proposed a theory of a World Ensemble, in which many parallel universes were created at the time of the Big Bang, with all possible combinations of physical constants, and only our universe has the right combination of constants to make life possible. Tipler's book, *The Physics of Immortality: Modern Cosmology, God and the Resurrection of the Dead* (1994), went even further, proposing that artificial life would achieve absolute power and absolute knowledge at the Omega Point, the culmination of the evolution of the universe. At that point, simulations of all intelligent life-forms would be resurrected. The anthropic principle has had little impact on mainstream physics, and these speculations, as with most esoteric cosmology, have had little impact on religious discourse.

Advances in medicine have produced ethical issues that religious traditions have adapted to. Inexpensive and effective methods of birth control have forced religious people to revisit their assumptions about sexual activity and choices about having children. The process of in vitro fertilization, perfected by Patrick Steptoe (1913–1988) and Robert Edwards (1925–) in 1978 with the first human "test-tube baby," faced considerable opposition from religious authorities before they were successful. Other issues of bioethics, such as the right to die, assisted suicide, and abortion, have engaged theologians, and arguments based on religious beliefs are often the criteria that individuals use when facing the difficult decisions involved.

With the successful cloning of Dolly the sheep in 1996 from an adult somatic cell, people immediately asked themselves whether humans could be cloned. The idea struck at the core of personal identity, and most religious leaders came out against the idea of human cloning. The sociobiologist Edward O. Wilson (1929–) is among those who have welcomed cloning and other genetic engineering advances as ways for humans to take conscious control of human evolution, but his position is apparently in the minority. Bioengineering holds the promise of eventually being able to create babies with designed characteristics and will prompt considerable controversy if the techniques are ever developed.

A revived interest in the psychology of religion during the 1950s accompanied the adoption of secular psychotherapeutic training and techniques by mainline Protestant clergy. Attitudes of psychologists of religion on how to best pursue their studies have varied quite a bit. There are those who follow the classic scientific technique of reductionism, but others have opposed reductionism in any form, insisting that religion is unique in all essential respects.

See also Big Bang Theory and Steady State Theory; Bioethics; Birth Control Pill; Cloning; Creationism; Dawkins, Richard; Edwards, Robert; Evolution; Medicine; Philosophy of Science; Psychology; Search for Extraterrestrial Intelligence; Steptoe, Patrick

References
Barbour, Ian G. *When Science Meets Religion: Enemies, Strangers, or Partners?* San Francisco: HarperSanFrancisco, 2000.
Brooke, John Hedley. *Science and Religion: Some Historical Perspectives.* New York: Cambridge University Press, 1991.
Lindberg, David C., and Ronald L. Numbers, editors. *God and Nature: Historical Essays on the Encounter between Christianity and Science.* Berkeley: University of California Press, 1986.
Margenau, Henry, and Roy Abraham Varghese. *Cosmos, Bios, Theos: Scientists Reflect on Science, God, and the Origins of the Universe, Life and Homo Sapiens.* La Salle, IL: Open Court, 1992.
Wilson, David Sloan. *Darwin's Cathedral: Evolution, Religion, and the Nature of Society.* Chicago: University of Chicago Press, 2002.
Wright, Robert. *Three Scientists and Their Gods: Looking for Meaning in an Age of Information.* New York: Random House, 1988.
Wulff, David M. *Psychology of Religion: Classic and Contemporary Views.* New York: Wiley, 1991.

Rogers, Carl R. (1902–1987)

The American psychologist Carl R. Rogers promoted an influential approach within humanistic psychotherapy, an approach that he called client-centered therapy. Carl Ransom Rogers was born in Oak Park, Illinois, where his father, Walter A. Rogers, was a successful businessman who bought a farm for the family to live on. His father ran the farm based on the latest scientific principles, prompting Rogers to enroll at the University of Wisconsin with a major in agriculture. Having been raised in a religious household, he then decided to become a Protestant minister instead and changed his major to history. While in college, he visited China as part of the World Student Christian Federation Conference, and the experience began his personal process of breaking away from the theology that his parents had taught him. After graduating from Wisconsin in 1924, Rogers enrolled in the Union Theological Seminary. The psychology courses at the seminary inspired him to begin to take similar courses at Columbia University Teachers College, and he decided to abandon his plan of becoming a minister. He received his M.A. degree in 1928 and his Ph.D. in 1931 from Teachers College. Rogers married in 1924 and fathered a son and a daughter.

Rogers began working with children in 1928, first at the Society for the Prevention of Cruelty to Children in Rochester, New York, and later while teaching at the University of Rochester. He wrote two books on measuring personality in children and clinically treating children. In 1940, Rogers became a full professor in clinical psychology at Ohio State University, and as he prepared his lectures, he realized that his ideas about psychotherapy were not in line with the predominant methods. His optimistic opinion of people contrasted with the dour pessimism of Freudian psychoanalysis. Three books, *Counseling and Psychotherapy* (1942), *Client-Centered Therapy* (1951), and *On Becoming a Person* (1961), introduced and established his idea of nondirective, client-centered therapy, in

Carl R. Rogers, psychologist, 1979 (Roger Ressmeyer / Corbis)

chology at the University of Chicago from 1945 to 1957, then moved to the University of Wisconsin from 1957 to 1962, before finally moving to California and founding the Center for Studies of the Person in La Jolla. He remained active toward the end of his life conducting workshops and seminars, and writing books that advanced the cause of humanistic psychology. Elements of humanistic psychology and its goal of self-realization, as well as the philosophy of client-centered psychotherapy, became dominant influences in clinical psychology in the 1970s and remained so at the end of the century.

See also Psychology
References
Kirschenbaum, Howard. *On Becoming Carl Rogers.* New York: Delacorte, 1979.
Rogers, Carl R. *On Becoming a Person: A Therapist's View of Psychotherapy.* Boston: Houghton Mifflin, 1961.
Thorne, Brian. *Carl Rogers.* Thousand Oaks, CA: Sage, 1992.

Rowland, F. Sherwood (1927–)

The chemist F. Sherwood Rowland discovered, with his research associate Mario Molina (1943–), that chlorofluorocarbon (CFC) gases damage the ozone layer of the atmosphere. Frank Sherwood Rowland, nicknamed Sherry, was born in Delaware, Ohio, where his father was a professor of mathematics at Ohio Wesleyan University. He enjoyed naval history as a boy and played with naval miniatures. An accelerated path through school led to his high school graduation at age sixteen, and he went to college at Ohio Wesleyan. After two years, Rowland left to join the navy, but World War II ended before he completed basic training. He returned to school and graduated with a B.A. in 1948, with a triple major in chemistry, physics, and mathematics. Always drawn to competitive athletics, Rowland played college basketball and baseball, as well as playing a few summers with a Canadian semiprofessional baseball team.

accordance with which he referred to the person seeking help as a client, not a patient. Rogers's client-centered therapy always urged that clients should find their own paths to health, while psychotherapists act as facilitators and do not impose their own values. Rogers also came to believe that the empathy and the genuine nature of the psychotherapist had more to do with the effectiveness of the treatment than the therapeutic techniques that the psychotherapist employed.

Client-centered therapy became a major element of humanistic psychology, and Rogers became a prominent leader among clinical psychologists. He served as president of the American Psychological Association (APA) in 1946–1947 and received the Distinguished Scientific Contribution Award from the APA in 1956. He was a professor of psy-

In 1948 Rowland began graduate studies in chemistry at the University of Chicago under the tutelage of Willard F. Libby (1908–1980), who had just developed the method of carbon-14 dating that proved so useful in archaeology. In research sponsored by the Atomic Energy Commission (AEC), Rowland worked on radiochemistry. In 1952, he earned his doctorate and married. His wife and he later had a daughter and a son. Rowland accepted a position as an instructor at Princeton University and four years later moved to the University of Kansas as an assistant professor. In 1964, he moved to the Irvine campus of the University of California, which opened the following year. The involvement of his daughter, Ingrid, in the environmental movement and the general temper of the times prompted Rowland in 1970 to change his research from radiochemistry to environmental issues. In 1971, his research showed that current mercury levels in tuna and swordfish were consistent with levels of mercury from preserved specimens of fish, one hundred years old, found in museums.

At a workshop in 1972, Rowland listened to a presentation by the biophysicist James Lovelock (1919–) on recent measurements of the level of CFCs in the atmosphere. All CFCs are made by human industry, and they were used at that time mainly in aerosol cans and air conditioners. Working with Mario Molina, a native of Mexico and new postdoctoral fellow, Rowland began to investigate what happened to CFC molecules in the atmosphere. They found that when the CFCs reach the ozone layer, high in the atmosphere, ultraviolet radiation from the Sun breaks down the CFC molecules. Each CFC molecule consists of one atom of carbon, one of fluorine, and three of chlorine ($CFCl_3$). The released chlorine atoms rapidly react with ozone molecules, converting them to normal oxygen. Small amounts of CFCs have the potential to rapidly reduce the amount of ozone. The ozone layer protects life on Earth from the harmful effects of solar ultraviolet radiation. Molina and Rowland published their findings in the journal *Nature* in 1974, and their laboratory discovery was later confirmed by measurements of the ozone layer. The realization that continued use of CFCs would deplete the ozone layer, with catastrophic results for human civilization, led to the Montreal Protocol international agreements of 1987, 1990, and 1992 to phase out the use of these gases.

Rowland actively campaigned for CFC reduction and he continues to conduct research on atmospheric chemistry with his students. When his regular AEC funding ended, he turned to the National Aeronautics and Space Administration (NASA) for funding. Rowland has received many awards for his scientific work, including being elected to the National Academy of Sciences in 1978. Paul Crutzen (1933–), Molina, and Rowland were awarded the 1995 Nobel Prize in Chemistry.

See also Chemistry; Crutzen, Paul; Environmental Movement; Libby, Willard F.; Lovelock, James; Meteorology; Molina, Mario; National Aeronautics and Space Administration; Nobel Prizes; Ozone Layer and Chlorofluorocarbons

References

"F. Sherwood Rowland—Autobiography." http://nobelprize.org/medicine/laureates/index.html (accessed February 13, 2004).

Newton, David E. *The Ozone Dilemma: A Reference Handbook.* Santa Barbara, CA: ABC-CLIO, 1995.

S

Sabin, Albert (1906–1993)

Although the virologist Albert Sabin failed to develop the first effective vaccine for polio, he did develop the live-virus form of vaccine used today. Albert Bruce Sabin was born in Bialystok, Russia, which later became part of Poland. His family moved to America in 1921, and Sabin graduated from high school two years later. Working odd jobs, he put himself through New York University, earning a B.S. degree in 1928 and an M.D. in 1931. He became a medical researcher, spending time in England and on the staff of the Rockefeller Institute for Medical Research, before settling at the University of Cincinnati College of Medicine in 1939. Sabin married in 1935 and fathered two daughters. During World War II, he served with the U.S. Army Medical Corps, developing vaccines for use against dengue fever and Japanese encephalitis. In 1946, he returned to Cincinnati and applied himself to the problem of finding a polio vaccine.

The private National Foundation for Infantile Paralysis funded Sabin's research to find a vaccine for poliomyelitis (infantile paralysis). Polio usually did not result in paralysis or death, but its possible effect on children, when most childhood diseases had finally been controlled in the United States, particularly terrified parents. Work by the microbi-ologist John Enders (1897–1985) on a new way to culture mammalian viruses allowed virologists to create enough material to work with, which helped the polio researchers. Sabin became the leading polio researcher of the day after demonstrating that the polio virus entered the body through ingestion, rather than through the respiratory system. His approach was similar to that which had succeeded with other viruses—to develop a vaccine made of viruses so weakened that they were not virulent, but would provoke the human body to produce antibodies.

The virologist Jonas Salk (1914–1995) developed a killed-virus polio vaccine in 1952 and began to test it. Sabin strongly objected to Salk's vaccine, fearing the long-term consequences of such an approach. Testing of Salk's vaccine was expanded and proved effective, and Salk became an international celebrity. Sabin continued to work on his own version of the vaccine, and after successful tests with chimpanzees, he tested the vaccine in 1955 on volunteer prisoners. Later tests conducted internationally, including in the Soviet Union, proved effective. Sabin vaccine could be given orally, maintained its immunization effect longer, was cheaper to manufacture, and was more effective against the different polio strains. The Sabin polio vaccine eventually became the more commonly used vaccine in the

United States and internationally. Sabin went on to work on vaccines for other viral diseases and received numerous honors before his retirement from the University of Cincinnati in 1971. He served as a research professor at the University of South Carolina in Charleston from 1974 to 1982.

See also Enders, John; Foundations; Medicine; Pharmacology; Salk, Jonas

References

Chanock, Robert M. "Reminiscences of Albert Sabin and His Successful Strategy for the Development of the Live Oral Polio Virus Vaccine." *Proceedings of the Association of American Physicians* 108, no. 2 (March 1996): 117–126.

Melnick, Joseph L. "Albert B. Sabin." *Biologicals* 21, no. 4 (December 1993): 299–303.

Paul, John R. *A History of Poliomyelitis.* New Haven: Yale University Press, 1971.

Sagan, Carl (1934–1996)

A prominent planetary astronomer, Carl Sagan also gained fame as a science popularizer. Carl Edward Sagan was born in Brooklyn, New York, the son of a Jewish Ukrainian immigrant who worked in the garment industry, eventually becoming a factory manager. As a child, Sagan was intensely interested in science, with a special love of astronomy, and he was excited by the ideas of space travel and science fiction. He graduated from high school at the age of sixteen and went to the University of Chicago on a scholarship. His undergraduate education was unusually broad in the sciences and humanities. He earned a general and specific honors B.A. degree in 1954 and a B.S. in physics in 1955. He remained at Chicago to earn an M.S. in physics in 1956 and a Ph.D. in astronomy and astrophysics in 1960. After a fellowship at the University of California at Berkeley, Sagan joined the faculty at Harvard University in 1962. After being denied tenure at Harvard, he moved to Cornell University in 1968.

Sagan followed his interests and specialized in planetary astronomy at a time when it was a backwater in the larger field of astronomy. Gerard Kuiper (1905–1973), a pioneer in planetary astronomy, mentored Sagan in this field. The field exploded in importance with the coming of the space age and the launching of robotic spacecraft to explore other planets. Sagan served as a key consultant with the Jet Propulsion Laboratory and the National Aeronautics and Space Administration (NASA) from the 1960s through 1990s.

Early in his career, Sagan developed the argument (based on prior work by the astronomer Rupert Wildt in 1940) that Venus was heated by a greenhouse effect created by the high concentration of carbon dioxide in its atmosphere, a prediction later borne out by spacecraft sensors. Sagan and his first graduate student, James Pollack (1938–1994), also argued that the changing features of Mars visible in telescopes from Earth were the result of dust storms, not biological changes in the form of plants. Later discoveries by the Mariner spacecraft confirmed the correctness of this theory as well. Sagan also worked on understanding the possible prebiotic conditions on Earth that promoted the creation of life, and he is acknowledged as one of the founders of the field of astrobiology.

In the 1960s, Sagan began a parallel career as a popularizer and strong advocate of science, even while continuing to produce a significant amount of scientific work. He proved to be a talented writer, and his 1977 book, *The Dragons of Eden: Speculations on the Evolution of Human Intelligence,* sold a million copies worldwide and won a Pulitzer Prize. Sagan cofounded The Planetary Society in 1980 to promote the exploration of space by advocating continued government and private efforts. Sagan's 1980 public television series, *Cosmos,* introduced viewers to a wonder-filled universe and proved immensely popular. Sagan appeared as a guest on popular late-night television talk shows and wrote columns for popular magazines. Among his other books were ruminations on astronomy, comets, space travel, evolution, asteroid hazards, and opposition to the Strategic Defense Initiative (commonly called "Star Wars").

Carl Sagan, planetary astronomer and science writer, holds the plaque he helped design. (Noel M. McKinnell/Corbis)

In the 1980s, Sagan and other researchers concluded that a massive exchange of nuclear weapons between the Soviet Union and United States would not only kill millions, but would result in so much dust being raised by the explosions that the world's temperature would drop and Earth would experience a "nuclear winter." Sagan came to this conclusion because he noticed that temperatures on Mars had dropped during a massive dust storm observed by an orbiting Mariner spacecraft.

Sagan also advocated an intensive search for extraterrestrial intelligence (SETI). In his role as a NASA consultant, Sagan, with others, created plaques of greetings, carried by the *Pioneer 10* and *11* spacecraft in case an alien intelligence encountered the craft at some unknown time in the future. He also coordinated something similar for the Voyager spacecraft, each of which carried a 12-inch, gold-plated copper disk with 115 images from Earth on it and a selection of sounds, including sounds of nature and voice greetings in fifty-five languages. Sagan's novel, *Contact,* published in 1985, described the results of first contact with an alien species via radio transmissions. The novel became a movie shortly after his death.

As a young man, Sagan believed that UFOs might be extraterrestrial vehicles and that Mars had canals on its surface. After disabusing himself of these ideas, he became a strong advocate of what he characterized as rational thinking, and an effective critic of UFOs and creationism. His last book, *The Demon-Haunted World: Science as a Candle in the Dark* (1996), debunked many pseudoscientific notions.

Sagan married Lynn Alexander in 1957. They had two sons before divorcing in 1963. Lynn remarried, and she retained her new married name of Lynn Margulis (1938–) even after her divorce. Margulis became

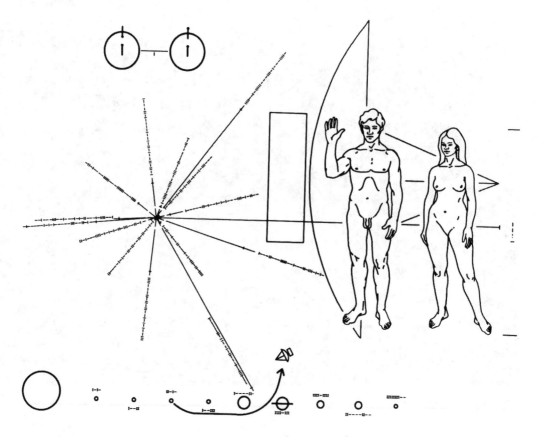

The plaque of greeting carried by both Pioneer 10 *and* 11 *spacecrafts (Courtesy of NASA)*

well-known as an advocate of symbiosis and the Gaia hypothesis and for her popular science writings, coauthored with her son, Dorion Sagan. Sagan married again in 1968 and had another son before divorcing in 1981. He then married Ann Druyan, a collaborator on *Cosmos* and some of his science writings, and had three more children. Sagan died in Seattle, Washington, after a two-year struggle with myelodysplasia.

See also Astronomy; Cold War; Creationism; Jet Propulsion Laboratory; Margulis, Lynn; Media and Popular Culture; National Aeronautics and Space Administration; Search for Extraterrestrial Intelligence; Space Exploration and Space Science; Voyager Spacecraft

References

Davidson, Keay. *Carl Sagan: A Life.* New York: Wiley, 1999.

Poundstone, William. *Carl Sagan: A Life in the Cosmos.* New York: Henry Holt, 1999.

Sakharov, Andrei (1921–1989)

Andrei Sakharov is credited with designing the Soviet hydrogen bomb and later became famous for his dissident activities in the causes of peace and human rights. Andrei Dmitrievich Sakharov was born in Moscow, where his father was a professor of physics at the Lenin Pedagogical Institute. He attended Moscow State University, and upon his graduation in 1942, he worked as a munitions engineer in a factory during World War II. Sakharov married in 1943 and eventually fathered three children.

After the war, Sakharov studied at the Lebedev Institute of the Soviet Academy of Sciences and earned a doctorate in the physical and mathematical sciences at age twenty-six. His mentor, Igor Yevgenyevich Tamm (1895–1971), brought Sakharov into the Soviet nuclear weapons program. The Soviets exploded their first atomic bomb in 1949; Sakharov developed a design like a layer cake for a hydrogen bomb, which used the fission of an atomic bomb as a trigger to create a thermonuclear explosion. The Americans exploded their first hydrogen bomb in 1952, and the Soviets exploded their first hydrogen bomb nine months later in 1953. Sakharov was credited with the major scientific advances of the Soviet hydrogen bomb and elected to the Soviet Academy of Sciences. At age thirty-two, he was the youngest person ever elected to the academy. He continued theoretical and practical work on nuclear weapons and nuclear reactors while living a privileged life in the Soviet Union.

In the late 1950s, Sakharov opposed continued nuclear testing because of the environmental effects on the atmosphere, but kept his objections confined to official channels. In 1963, he joined other scientists in discrediting Trofim Lysenko (1898–1976), a geneticist favored by Josef Stalin (1879–1953) who damaged Soviet agriculture and the Soviet practice of genetics with his unscientific theories. As the 1960s progressed, Sakharov became a public dissident, opposed to the repression inherent in Soviet society. In 1968, he published an essay, "Thoughts on Progress, Peaceful Coexistence, and Intellectual Freedom," calling for an end to the cold war and advocating human rights. The Soviet government removed him from nuclear weapons research and assigned him to the lowest possible position at the Lebedev Institute that a member of the Soviet Academy of Sciences could hold. He continued his scientific inquiries into quantum mechanics, quarks, and the idea of antiquarks.

After the death of his first wife, Sakharov married a fellow dissident, Yelena G. Bonner, in 1971. He became ever more vocal as a dissident, and his status as a scientist transformed him into an important symbol. When he received the Nobel Peace Prize in 1975 for his dissident activities, Soviet authorities refused to allow him to leave the Soviet Union to accept the prize. In 1980, when Sakharov

publicly condemned the Soviet invasion of Afghanistan, the authorities exiled him to the closed city of Gorky. In 1986, his exile ended, thanks to the liberalization program of Mikhail Gorbachev (1931–), and he was elected to the Congress of People's Deputies shortly before his death.

See also Big Science; Cold War; Ethics; Hydrogen Bomb; Nobel Prizes; Nuclear Physics; Soviet Academy of Sciences; Soviet Union; Teller, Edward
References
Lourie, Richard. Sakharov: A Biography. Hanover, NH: Brandeis University Press, 2002.
Sakharov, Andrei. Memoirs. New York: Knopf, 1990.

Salam, Abdus
See Third World Science

Salk, Jonas (1914–1995)
The virologist Jonas Salk developed the first effective polio vaccine, becoming an international celebrity and founding the institute that bears his name. Jonas Salk was born in New York City, where his intelligence was recognized early. He graduated from a high school for gifted children at age fifteen and entered the City College of New York. With a B.S. degree awarded in 1934, he entered New York University's School of Medicine. Salk obtained his M.D. in 1939 and followed his mentor, the virologist Thomas Francis Jr. (1900–1969), to the University of Michigan. During World War II, Francis headed the Army Influenza Commission. Francis believed that vaccines from dead viruses could work as well as the more common vaccines derived from live viruses. Francis and Salk worked on creating flu vaccines using killed viruses and found some success.

In 1947, Salk moved out on his own to the University of Pittsburgh School of Medicine, where he headed the Virus Research Laboratory. He continued to work on influenza vaccines, which the army funded, but his interest

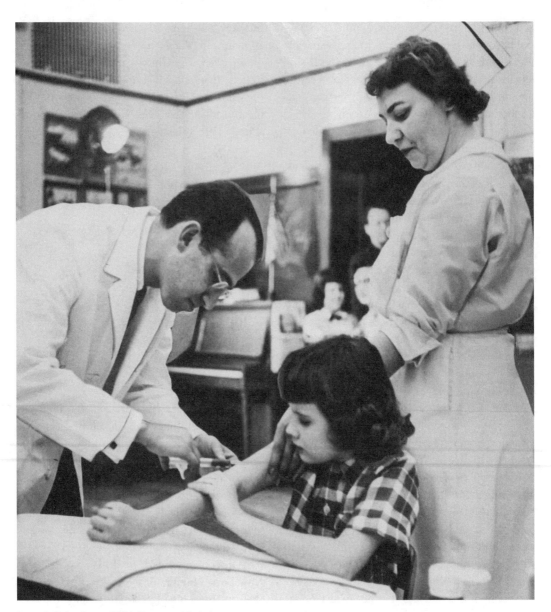

Jonas Salk, virologist, 1954 (Bettmann / Corbis)

soon turned to poliomyelitis (infantile paralysis). Polio usually did not result in paralysis or death, but its effect on children sometimes included paralysis or death, which particularly terrified parents at a time when most childhood diseases had finally been controlled in the United States. The country considered itself in the midst of an epidemic, and a National Foundation for Infantile Paralysis raised private money to combat the disease. Salk ob-

tained funding from the foundation to support his research.

In 1948, John Enders (1897–1985) and his colleagues succeeded in culturing the polio virus in normal human cells. Up to that time, the polio virus had only been cultured in human nerve cells and in live monkeys. Salk used this discovery to more quickly culture viruses. Other researchers established that there existed three types of polio, which helped Salk isolate the viruses. Experiments with his viruses on monkeys in 1952 created high levels of antibodies. Salk tested the virus on himself, his family, and his staff, before starting the first trials on children. The trials demonstrated higher levels of polio antibodies in the subjects. The use of a killed-virus vaccine proved controversial, and Albert Sabin (1906–1993), the leading polio researcher of the day, objected to such an approach, fearing long-term dangers.

In 1953, the National Foundation for Infantile Paralysis supported Salk in a larger trial on children. When word of the trial leaked to the press, Salk became an instant celebrity. The trials expanded to include nearly 2 million children, with the results analyzed by a team headed by Francis. Pronounced a success in 1955, the Salk vaccine began to be used in mass immunizations. Salk received a Congressional Gold Medal in 1955, as well as the 1956 Albert Lasker award, and he became the Commonwealth Professor of Medicine at the University of Pittsburgh. Sabin eventually developed a live-virus vaccine, and after successful international tests, Sabin's vaccine began to be used in the United States in 1962, eventually superseding Salk's vaccine.

Salk used his fame to set up an independent laboratory, the Salk Institute for Biological Studies in La Jolla, California. He attracted other luminaries, such as Francis Crick (1916–2004), to relocate to the institute. By the end of the century, the institute employed over nine hundred researchers.

Salk married the day after receiving his M.D. in 1939 and fathered three sons. The marriage ended in divorce in 1968, and he remarried three years later. Starting with *Man Unfolding* in 1972, Salk published several optimistic books for popular audiences on the implications of biology for humanity. In the 1980s, Salk returned to directing research; he unsuccessfully tried to develop a vaccine to combat the new scourge of acquired immunodeficiency syndrome (AIDS).

See also Acquired Immunodeficiency Syndrome; Crick, Francis; Enders, John; Foundations; Medicine; Pharmacology; Sabin, Albert
References
Carter, Richard. *Breakthrough: The Saga of Jonas Salk.* New York: Trident, 1966.
Salk Institute for Biological Studies. http://www.salk.edu/ (accessed February 13, 2004).
Smith, Jane S. *Patenting the Sun: Polio and the Salk Vaccine.* New York: William Morrow, 1990.

Sanger, Frederick (1918–)

The British biochemist Frederick Sanger received two Nobel prizes in chemistry. Frederick Sanger was born in Rendcombe, Gloucestershire, England, where his father was a physician. Before arriving at Saint John's College, Cambridge University, Sanger decided to change his ambitions from medicine to science because he wanted to be single-minded in his activities. In 1939, he graduated with a B.A. in biochemistry. His Quaker beliefs led him to avoid military service as a conscientious objector during World War II, though he did not remain committed to Quakerism later in life. The British government allowed him to continue to study for his Ph.D. degree, and he graduated in 1943. Sanger married in 1940 and fathered two sons and a daughter. Sanger remained at Cambridge as a researcher until 1951, when he joined the staff of the Medical Research Council (MRC). A researcher his entire life, he never took a position that included extensive teaching duties and only later gained administrative duties. Private income from his mother's inheritance helped support his commitment to research.

A passion for understanding the sequence and structure of organic molecules dominated his life. Mostly working alone, Sanger developed new techniques for sequencing amino acids and eventually described the complete sequence of insulin in 1953. He chose insulin when other researchers looked at other proteins because he could easily buy insulin in its pure form. Sanger received the

1958 Nobel Prize in Chemistry for this work, and he found the award opened more opportunities to him in terms of funding, staff, and resources. In 1962, Sanger moved to the newly built Laboratory of Molecular Biology at Cambridge, still part of the Medical Research Council.

Sanger became interested in sequencing nucleic acids, which form the basis of deoxyribonucleic acid (DNA) and ribonucleic acid (RNA) molecules. Unlike his earlier solitary work, Sanger's later work profited from considerable collaboration with students and postdoctoral students. Sanger and his colleagues developed dideoxy, sometimes called Sanger sequencing, a method still used to rapidly sequence genes. In 1979, Sanger's team produced the first complete genome sequence of an organism, the 10 bases and 5,000 base pairs of the bacteriophage phi-X174. For this work, Sanger shared the 1980 Nobel Prize with Paul Berg (1926–) and Walter Gilbert (1932–). (Berg and Gilbert had made important advances in the creation of the science and technology of genetic engineering.) Only three people have been awarded two Nobel Prizes. Sanger's innovations made possible the Human Genome Project and further advances in genetic engineering. Other awards for Sanger included becoming a fellow of the Royal Society in 1954, a member of the Order of Merit in 1986, and a Companion of Honour in 1981.

> **See also** Biotechnology; Chemistry; Genetics; Gilbert, Walter; Human Genome Project; Nobel Prizes
>
> *References*
> "Frederick Sanger—Autobiography." http://nobelprize.org/medicine/laureates/index.html (accessed February 13, 2004).
> Jones, Glyn, and Kate Douglas. "The Quiet Genius Who Decoded Life." *New Scientist* 144, no. 1946 (October 8, 1994): 32–35.

Sanger, Margaret (1879–1966)
See Birth Control Pill

Satellites

The Russian physicist Konstantin Tsiolkovksy (1857–1935) developed the principles that later led to rockets and spacecraft. He proposed the artificial satellite, a man-made analogue to natural satellites like the Moon, in an 1895 book. This vision inspired early rocket inventors, and the impetus of World War II led the Germans to develop sophisticated rocket technology. After the war, the Soviet Union and United States used captured German technology, scientists, engineers, and technicians to create their own rocket technology. Although rocket development became an important part of the cold war with the development of nuclear-tipped missiles, spacecraft enthusiasts had not forgotten the dream of artificial satellites. A 1956 book edited by the U.S. physicist James A. Van Allen (1914–), *Scientific Uses of Earth Satellites,* described the many possible uses of satellites. Launching an artificial satellite became one of the goals of the International Geophysical Year (1957–1958). The Soviets surprised the world by launching *Sputnik I* on October 4, 1957. In their haste to beat the Americans, they made the first satellite very simple, with few scientific instruments. On November 3, 1957, the much larger *Sputnik II* was launched with a dog, Laika, aboard, as well as many more scientific instruments. Laika died in orbit.

The embarrassed Americans compounded their embarrassment on December 6, when the launcher of the first American satellite, *Vanguard 1,* exploded on launch. The Americans launched the first successful American satellite, *Explorer 1,* on January 31, 1958. The satellite discovered zones of radiation around Earth, later named the Van Allen Radiation Belts in honor of the creator of the scientific instrument that detected the radiation. Scientists later learned that Earth's magnetic field created these belts of radiation and that the belts helped protect Earth from some of the more harmful effects of solar radiation.

The cold war drove competition between the Soviets and Americans, and science bene-

A painting of one of the twenty-four satellites that make up the Global Positioning System (AP/WideWorld Photos)

fited. Arthur C. Clarke (1917–) had proposed the idea of communications satellites in 1945, but his idea was lost and only belatedly rediscovered as the idea became a reality. Early efforts relied on radio signals bouncing off experimental communications satellites made of aluminized Mylar balloons that expanded in space after delivery into orbit. *Telstar 1,* launched in 1962, was paid for by the American Telephone and Telegraph corporation, beginning commercial exploitation of space. Later satellites provided television and telephone relays between ground stations. By 1964, satellites were being placed in geosynchronous orbit, 22,300 miles (35,880 kilometers) above the equator, where their orbit kept them in station over the same place on Earth. Also in 1964, Intelsat (International Telecommunications Satellite Organization)

was formed to launch and maintain communications satellites for member nations of the free world. Satellite data links, satellite phones, and satellite television revolutionized how scientists communicated and provided a way for scientific knowledge to be more effectively broadcast and popularized. Since the beginning of the space age, over 5,000 satellites have been launched.

Satellites also revolutionized the practice of meteorology. The first high-altitude photograph to reveal new cloud formations was taken from a V-2 rocket captured from the Germans and launched in 1947. In 1959, the American satellite *Explorer 6* took pictures of Earth from orbit. In 1960, the National Aeronautics and Space Administration (NASA) launched *Tiros 1,* the first satellite of many designed for meteorology. The first ten satellites of the Tiros (Television and Infra-Red Observation Satellite) series developed the technology, and the following nine Tiros satellites formed an operational system. *Tiros 3* tracked Hurricane Carla in 1963, allowing hundreds of thousands to evacuate before Carla reached land and thus proving the value of weather satellites. The satellites also revealed unknown cloud formations and allowed scientists to readily follow the development of storm systems. More advanced satellites from the United States, the Soviet Union, and the European Space Agency followed the Tiros series. Japan, China, and India have also added their own satellites to the system. The World Meteorological Organization (WMO), part of the United Nations, coordinates exchanging information between different countries and their weather satellites. No tropical storm has escaped detection and tracking since 1966, and even though satellites have only made weather forecasting better, not perfect, meteorologists now know exactly what is happening anywhere in the world.

Satellites containing telescopes and other sensors have completely revolutionized solar, interstellar, and interplanetary astronomy. NASA launched in 1990 the most famous of these satellites, the Hubble Space Telescope. A manufacturing flaw in the main telescope's mirror required a 1993 space shuttle visit for space-walking astronauts to repair the defect.

NASA's Landsat program changed the way that we look at our planet. The first three Landsat satellites, launched in 1972, 1975, and 1978, transmitted back data to be converted into colored pictures. The pictures revealed topological features, geological fault lines, crop and vegetation patterns, hidden archaeological sites, pollution concentrations, ocean and surface temperature patterns, and tropical rain forest depletion. Other Earth observation satellites followed from the Soviet Union, France, the European Space Agency, Japan, India, Canada, and China, as well as further U.S. satellites in the Landsat series and other efforts. In the 1980s and 1990s, commercial satellites from Russia, France, and the United

States began to take images for commercial sale to all customers, offering resolution of ground objects smaller than two meters.

The same technology used for weather satellites and Earth observation satellites was refined by the U.S. and Soviet militaries to create military reconnaissance satellites. The requirement for the U.S. military to always be able to precisely locate their personnel and equipment anywhere on Earth's surface led to the U.S. Global Positioning System (GPS). An experimental system of ten satellites was launched between 1978 and 1985. An operational system of twenty-four satellites, launched between 1989 and 1994, now allows those with a GPS receiver to locate themselves on the surface of Earth with up to 10 meters' accuracy. U.S. military users have better resolution. GPS changed the practice of scientific fieldwork by allowing exact measurements of continental drift, pinpointing the location of geological formations, archaeological sites, animal populations, and any other application requiring the precision that GPS offers.

See also Astronomy; Clarke, Arthur C.; Cold War; European Space Agency; International Geophysical Year; Space Exploration and Space Science

References

Chaisson, Eric J. *The Hubble Wars: Astrophysics Meets Astropolitics in the Two-Billion-Dollar Struggle over the Hubble Space Telescope.* New York: HarperCollins, 1994.

Curtis, Anthony R. *Space Satellite Handbook.* Third edition. Houston: Gulf, 1994.

Mack, Pamela E. *Viewing the Earth: The Social Construction of the Landsat Satellite System.* Cambridge: MIT Press, 1990.

Tedeschi, Anthony Michael. *Live via Satellite: The Story of COMSAT and the Technology That Changed World Communication.* Washington, DC: Acropolis, 1989.

Tucker, Wallace, and Karen Tucker. *Revealing the Universe: The Making of the Chandra X-Ray Observatory.* Cambridge Harvard University Press, 2001.

Williamson, Mark. "'And Now the Weather': The Early Development of the Meteorological Satellite." *Transactions of the Newcomen Society* 66 (1995): 53–76.

Schwinger, Julian (1918–1994)

The American physicist Julian Schwinger made important contributions to the theory of quantum electrodynamics (QED). Julian Seymour Schwinger was born in New York City, where his father worked as a dress designer and manufacturer. A child prodigy, he quickly moved through the public school system and graduated at age fourteen. He entered the City College of New York and began to work on quantum mechanics. Schwinger published his first paper in the *Physical Review* when only sixteen years old. I. I. Rabi (1898–1988), a Polish-born physicist who later won the 1944 Nobel Prize in Physics, took notice of the young man and arranged for a scholarship to bring Schwinger to Columbia University. He earned his B.S. in physics in 1936 and spent time studying at the University of Wisconsin at Madison and Purdue University before earning his Ph.D. in physics at Columbia in 1939, while still only twenty-one years old. After two years at the University of California at Berkeley, conducting research with J. Robert Oppenheimer (1904–1967), Schwinger joined the faculty at Purdue.

During World War II, Schwinger worked on the Manhattan Project at the University of Chicago for a time before spending the rest of the war at the Radiation Laboratory at the Massachusetts Institute of Technology, where he worked on radar. When the war ended, he joined the faculty of Harvard University and became a full professor in 1947, one of the youngest ever appointed to that rank at Harvard. Also in 1947, Schwinger married; the marriage produced no children.

The English mathematician and theoretical physicist P. A. M. Dirac (1902–1984) developed the original theory describing the relationship between electromagnetic fields and charged particles, though it was widely recognized as flawed. Schwinger was part of the second generation that refined the theory. Schwinger developed a new theory of QED to renormalize Dirac's theory and fixed flawed mathematical equations that led to infinite results and problems with how to handle the self-energy of particles. Richard P. Feynman (1918–1988), a fellow New Yorker, independently developed a similar theory, as did Shinichiro Tomonaga (1906–1979), working in war-imposed isolation in Japan. Freeman Dyson (1923–) showed that the three theories were mathematically equivalent, and the 1965 Nobel Prize in Physics was shared among Schwinger, Feynman, and Tomonaga. Feynman's mathematical notation system for QED was accepted by the physics community over the systems created by Schwinger and Tomonaga.

Schwinger proposed in 1957 that two forms of neutrinos existed, one associated with electrons and the other with muons. A neutrino associated with the formation of a muon particle was discovered in 1962. Schwinger also mentored Sheldon Glashow (1932–), who shared in the 1979 Nobel Prize in Physics for his work on quarks. Schwinger's later research pursued a variation of the search for a grand unified theory. In 1972, Schwinger accepted a position at the University of California at Los Angeles, where he died in 1994.

See also Dyson, Freeman; Feynman, Richard; Glashow, Sheldon L.; Grand Unified Theory; Neutrinos; Nobel Prizes; Oppenheimer, J. Robert; Particle Physics; Physics

References

Martin, Paul C., and Sheldon L. Glashow. "Julian Schwinger: Prodigy, Problem Solver, Pioneering Physicist." *Physics Today* 48, no. 20 (1995): 40–46.

Mehra, Jagdish, and Kimball A. Milton. *Climbing the Mountain: The Scientific Biography of Julian Schwinger.* New York: Oxford University Press, 2000.

Science Fiction

Science fiction is the literature of the scientific age. Born in adventure pulp magazines during the first half of the twentieth century, the science fiction genre began to mature in the 1940s and 1950s. Although many of the early stories suffered from simplistic characters and juvenile plots, speculations on the impact of science and technology, and extrapolations of possible future directions of science and technology, have proved to be insightful and useful. Isaac Asimov (1920–1992), prolific writer of both popularized science and science fiction, thought the genre was dominated by adventure themes up to 1939 and by technology themes until 1950, with sociological themes becoming dominant after 1950. Although predominant themes are apparent, the genre has been characterized by wide variation in quality and diversity in content.

In the 1950s, writers in the Soviet Union and communist Eastern European countries used science fiction as a way to break away from the confines of government-encouraged socialist realism in literature. The works of the Polish Stanislaw Lem (1921–) and the Soviet brothers Arkady Strugatski (1925–1991) and Boris Strugatski (1931–) effectively delved into philosophical and ethical elements of science. Although often viewed as a quintessentially American genre, science fiction has found practitioners in many countries, and the genre has international appeal.

In the 1960s a New Wave movement out of Britain expanded the bounds of science fiction from conventional space travel stories. Traditional science fiction was optimistic about the future, whereas New Wave authors tended toward pessimism, had a taste for dystopian societies, and enjoyed breaking taboos. The New Wave forced the genre into more complex stories, more adult discussions of sexuality, and a willingness to examine every possible sort of question about the future.

In the early 1980s, computers inspired the cyberpunk movement in the genre. Oddly enough, the rise of personal computers was almost completely absent from science fiction prior to the 1970s, but cyberpunk fiction made up for that, describing worlds where computers and the flow of data completely transformed human relationships and the common reality that everyone experienced. The exciting possibilities of nanotechnology at times turned the cyberpunk world into a place of seeming magic. Often the cyperpunk vision turned depressing and pessimistic, though the 1990s saw cyberpunk themes absorbed into the more optimistic mainstream of science fiction, just as the earlier New Wave movement was absorbed.

Commentators have noted a difference between "soft" science fiction and "hard" science fiction. Soft science fiction usually concentrates on psychological or sociological themes. Hard science fiction is informed by a sustained effort to make the proposed science and technology plausible. Noted authors of hard science fiction include Asimov, Stephen Baxter (1957–), Greg Bear (1951–), Gregory Benford (1941–), David Brin (1950–), Arthur C. Clarke (1917–), Michael Crichton (1942–), James P. Hogan (1941–), Larry Niven (1938–), Frederik Pohl (1919–), Brian M. Stableford (1948–), and Olaf Stapledon (1886–1950). Poul Anderson's *Tau Zero* (1970) follows the story of a starship that accelerates so close to the speed of light that they witness the end of the universe and a new Big Bang. Gregory Benford's *Timescape* (1981) is a time travel novel that strives to maintain scientific plausibility. Robert L. Forward's *Dragon's Egg* (1980) portrays how life might exist on a neutron star. Impending ecological collapse and global warming are portrayed in David Brin's *Earth* (1987). Practicing scientists such as Carl Sagan (1934–1996), Fred Hoyle (1915–2001), and Robert L. Forward (1932–2002) have also written hard science fiction. The best science fiction remains in the written format, with most movies and television series in the genre engaging in little scientific rigor.

Science fiction has effectively examined many of the possibilities that arise from space exploration, the possible success of the search for extraterrestrial intelligence (SETI), quantum reality, and the implications of extrasolar planets containing life. Even before scientists and the general population became concerned about environmental issues, science fiction often incorporated environmental themes, particularly overpopulation and ecological

A still photograph from the science fiction movie 2001: A Space Odyssey *(Underwood & Underwood/Corbis)*

collapse. Although successful predictions of future science or technology are easy enough to come by, science fiction is at its best when portraying the scientific method as a way to knowledge and the implications of scientific and technological advances. Science fiction has also excited many young people to pursue science or engineering as a career choice.

See also Asimov, Isaac; Big Bang Theory and Steady State Theory; Clarke, Arthur C.; Environmental Movement; Extrasolar Planets; Hoyle, Fred; Media and Popular Culture; Nanotechnology; Sagan, Carl; Search for Extraterrestrial Intelligence; Space Exploration and Space Science

References

Adliss, Brian, and David Wingrove. *Trillion Year Spree: The History of Science Fiction.* New York: Atheneum, 1986.

Clute, John, and Peter Nicholls. *The Encyclopedia of Science Fiction.* New York: St. Martin's Griffin, 1993.

Disch, Thomas M. *The Dreams Our Stuff Is Made Of: How Science Fiction Conquered the World.* New York: Free Press, 1998.

Goswami, Amit, and Maggie Goswami. *The Cosmic Dancers: Exploring the Physics of Science Fiction.* New York: Harper and Row, 1983.

Hartwell, David. *Age of Wonders: Exploring the World of Science Fiction.* Revised edition. New York: Tor, 1996.

Nicholls, Peter. *The Science in Science Fiction.* New York: Knopf, 1983.

Search for Extraterrestrial Intelligence

The search for extraterrestrial intelligence (SETI) is the attempt to find other intelligent life in the universe. Although science fiction writers made intelligent extraterrestrial aliens part of their staple fare from the beginning, scientists only occasionally speculated on this area. In 1950, the nuclear physicist and Nobel laureate Enrico Fermi (1901– 1954) asked the obvious question: if extraterrestrial beings are common, where are they? The physicists Giuseppe Cocconi (1914–) and Phillip Morrison (1915–) published a paper in the journal *Nature* in 1959 proposing that the new radio technology allowed us to listen for deliberate radio transmissions from extraterrestrial civilizations around other stars. The astronomer Frank D. Drake (1930–), who worked at the National Radio Astronomical Observatory in Green Bank, West Virginia, had already been formulating plans to do just that. He proposed to listen for extraterrestrial transmissions from two nearby stars, Tau Ceti and Epsilon Eridani. Each star was similar to our own Sun. Drake called the effort Project Ozma, after the queen of Oz in the Frank Baum novels.

On April 8, 1960, Drake and his colleagues

began to listen at the 1,420-megahertz frequency, based on the idea that this frequency is emitted by hydrogen, the most basic atom. Although this first effort did not bear fruit, Drake continued to emphasize the subject in his research. He developed a mathematical equation, known as the Drake formula, that attempted to calculate the chances of eventual success. The values for several factors in the equation are completely unknown. How often are planets formed around stars? How many of those planets will develop conditions hospitable to life? How many of those planets will develop life? How often will life eventually evolve intelligence and develop the technology to communicate via radio through interstellar space? What is the average lifetime of an extraterrestrial technological civilization?

The next efforts in SETI occurred in the Soviet Union, where Nikolai Kardashev (1932–) and Vsevolod Troitsky participated in several listening efforts. Instead of concentrating on individual stars, the Soviets searched large sections of the sky, hoping to find extraterrestrial civilizations that broadcast powerful signals. Radio telescopes bloomed around the world in the 1960s and afterward, but SETI had such a low probability of success that time on radio telescopes remained hard to obtain. The regular radio pulse of the first pulsar, discovered in 1967 by Jocelyn Bell Burnell, made some wonder if she had actually stumbled across a transmission from an extraterrestrial intelligence. The discovery of other pulsars and a natural explanation removed that tantalizing possibility. Other efforts at SETI in the United States eventually resumed. The National Aeronautics and Space Administration (NASA) proposed a Project Cyclops in the mid-1970s to build a large radio telescope array dedicated to SETI. The proposed effort, costing billions of dollars, was not funded, but the research became a basis for many other efforts.

The All Sky Survey at Ohio State University has regularly used their radio telescope as part of a long-running SETI effort since the

1970s. In 1977, they received an intense signal, dubbed the "wow" signal, that was never repeated and entered the folklore as an intriguing mystery. These efforts around the world expanded their search beyond the original frequency of 1,420 megahertz. Eventually electronic devices were developed to simultaneously record many frequencies and automatically record frequencies while radio telescopes performed other research. An effort in the 1990s at the radio telescope at Arecibo, Puerto Rico, was called SERENDIP III (Search for Extraterrestrial Radio Emissions from Newly Developed Intelligent Populations); it recorded 10,000 hours of time. The Planetary Society, cofounded by Carl Sagan, funded the SERENDIP projects. Analyzing the data from SERENDIP III required massive amounts of computer processing power, and an innovative project called SETI@home allowed personal computers in homes to be used in this effort. The SETI@home program ran as a screen saver, periodically downloading over the Internet new units of data to work on and returning the results of data that had been processed. Over a million people worldwide have downloaded and run this program.

NASA periodically funded SETI efforts until congressional action suspended this support in 1993. SETI advocates turned to private donors to fund their continued efforts. A major organization in this private effort was the SETI Institute at Mountain View, California, founded in 1984 by Drake, Carl Sagan, and others. In 2003, NASA resumed funding SETI efforts.

In 1974, Drake and others engaged in an obvious extension of SETI. Using the dish at Arecibo, a message in binary code was transmitted toward the Great Cluster in the constellation Hercules. The message included a visual representation of the formulas for the four chemicals that form DNA molecules, a description of the solar system, and other information. The NASA Pioneer spacecraft, launched to visit Jupiter and Saturn, eventually left the solar system, and each carried a

The 1,000-foot-wide dish of the Arecibo Radio Telescope as viewed from its cable-suspended receiver (Roger Ressmeyer/Corbis)

plaque of greeting in case an alien intelligence encountered the craft at some unknown time in the future. NASA's Voyager space probes each carried a 12-inch, gold-plated copper disk with 115 images from Earth on it and a selection of sounds, including natural sounds and voice greetings in fifty-five languages. The astronomer Carl Sagan coordinated the selection of the material.

The discovery of extrasolar planets in 1995 added enthusiasm to the SETI effort, beginning an effort to determine a value for one of the unknowns in Drake's formula. In the 1990s, the new interdisciplinary discipline of astrobiology was born. Astrobiology, also called exobiology or bioastronomy, tried to understand the processes of life on Earth better in order to understand how life might emerge and persist on other planets. NASA formed an Astrobiology Institute in 1998, part of a surge in international efforts that included the Centro de Astrobiologia in Spain, the French Groupement de Recherche en Exobiologie, and the Australian Centre for Astrobiology. The University of Washington began a graduate program in astrobiology, funded by a grant from the National Science Foundation. Astrobiologists have drawn encouragement from the discovery of life in deep-sea hydrothermal vents, on the fringes of volcanoes, and even in the frigid waters of Lake Vostok near the South Pole in Antarctica. Landings by robotic craft are planned to seek primitive life in possible liquid oceans under ice mantles on large moons of Jupiter and Saturn.

If any extraterrestrial intelligence were found, it would certainly have a dramatic religious, social, and scientific impact on humanity. Socially marginal movements of quasi-religious advocates claim that extraterrestrial aliens have already visited Earth. They point to unidentified flying objects (UFOs), alien abduction accounts, and the notion that ancient alien astronauts taught science and technologies to the ancient civilizations of Earth.

See also Astronomy; Bell Burnell, Jocelyn; Deep-Sea Hydrothermal Vents; Extrasolar Planets; Internet; Lovelock, James; National Aeronautics and Space Administration; National Science Foundation; Sagan, Carl; Science Fiction; Space Exploration and Space Science; Telescopes; Volcanoes; Voyager Spacecraft

References

Asimov, Isaac. *Extraterrestrial Civilizations.* New York: Crown, 1979.

Darling, David. *Life Everywhere: The Maverick Science of Astrobiology.* New York: Basic, 2001.

Dick, Steven J. *The Biological Universe: The Twentieth-Century Extraterrestrial Life Debate and the Limits of Science.* New York: Cambridge University Press, 1996.

Parker, Barry. *Alien Life: The Search for Extraterrestrials and Beyond.* New York: Plenum, 1998.

SETI Institute. http://www.seti-inst.edu (accessed February 13, 2004).

Shklovskii, Iosef Shumeulovich, and Carl Sagan. *Intelligent Life in the Universe.* New York: Holden-Day, 1966.

Sullivan, Walter. *We Are Not Alone: The Continuing Search for Extraterrestrial Intelligence.* Revised edition. New York: Dutton, 1993.

Selye, Hans (1907–1982)

The endocrinologist Hans Selye demonstrated how physiological stress affects the human body, fundamentally changing the perspective of scientists on stress. Hans Hugo Bruno Selye was born in Vienna, Austria, to a wealthy family; his father was a surgeon. Educated at first by governesses, Selye attended a private secondary school in Czechoslovakia, then entered the German University of Prague in 1924. Studies at the University of Paris and University of Rome supplemented his study at the German University, where he obtained his M.D. in 1929. Two years later he received a Ph.D. in organic chemistry from the same university. Drawn to research rather than clinical practice, he obtained a Rockefeller fellowship to study at Johns Hopkins University for a year, and then for another year at McGill University in Montreal, Quebec. Fluent in ten languages, he chose to remain at McGill, where he became a lecturer in biochemistry in 1933. He moved up the academic ranks to become a professor and director of the Institute of Experimental Medicine and Surgery at the University of Montreal.

In 1936, Selye injected laboratory rats with sex hormones and found that the hormones caused ulcers, damaged the outer tissue of the adrenal glands, and shrunk the lymphatic structures, leading to eventual death. Further research showed that injecting any toxic substance caused the same reaction. Selye concluded that physiological stress damaged the body's endocrinologic system. He developed a theory that stress caused an alarm reaction, mobilizing the body's defenses to a state of resistance, and finally led to a stage of exhaustion that might result in death. He published an extensive monograph, *Stress,* in 1950 expounding and defending his theory. Though he faced resistance to this theory from other physicians, his experiments and consistent results eventually changed their minds. Physicians came to accept that physiological stress could contribute to many problems, including heart attacks, arthritis, allergies, kidney disease, and inflammatory tissue diseases. Germ theories and other basic physiological explanations had to be supplemented by the idea of stress.

Selye retired from the University of Montreal in 1977, but continued to direct the International Institute of Stress, which he had founded the previous year. A skilled and prolific writer, author of over forty books and over a thousand articles, Selye wrote for both his colleagues and lay audiences. In his later years, he promoted a code of ethics designed to minimize negative stress in the lives of people; the code emphasized self-direction of goals and observation of the Golden Rule. Selye married three times, fathering five children, and received numerous honors, including the Companion of the Order of Canada. A well-deserved Nobel prize eluded him.

Eugene Shoemaker, planetary geologist, before the Barringer Crater in Arizona, 1988 (Roger Ressmeyer/Corbis)

See also Medicine; Nobel Prizes
References
Selye, Hans. *The Stress of My Life: A Scientist's Memoirs.* Toronto: McClelland and Stewart, 1977.
Viner, Russell. "Putting Stress in Life: Hans Selye and the Making of Stress Theory." *Social Studies of Science* 29, no. 3 (June 1999): 391–410.

Shoemaker, Eugene (1928–1997)

Eugene Merle Shoemaker, a pioneering planetary geologist, specialized in lunar geology, impact craters, and searching the sky for comets and asteroids. Born in Los Angeles, Shoemaker moved around the United States, where his father held many different types of jobs, including being a teacher, before his family moved back to Los Angeles. In 1944, an early interest in geology led him to the California Institute of Technology (Caltech) at the age of sixteen. He graduated with honors in 1947 and received his

M.S. the following year. He found employment with the United States Geological Survey (USGS) in Grand Junction, Colorado, where he worked with the uranium industry in that area. Later work evaluating the craters formed by nuclear testing in Nevada gave him an interest in understanding impact craters. He earned his Ph.D. from Princeton in 1960 while still working at the USGS. His dissertation proved that the enigmatic Barringer Crater in Arizona was actually the result of a meteorite impact, not volcanic action.

In 1960, Shoemaker created an astrogeological section in the USGS to support the U.S. space program. During the Apollo project, Shoemaker was involved as a senior scientist and administrator. His work included analyzing images from the Ranger, Surveyor, and Lunar Orbiter spacecraft and Moon landers, and determining where it was safe for the astronauts to land. He wanted to be one of the Apollo astronauts, but a diagnosis of Addison's disease prevented him from achieving this dream.

Retaining his association with the USGS, Shoemaker became a professor of geology at his alma mater, Caltech, in 1969. Using the 18-inch (0.46-meter) Schmidt wide-view telescope at Palomar Observatory, his colleagues and he developed a program to search for comets and asteroids that passed close to Earth. This effort led to numerous discoveries, including 32 comets and over a thousand asteroids. The discovery in 1993 of the Shoemaker-Levy 9 comet led to a singular event watched and analyzed by astronomers, as the comet subsequently broke up and impacted the planet Jupiter. The implications of impact craters led Shoemaker and others to express concern about future impacts on Earth, with possibly catastrophic results for human civilization. He helped found the Spaceguard Foundation to coordinate international study of this threat.

Shoemaker married in 1951 and fathered three children, and his wife, Carolyn, in later years became a full partner in his scientific research. Shoemaker died in 1997 in a car accident near Alice Springs, Australia, while researching impact craters. The Near Earth Asteroid Mission (NEAR), sent to orbit an asteroid, was renamed the NEAR Shoemaker spacecraft in his honor. The Lunar Prospector spacecraft that impacted the Moon in 1999 carried an ounce of his cremated ashes.

See also Apollo Project; Astronomy; Geology; National Aeronautics and Space Administration; Shoemaker-Levy Comet Impact; Space Exploration and Space Science; Telescopes

References
Levy, David H. *Shoemaker by Levy: The Man Who Made an Impact*. Princeton: Princeton University Press, 2000.
Marsden, Brian. "Eugene Shoemaker (1928–1997)." http://www2.jpl.nasa.gov/sl9/news81.html (accessed February 13, 2004).

Shoemaker-Levy Comet Impact

On March 25, 1993, Carolyn Shoemaker discovered a comet on film from the 18-inch (0.46-meter) Schmidt wide-view telescope at Palomar in southern California. She was part of a team that included her husband, the noted planetary geologist Eugene Shoemaker, and their colleague, David Levy. Their effort was part of a long-term sky survey to find comets and asteroids, an effort started by Eugene in the early 1970s. As was traditional, the comet was named in honor of its discoverers. They soon learned that the comet was really multiple objects, not a single ball of ice and debris. After the trajectory of the comet was calculated, astronomers realized that the comet had approached to within 50,000 miles of Jupiter the previous year and that the gas giant's tidal forces had probably fragmented the comet at that time. Even more exciting, Jupiter's gravity had altered the comet's course so that it was coming back to impact the planet on its next orbit.

Astronomers pointed the Hubble Space Telescope for the first time at the comet on July 1, 1993, and confirmed the twenty-one separate fragments that ground-based astronomers had already identified. From July 16, 1994, until July 22, 1994, the fragments hit Jupiter in a series of spectacular events. Astronomers in observatories all over the world watched and analyzed these impacts. The Hubble Space Telescope and space-based telescopes also provided observations. The spacecraft *Galileo* was en route to the planet and observed the impacts with its instruments. When *Galileo* entered orbit around Jupiter in December 1995, it provided close-up analysis of the continuing effects of the impacts. Astronomers used the Internet to quickly send e-mail messages, data files, and pictures to each other. Pictures and other information were released to the world media as soon as they were received.

Astronomers were astonished by the dramatic effects on the planet. Great fireballs erupted on some impacts, one 3,000 kilometers high, leaving dark smudges on Jupiter's atmosphere, while other fragments were swallowed by Jupiter with barely a trace. The fragments are thought to have been generally 1 to 2 kilometers in diameter, with fragment G being perhaps 3 kilometers wide. Scientists still disagree over how large the comet fragments were and exactly what happened in the atmosphere. The results of the impacts were visible for years afterward as disturbances in Jupiter's atmosphere.

The lesson of Shoemaker-Levy 9 is that the solar system is filled with asteroids and comets that sometimes collide with planets. Such an impact on Earth would be catastrophic for human civilization and maybe even for all higher life-forms on the planet. Astronomers believe that the mysterious 1908 explosion in Tunguska, Siberia, was either a small comet or an asteroid hitting Earth or exploding in the atmosphere. The remote area was devastated. Tunguska was not an isolated event, since Earth is covered with craters from past impacts.

> See also Astronomy; Internet; Satellites; Shoemaker, Eugene; Space Exploration and Space Science; Telescopes

References

Levy, David H. *Impact Jupiter: The Crash of Shoemaker-Levy 9.* New York: Plenum, 1995.

Noll, Keith S., Harold A. Weaver, and Paul D. Feldman, editors. *The Collision of Comet Shoemaker-Levy 9 and Jupiter.* Cambridge: Cambridge University Press, 1996.

Spencer, John R., and Jacqueline Mitton, editors. *The Great Comet Crash: The Impact of Comet Shoemaker-Levy 9 on Jupiter.* Cambridge: Cambridge University Press, 1995.

Verschuur, Gerrit L. *Impact! The Threat of Comets and Asteroids.* New York: Oxford University Press, 1996.

Skinner, B. F. (1904–1990)

The psychologist B. F. Skinner became famous for promoting a theory of strict behaviorism that rejected free will or any form of unconscious motivation. Burrhus Frederic Skinner was born in Susquehanna, Pennsylvania, where his father was a lawyer. He attended Hamilton College in Clinton, New York, where he earned an A.B. in 1926, majoring in English. He intended to be a fiction writer, but gave up after a year. After reading psychology articles by the British philosopher Bertrand Russell (1872–1970) and U.S. psychologist John B. Watson (1878–1958) on behaviorism, he found a new path in life. In 1928, he entered Harvard University, earning an M.A. in 1930 and a Ph.D. in 1931, both in psychology. Skinner remained at Harvard for five years as a researcher before accepting an academic appointment at the University of Minnesota in 1936. Skinner believed that only observed actions could explain human behavior. As a child, he often built things, and as a psychologist he built what became known as the Skinner box, which provided a controlled environment where rat behavior could be carefully monitored with only a few variables at a time. He developed a theory of operant conditioning, according to which positive rewards or negative rewards created habits of behavior in rats, in other animals, and by extension, in humans.

Skinner married in 1936, fathered two daughters, and for the second daughter he built a special mechanical crib that she spent much of her first two years in. The crib stimulated her with operant conditioning in a germ-free environment. When he published a 1945 article in the popular *Ladies' Home Journal* describing the crib, some people reacted with interest, trying to market the device, while others were horrified.

During World War II Skinner received a contract to train pigeons with operant conditioning to guide missiles toward a target by pecking at the center of the target's image. The idea worked, but was never actually implemented. After the war, Skinner became a professor of psychology at Indiana University. His strict behaviorism had established his scientific reputation, and in 1948 he returned to Harvard. That same year he published *Walden Two,* a novel about a utopian community based

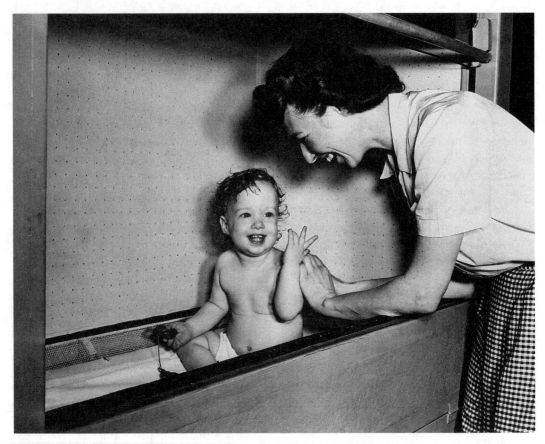

Yvonne Skinner, wife of B. F. Skinner, with their daughter and a mechanical crib that he designed to demonstrate his psychological theories (Bettmann/Corbis)

on behaviorist principles. This novel attracted a lot of attention and helped transform Skinner into a cultural icon. He used the publicity to promote what he called the "technology of behavior," which he equated with the laws of behavioral science being used to create a better world. His *Science and Human Behavior* (1953) became a standard text. Later books by Skinner argued that language came from operant conditioning, a view rejected by linguists in favor of the theories of Noam Chomsky (1928–), and Skinner promoted using operant conditioning principles and learning machines in education. His 1971 book, *Beyond Freedom and Dignity,* argued that the ideas of freedom and dignity were illusions and would disappear if people followed the principles of his behaviorism.

Skinner remained at Harvard until his re-tirement in 1974. He received numerous honors, including a Distinguished Scientific Contribution Award from the American Psychological Association (APA) in 1958, a National Medal of Science in 1968, and the Gold Medal of the APA in 1971. Although the most radical ideas of his behaviorism are rejected by mainstream psychology, his name has become synonymous with a stance that remains useful when defining different theories and schools of thought in modern psychology.

See also Chomsky, Noam; Psychology
References

Bjork, Daniel W. *B. F. Skinner: A Life.* New York: Basic, 1993.
Skinner, B. F. *Particulars of My Life.* New York: Knopf, 1976.
Wiener, Daniel N. *B. F. Skinner: Benign Anarchist.* Boston: Allyn and Bacon, 1996.

Smallpox Vaccination Campaign

Modern medicine completely eradicated the deadly disease of smallpox in an international campaign directed by the World Health Organization (WHO). Smallpox is caused by several strains of the variola virus, leading to a fever, sores, and possible death. Pockmarks often scarred survivors for life. Ancient manuscripts from China and India refer to the disease, and the mummy of the ancient pharaoh Ramses V is thought to show evidence of the disease. Survivors of the infection are immune to further infection. Although not strongly infectious, the virus itself is able to survive for up to eighteen months outside of the human body and remain viable. Smallpox was one among many diseases that decimated the native populations after Christopher Columbus (1451–1506) opened the New World to settlement and exploitation by the Old World. What made these diseases so devastating was not an innate genetic weakness among the American natives, but the fact that the populations had not been exposed to the diseases before. In a so-called virgin territory epidemic, everyone gets sick at the same time, and there is no one to care for the ill, unlike an epidemic in Europe or Asia, where many of the adults would possess immunity from earlier epidemics. The result is a much higher mortality rate.

Folk remedies for smallpox contained considerable wisdom, including transferring pus from the sore of an infected person into a small cut on the skin of an uninfected person. The recipient got a mild form of the disease that gave immunity but did not kill. In a 1721 smallpox epidemic in Boston, a variation of this folk remedy, copied from Turkish practices, helped stem the disease. The English physician Edward Jenner (1749–1823) noticed that milkmaids who contracted cowpox, resulting in minor illness and skin lesions, gained immunity to smallpox. He developed the technique that he called vaccination to use cowpox to immunize humans to smallpox.

Vaccination laws in the late nineteenth and early twentieth centuries made smallpox and other diseases a distant threat in the industrialized West, but smallpox still continued to kill millions of people a year in Third World countries. The Soviet Union, in a period of relaxed tensions during the cold war, proposed to WHO in 1958 that the international organization launch a program to eradicate smallpox. WHO created a Smallpox Eradication Program in 1958, and in 1967, the ambitious effort started in earnest, led by the U.S. physician Donald A. Henderson (1928–). Health care workers fanned out around the world, inoculating millions. Under the original plan, every single person received a vaccination, but when the program in Nigeria ran short of vaccine, the medical advisor to the program found that targeting the vaccine in a "surveillance-containment" strategy of vaccinating only those people recently exposed to smallpox succeeded in eliminating the disease within the local population. Not everybody on the planet had to be vaccinated.

Smallpox was the ideal candidate for absolute eradication. A single vaccination offered protection from all strains of the variola virus, and the virus only lived in human beings. Most other viruses and bacteria also infect animal hosts and can form a reservoir for future infection even if no humans are available as carriers. Although humanitarian motives were important in providing the billions of dollars necessary to fund the eradication effort, the industrialized nations also wanted to stop vaccinating their own populations, and could not do this if smallpox remained on the planet to cause a possible epidemic.

Ali Maow Maalin of Somalia contracted smallpox in October 1977 and gained the distinction of being the last person to naturally acquire smallpox. A year later at the Birmingham University Medical School in Alabama, a worker died after contracting the disease in a laboratory. Other laboratory infections also occurred in England. After monitoring for further outbreaks, WHO officially declared smallpox eradicated in 1979. Henderson felt

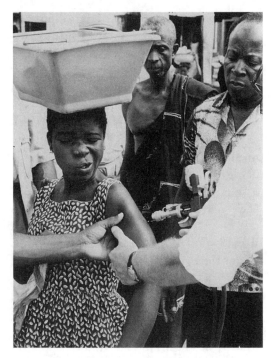

In this 1969 photo a Nigerian girl winces as she gets one of the 100 million vaccinations against smallpox. (Bettmann / Corbis)

fare development program conducted by the Soviet Union for decades, which may well have ended only in the late 1990s, scientists have remained concerned that smallpox may return as a weapon of war or a tool of deliberate terrorism.

See also Centers for Disease Control and Prevention; Medicine; Pharmacology; World Health Organization

References

Bazin, H. *The Eradication of Smallpox: Edward Jenner and the First and Only Eradication of a Human Infectious Disease.* San Diego, CA: Academic, 2000.

Hopkins, Donald R. *Princes and Peasants: Smallpox in History.* Chicago: University of Chicago Press, 1983.

Tucker, Jonathan B. *Scourge: The Once and Future Threat of Smallpox.* New York: Atlantic Monthly, 2001.

that the effort succeeded in spite of the bureaucracy of WHO, not because of it. After its success with smallpox, WHO turned its attention to polio, coming close to eradicating it by the end of the twentieth century. Other candidates proposed for future eradication included the guinea worm, measles, rubella, and hepatitis B.

As the scars on shoulders from jet-injected shots have faded in the world population and new generations are born, the global population has gradually lost its immunity to smallpox. Smallpox officially remains in laboratories at the Centers for Disease Control and Prevention in Atlanta, Georgia, and the Institute for Viral Preparations in Moscow, Russia, with concerns that unofficial samples may also remain in other laboratories. Scientists and various public health organizations have called for the destruction of remaining samples of smallpox to prevent accidental release from the high-containment laboratories. Especially in light of the secret biological war-

Social Constructionism

The scientific method is followed by scientists as a unique method for finding out truths about the physical world and reality. Postmodern theorists have pointed out the human fallibility inherent in scientific endeavors. Social constructionists, present on the cutting edge of sociology, anthropology, and history in the 1980s and 1990s, advocate a position of cultural and cognitive relativism. Critics of social constructionism have argued that postmodernism and social constructionism only lead to intellectual anarchy and surrender the best-known technique for finding truth.

In the early 1970s, the "strong programme" in the sociology of scientific knowledge emerged from a group of sociologists at the University of Edinburgh, a movement that is now also called social constructionism. In the age-old nature vs. nurture debate, social constructionists definitely came down on the side of nurture, or environment. Social constructionists believe that all knowledge is subjective, a result of social and cultural conditioning. Social constructionists make the argument that cognitive relativism should be as acceptable as moral relativism. Moral relativism assumes that there are no absolute moral truths, and thus cognitive relativism would have to argue that there is not a physical reality that all humans share and can objectively evaluate. In its most radical form, social constructionism denies the possibility of objective scientific knowledge. At its best, social constructionism brings out the way social privilege promotes certain scientific ideas and pseudoscientific prejudices. The best work by social constructionists has examined the softer sciences; these sciences, with their focus on human beings and animals, have such a difficult time attaining objective knowledge that they are easy targets.

In the 1980s, social constructionism, postmodernism, and certain developments in the philosophy of science sparked what became known as the science wars. The questioning of social constructionists was not welcomed by practicing scientists, who vigorously defended the scientific method as a way of objectively obtaining knowledge and defended the body of scientific knowledge as providing a correct view of the universe.

The intellectual vacuum that social constructionism throve in was vividly demonstrated in 1994 when Alan D. Sokal, a theoretical physicist at New York University, published an article, "Transgressing the Boundaries: Toward a Transformative Hermeneutics of Quantum Gravity," in an issue on "Science Wars" of the cultural studies journal *Social Text*. The editors admitted that his credentials as an actual physicist encouraged them to publish his article. The article was a nonsensical parody, filled with absurd statements, using the rhetoric of postmodernism and social constructionism. The editors of the journal were not in on the joke when Sokal publicly revealed his hoax three weeks later. A storm of controversy erupted, with critics condemning Sokal for his duplicity and others applauding his intellectual exposure of extreme social constructionism, radical feminist epistemology, deconstructive literary theory, and New Age ecology.

See also Philosophy of Science
References
Bloor, David. *Knowledge and Social Imagery*. Second edition. Chicago: University of Chicago Press, 1991.
Gergen, Kenneth J. "The Social Constructionist Movement in Modern Psychology." *American Psychologist* 40, no. 3 (March 1985): 266–275.
Koertge, Noretta, editor. *A House Built on Sand: Exposing Postmodernist Myths about Science*. New York: Oxford University Press, 1998.
Latour, Bruno. *Science in Action: How to Follow Scientists and Engineers through Society*. Cambridge: Harvard University Press, 1987.

Sociobiology and Evolutionary Psychology

The prominent biologist Edward O. Wilson (1929–) published *Sociobiology: The New Synthesis* in 1975. A recognized authority on ants and other social insects, Wilson considered his book to contain a logical evolution of the ideas in his 1971 book *The Insect Societies,* which combined entomology with population biology. In *Sociobiology,* Wilson argued that what biologists had learned about innate behavior in insect and animal societies, along with insights from the theory of natural selection, could also be applied to understanding human behavior and human society.

The idea was not new, since Charles Darwin (1809–1882) himself had suggested that human society was similar to animal societies, with instincts mediated by reason. The field of ethology, promoted by Konrad Lorenz (1903–1989) and others, had laid a strong foundation of knowledge about how animals behave. But how to explain altruistic behavior, in which an animal or human might sacrifice its own well-being for others? An important 1964 article by W. D. Hamilton at the University College at London, "The Genetic Evolution of Social Behavior," argued that kin selection was the answer. Altruistic behavior may not help pass on an individual's genes, but that behavior could help pass on the genes in relatives helped by the altruistic behavior, leading to reciprocal altruism.

Wilson drew these threads together and produced a well-written, exhaustively researched, powerful book. His comprehensive theory addressed issues of communication, altruism, aggression, sex, parental care, social hierarchy, dominance, and social behavior, applying sociobiology to social insects, birds, fish and other cold-blooded vertebrates, elephants, carnivores, primates, and finally humans. Wilson argued that even human ethics derived from genetic selection, and that sexual morality had a basis in what was best for the propagation of a set of genes. Reciprocal barter systems embodied altruism and kin selection, and perhaps certain genes led to people falling into certain classes. Wilson objected to extreme behaviorism, of which B. F. Skinner (1904–1990) is the most prominent example. He also objected to using only a few examples from a limited number of species to draw broad conclusions about animal and human traits, as in certain popular ethological books: Konrad Lorenz, *On Aggression* (1966); Robert L. Ardrey, *The Territorial Imperative: A Personal Inquiry into the Animal Origins of Property and Nations* (1966); and Desmond Morris, *The Naked Ape: A Zoologist's Study of the Human Animal* (1967).

Sociobiology caused a storm of controversy, and its critics worried that sociobiology seemed reminiscent of past pseudoscientific justifications that condoned racism and viewed humans as genetically determined rather than individuals more shaped by nurture than by nature. The paleontologist Stephen Jay Gould (1941–2002) and geneticist Richard C. Lewontin (1929–), colleagues at Harvard, were among the most vocal critics who rejected sociobiology. Much of the opposition came from leftist or Marxist ideologues. Wilson was even doused with water by political activists during a presentation at a meeting of the American Association for the Advancement of Science in 1978.

Wilson's next book, *On Human Nature* (1978), defended his ideas against his critics and won a Pulitzer Prize. He acknowledged that most human behavior was learned, not inborn, but also argued that the urge toward religious belief seemed innate. In his own life, he believed in scientific materialism, rather than any conventional religion. He also argued that altruistic behavior, which seems contrary to natural selection, is consistent with natural selection if an individual's altruism is directed toward close relatives. This makes the gene, rather than the individual, the mechanism of natural selection.

In his push for a science of evolutionary psychology, the zoologist Richard Dawkins (1941–) also promoted the emphasis on genes as the focus of evolution. His first book in 1976, *The Selfish Gene,* argued that natural selection occurs on the level of genes, rather than the level of individuals or species. Dawkins argued that individuals are just temporary carriers of genes and that physiology and behavior can best be explained by the need for genes to propagate. Not all scientists accept Dawkins's view of evolution, though evolutionary psychology and its kin, sociobiology, have become a powerful intellectual current supported by many scientists and continue to be controversial. In 1989, members of the Animal Behavior Society voted Wilson's *Sociobiology* the most important book on animal behavior ever published.

See also Biology and the Life Sciences; Dawkins, Richard; Evolution; Gould, Stephen Jay; Lorenz, Konrad; Religion; Skinner, B. F.; Wilson, Edward O.; Zoology

References

Alcock, John. *The Triumph of Sociobiology.* New York: Oxford University Press, 2001.

Pinker, Steven. *How the Mind Works.* New York: Norton, 1997.

Ruse, Michael. *Sociobiology: Sense or Nonsense?* Boston: D. Reidel, 1979.

Wilson, David Sloan. *Darwin's Cathedral: Evolution, Religion, and the Nature of Society.* Chicago: University of Chicago Press, 2002.

Wilson, Edward O. *Sociobiology: The New Synthesis.* Twenty-Fifth Anniversary Edition. Cambridge: Belknap Press of Harvard University Press, 2000.

Wright, Robert. *The Moral Animal: Evolutionary Psychology and Everyday Life.* New York: Random House, 1995.

Sociology of Science

The prominent sociologist Robert K. Merton (1910–2003) founded what became known as the Mertonian approach to the sociology of science. Merton advocated understanding science by examining the external social, political, religious, and economic contexts surrounding the scientific community. Merton's most influential books in this area were his *Science, Technology, and Society in Seventeenth Century England* (1938) and *The Sociology of Science: Theoretical and Empirical Investigations* (1973). Merton was also interested in the best ways for scientists to maximize their production of scientific knowledge, leading to his strong interest in scientific institutions and scientific careers. Merton and his followers, though recognizing the social context of science, did not dispute that genuine scientific knowledge could be achieved through the scientific method.

A good example of the work of a Merton-influenced sociologist is the work of Joseph Ben-David, who sought to find the social conditions that facilitate the rise of science and the practice of science. He found that decentralized organizations and a tradition of pluralism are the most ideal conditions yet invented. Even if the system of creating scientific knowledge decays, the knowledge remains behind. Ancient Greek science as a practice died after a brief period of flowering, but the knowledge continued to be accepted as dogma for the next two millennia.

In the early 1970s, the "strong programme" in the sociology of scientific knowledge emerged from a group of sociologists at the University of Edinburgh, a movement that is now also called social constructionism. Social constructionists believe that all knowledge is subjective, a result of our social and cultural conditioning. At its best, social constructionism points to the way that social privilege promotes certain scientific ideas and pseudoscientific prejudices. The social constructionists have also done good work in their examination of the softer sciences, since humans and animals prove to be difficult subjects about which to gain objective knowledge. Other work by social constructionists has examined the social relations within scientific laboratories, again concentrating on how scientific work is done, rather than on whether scientific knowledge is a valid way to examine physical reality. Some social constructionists have pursued a more controversial line of thought, demanding that the reflexivity applied to science should be applied to their own scholarship, questioning the methods and standards that sociologists use to arrive at their own flavor of knowledge.

As might be expected from a profession that arose in connection with the need for social activism, some sociologists have used their scholarship to advocate changes in how science is practiced. Some are concerned with only using science in a moral manner, and creating a system of moral responsibility for scientists to acknowledge. Others have demonstrated how the social structures of science and engineering can drive away women and minorities, and these thinkers have often recommended effective ways to change pedagogy in order to help women and minorities to succeed in science and other technical fields.

See also Feminism; Social Constructionism

References
Ben-David, Joseph. *The Scientist's Role in Society: A Comparative Study.* Upper Saddle River, NJ: Prentice-Hall, 1971.
Bloor, David. *Knowledge and Social Imagery.* Second edition. Chicago: University of Chicago Press, 1991.
Etzkowitz, Henry, Carol Kemelgor, and Brian Uzzi. *Athena Unbound: The Advancement of Women in Science and Technology.* Cambridge: Cambridge University Press, 2000.
Kneen, Peter. *Soviet Scientists and the State: An Examination of the Social and Political Aspects of Science in the USSR.* Albany: State University of New York Press, 1984.
Ravetz, Jerome R. *Scientific Knowledge and Its Social Problems.* Oxford: Clarendon Press, 1971.

Sokal Hoax

See Social Constructionism

Soviet Academy of Sciences

As part of his effort to modernize Russia along Western lines, Czar Peter the Great (1672–1725) founded in 1724 the Academy of Sciences in Saint Petersburg. The academy persisted under various names for two centuries. After coming to power, the communists restructured the academy into the Union of Soviet Socialist Republics (USSR) Academy of Sciences in the 1920s. Prior to the Soviet reorganization, most research in Russia had occurred in universities. The Soviet Academy created a system of institutes (eventually reaching at least 260 in number) to direct the practice of science, creating a centralized system of control not seen in other countries (save for imitators in other communist countries). The American National Academy of Sciences and other national societies in the Western world are primarily honorific organizations that promote scientific communication and play minor roles in actual research. The Soviet Academy directed not only the practice of science, but also technology, though in the late 1950s the Communist Party removed most of the more technologically directed institutes from academy control and assigned them to various industrial ministries. A new State Committee for Coordination of Scientific Research (later renamed the State Committee for Science and Technology) asserted overall direction over Soviet science, while the academy continued to do the actual work of research and development.

Especially during the rule of Josef Stalin (1879–1953), scientists who fell under political suspicion often died as a result. In 1936 and 1937 alone, one out of every five astronomers was arrested. Many of the Soviet Union's best scientists and engineers spent time at some point in their careers conducting research in prison laboratories. Sergei Vavilov (1891–1951) served as president of the Soviet Academy of Sciences from 1945 to 1951, yet his own brother, a famous geneticist, had starved to death in prison in 1943, victim of a Stalinist purge. Vavilov walked a fine line, as many have done in totalitarian societies, trying to remain moral and to survive at the same time, and managed to blunt some of the more extreme efforts by communist ideologues to twist the practice of science. The academy under Stalin proved that fruitful science can still be conducted in an atmosphere of fear, restricted communications, and imminent threat of arrest.

After Stalin, the situation for scientists became less hazardous, in that they no longer feared for their lives, but the government remained totalitarian. Contact with foreign scientists was always a difficult problem for the Soviets. In 1959, a joint program with the American National Academy of Sciences sponsored international workshops and started a period of visits to both countries by scientists. Soviet scientists were sometimes threatened with severe consequences to their families if they defected.

As the rest of Soviet society ossified into a bureaucratic stupor, critics accused the academy of being dominated by bureaucrats instead of researchers. When Mikhail Gorbachev (1931–) came to power in 1985, he sought to revitalize the Soviet economy in order to better compete in the cold war. His reform policies of glasnost and perestroika introduced more economic and social openness, as well as offering limited democracy. In 1987, as part of these efforts at revitalization, a new rule required directors of academy institutes to retire at the age of seventy, rather than holding their positions for life. Other researchers were forced to retire or recertify to prove that they were still capable in their specialties. Gorbachev's efforts failed in many ways, and the dissolution of the Soviet Union in 1991 led to the academy being renamed the Russian Academy of Sciences.

In 1991, the academy had 337 full members, 651 corresponding members, and tens of thousands of researchers and staff working in their institutes and other organizations. By 1995 the number of scientific articles and reports issued by the publishing arm of the academy had declined to about a fifth of its

Members of the Soviet Academy of Sciences, 1959 (Time Life Pictures / Getty Images)

former level. By 1996, financing for the academy had dropped to a tenth of its former level. Efforts at cooperative ventures with private corporations were resisted by academy bureaucrats. Many of the best scientists took advantage of the end of communism to move to Western countries and continue their careers there.

See also Associations; Cold War; Soviet Union
References
Fortescue, Stephen. "The Russian Academy of
    Sciences and the Soviet Academy of Sciences:
    Continuity or Disjunction?" *Minerva: Review of
    Science, Learning and Policy* 30 (1992): 459–478.
Lewis, Robert A. "Government and the
    Technological Sciences in the Soviet Union:
    The Rise of the Academy of Sciences." *Minerva:
    Review of Science, Learning and Policy* 15 (1977):
    174–199.
Schweitzer, Glenn E. "US-Soviet Scientific
    Cooperation: The Interacademy Program."
    Technology in Society 14 (1992): 173–185.

Soviet Union

The Union of Soviet Socialist Republics (USSR) gained superpower status in the aftermath of World War II, as the USSR acquired an Eastern European empire of satellite states and developed atomic weapons. The ideological power struggle of the cold war dominated global politics until the dissolution of the Soviet Union in 1991. Communist ideological conformity, enforced by spies for the secret police, permeated the Soviet practice of science. The economic and military needs of the state dictated the fields and goals of scientific and technological research. The Soviet Academy of Sciences and its numerous institutes (eventually reaching at least 260 in number) directed the practice of science, creating a centralized system of control not seen in other countries, save for imitators in other communist countries. Especially during the rule of Josef Stalin (1879–1953), scientists who fell under political suspicion often died as a result.

Stalin strongly supported the theory of non-Mendelian genetics promoted by Trofim Lysenko (1898–1976). Lysenko directed the Institute of Genetics of the Academy of Science of the USSR from 1940 to 1965 and served as president of the powerful V. I. Lenin All-Union Academy of Agricultural Sciences. As the effective dictator of Soviet agricultural science, Lysenko wielded his power ruthlessly. Soviet scientists who did not support his ideas found themselves at best unemployed, and some Soviet scientists who disagreed with Lysenko died from either execution or starvation after being arrested and taken into custody. Lysenko survived even though widespread adoption of his ideas led to agricultural failures. Lysenko's power declined after Stalin died in 1953. The new Soviet leader, Nikita S. Khrushchev (1894–1971), believed in Lysenko's Marxist science, but allowed scientific criticism of Lysenko to go unpunished. In 1964, after changes in the Soviet leadership, Lysenko was finally completely discredited and exiled to a remote agricultural station, though it took time for the effects of his regime to fade.

Communist ideology also threatened the integrity of other sciences. During World War II, Stalin relaxed restrictions in order to help the war effort; after the war, he reimposed control. Postwar conferences were organized in chemistry, ethnography, astronomy, and physiology, at which scientists were encouraged to draw back from Western science and only practice science in harmony with communism. Claims of Russian and Soviet priority in many common inventions were also promoted, making it seem that, oddly enough, everything of importance had been invented there first.

Physicists at Moscow University took the postwar opportunity to attack physicists in the institutes of the Soviet Academy of Sciences, accusing them of being enamored of Western culture and promoting science not in accordance with communist principles. Quantum mechanics and relativity were among the theories they thought should be banned. Stalin elected to support the Soviet Academy of Sciences because he wanted the Soviet atomic bomb project, run mostly by members of the academy, to succeed. Ironically, the atomic bomb is based on the theories that the Moscow University physicists proposed to ban.

The 1949 explosion of the Soviet Union's first atomic bomb, partially based on secrets stolen from the U.S. project at Los Alamos, shocked the world and established the Soviets as contenders in technological prowess. The United States reacted by developing the hydrogen bomb. The brilliant Andrei Sakharov (1921–1989) designed the first Soviet hydrogen bomb, exploded on August 12, 1953. Sakharov later gained international acclaim as a human rights activist and internal opponent to the Soviet regime.

In general, Soviet policy was indifferent to the well-being of individuals and the health of the environment. Medical doctors engaged in medical research on prisoners in the Siberian Gulag reminiscent of experiments conducted by Nazi doctors on prisoners in concentration camps. Industrial concerns poisoned rivers, scarred the Siberian forests with acid rain, and turned at least two towns into radioactive wastelands. Although environmental concerns held a low rank among Soviet priorities, scientific studies of the danger of airborne lead pollution did lead the Soviets to ban leaded gasoline in major cities in 1967, earlier than similar efforts in Western nations.

When the Soviets launched the first artificial satellite, *Sputnik I*, in 1957, they seemed to take the lead in the cold war technological contest to show which political and economic ideology was most effective. The second Sputnik carried a dog named Laika into orbit, and the Soviets orbited the first man in April 1961. In response, the Americans launched the Apollo project a couple of months later, with the intent of landing a man on the Moon and returning him safely to Earth before the end of the decade. The Soviets launched a secret program to beat the Americans, but suffered a crippling setback when a rocket exploded on the pad and killed dozens of key engineers and technical personnel. The Soviets announced that they were not interested in following the U.S. lead. During a thaw in the cold war, a period of détente, the United States and the USSR cooperated on the Apollo-Soyuz mission of 1975, during which American astronauts and Soviet cosmonauts rendezvoused in orbit and shook hands inside a temporary tunnel connecting the two spacecraft.

The collapse of the Soviet Union, with the Russian Federation and other republics dividing up the nation, did not reverse the economic decline that had begun in the 1970s. The practice of science suffered from lack of funding, despite the large number of well-educated scientists. Many scientists fled to Western countries to continue their careers there. Large U.S. subsidies allowed the Russian space program to continue to function and participate in the International Space Station project. Other U.S. and European subsidies helped keep scientists in sensitive career fields employed, in order to encourage them not to leave for totalitarian nations and work on weapons of mass destruction. Some Russians, benefiting from a strong mathematical education system, found employment working as programmers for Western companies.

See also Apollo Project; Big Science; Cold War; Environmental Movement; Hydrogen Bomb; Lysenko, Trofim; Sakharov, Andrei; Soviet Academy of Sciences

References

Berstein, Vadim J. *The Perversion of Knowledge: The True Story of Soviet Science.* Boulder, CO: Westview, 2001.

Graham, Loren R. *Science in Russia and the Soviet Union: A Short History.* Cambridge University Press, 1993.

———. *What Have We Learned about Science and Technology from the Russian Experience?* Stanford, CA: Stanford University Press, 1998.

Josephson, Paul R. *Red Atom: Russia's Nuclear Power Program from Stalin to Today.* San Francisco: W. H. Freeman, 2000.

Kneen, Peter. *Soviet Scientists and the State: An Examination of the Social and Political Aspects of Science in the USSR.* Albany: State University of New York Press, 1984.

Space Exploration and Space Science

Early efforts at space exploration used high-altitude balloons and sounding rockets that briefly left the atmosphere, taking photographs and sensor readings before falling back to Earth. On October 4, 1957, the Soviet Union launched the first artificial satellite, *Sputnik I.* The effort by both the Soviets and the Americans to launch satellites was a focus of the International Geophysical Year (IGY) of 1957–1958, an international project organized to understand Earth better. The second Sputnik carried a dog named Laika into orbit, and instruments monitored her condition until the air ran out. The first U.S. effort to launch a satellite exploded on the launch pad on December 6, 1957. The second effort, *Explorer 1,* launched on January 31, 1958, carried instruments that detected cosmic rays. The scientific team led by James Van Allen discovered that radiation belts, created by Earth's magnetic field, surrounded our planet. This phenomenon was named the Van Allen radiation belts.

Both the Americans and the Soviets launched dozens of satellites during the first three decades of the space age with scientific instruments for studying Earth, the local region of space, and astronomical objects. Weather satellites allowed major advances in meteorology and the ability to make weather predictions. Efforts by the National Aeronautics and Space Administration (NASA) to miniaturize electronic components became an important contribution in the U.S. computer industry.

The Ranger, Surveyor, and Lunar Orbiter projects sent unmanned robotic craft to the Moon to scout the way for the astronauts of the Apollo project. Though most Ranger probes experienced technical failures, the last three smashed into the Moon while sending back television pictures; Surveyor probes actually landed; and the Lunar Orbiters used radar and photography to chart the surface of the Moon, searching for safe landing sites for the Apollo astronauts. The Luna series of probes by the Soviets included numerous failures, but also included the first image of the far side of the Moon, taken by *Luna 3* in October 1959, and the first successful soft landing on the Moon with *Luna 9* in 1966.

Six pairs of Apollo astronauts visited the Moon from 1969 to 1972, gathering rock samples and exploring. Three Soviet probes successfully landed on the Moon, took soil samples, and returned them to Earth. Other Soviet landers brought robotic lunar rovers that moved about, sending back television pictures. All these efforts revealed a wealth of information about the Moon. They found that craters on the Moon extended in size all the way down to microscopic size, since there was no atmosphere to burn up smaller meteors. The Moon possesses no magnetic field to protect it from the solar wind and cosmic rays. The maria of the Moon, which look like plains in telescopes, were formed by volcanic lava flows over three billion years ago. The oldest rocks found on the Moon are 4.3 billion years old, comparable to the oldest rocks found on Earth. The question of the Moon's origin, whether it was formed at the same time as Earth or later, has not been resolved. Evidence of past water or ice has not been found, though tantalizing hints of ice in craters at the Moon's poles were found in the 1990s by the Clementine and Lunar Prospector orbiting scientific satellites, perhaps deposited by the impact of comets.

The Americans and Soviets also sent scientific spacecraft to other planets. On December 14, 1962, the U.S. *Mariner 2,* the first successful interplanetary probe, flew by Venus. The probe determined that Venus possessed neither radiation belts nor a magnetic field. Four days after the spacecraft passed Venus, the electronics on *Mariner 2* failed from the heat of the Sun. The Soviets achieved their first success with interplanetary probes when *Venera 4* flew by Venus on October 18, 1967. It dropped a probe into the atmosphere. Higher than expected atmospheric pressure crushed the probe before it reached the planet's surface. The Soviets launched

many more probes than the United States, but were plagued by persistent failures at launch or during the long journey to the other planets. Soviet probes were mass-produced and attempted to maintain Earth-like conditions within their large spacecraft. The more expensive American spacecraft were produced individually, with each component rigorously tested and designed to work in the vacuum of space.

Mariner 5 arrived at Venus only a few days after *Venera 4.* Spacecraft bound for other planets were often launched during a short window of time when Earth and the target planet were in the right positions, meaning that Soviet and U.S. probes often arrived at their destinations together. When *Mariner 5* passed behind Venus, its radio signal passed through the planet's atmosphere. U.S. astronomers measured the strength of the returning signal to learn the density of the atmosphere. Venus possessed a surface atmospheric pressure one hundred times greater than Earth's surface. The surface temperature was a scorching 800 degrees Fahrenheit (427 degrees Celsius). Venus became the textbook example of a runaway greenhouse effect, where carbon dioxide clouds retain the Sun's heat. Later Soviet Venera probes succeeded in reaching the surface of Venus and transmitting back scientific readings, as well as a limited number of television pictures. Venera probes were also the first to orbit Venus in the 1970s. The American Magellan orbiter mapped Venus from 1990 to 1994 with radar, penetrating the mystery of the ever present clouds. The lack of many craters led scientists to conclude that widespread volcanism had resurfaced most of the planet about 500 million years ago. Magellan also found no evidence of plate tectonics.

Mariner 4 flew by Mars on July 15, 1965, coming within 6,200 miles of the surface. A television camera took 21 pictures. The murky pictures transmitted back to Earth surprised astronomers by revealing a cratered surface much like Earth's Moon. By analyzing the strength of *Mariner 4*'s radio signals as they passed through the Martian atmosphere, scientists determined the surface pressure of the atmosphere. This measurement revealed a very thin atmosphere, with a surface pressure equal to less than 1 percent of Earth's atmosphere at sea level. The atmosphere contained mostly carbon dioxide, with only a tiny amount of water vapor. *Mariner 6* and *Mariner 7* passed by Mars on July 31, 1969, and August 5, 1969, respectively. Passing as close as 2,100 miles from the surface, *Mariner 6* returned 75 pictures, and *Mariner 7* returned 126 pictures. Even with these pictures, only a tenth of the planet's surface had been photographed by spacecraft. The probes confirmed the low pressure of the atmosphere and determined that Mars did not possess a magnetic field. A temperature reading of the southern polar cap returned a measurement of −190 degrees Fahrenheit (−124 degrees Celsius). Despite the cold, cratered surface, scientists held out hope that Mars had possessed a denser atmosphere during an earlier epoch millions of years ago and that perhaps life had thriven then.

Mariner 9 entered orbit around Mars on November 13, 1971. A great dust storm obscured the surface of the planet, and the probe orbited for a month before clear pictures were possible. The Soviets made numerous attempts to send probes to Mars, all of which failed for various reasons until *Mars 2* and *Mars 3* reached Mars in 1971 shortly after *Mariner 9*. Both Soviet probes dropped descent modules to examine the surface, then went into orbit around the planet. Both descent modules failed, and the Soviet orbiters circled the planet, operating on automatic programs, taking futile pictures of the dust storm. Ground controllers on Earth reprogrammed *Mariner 9* to wait until the dust storm cleared before beginning research in earnest. Among the 7,239 pictures returned were what looked like ancient dry riverbeds and runoff channels, suggesting that free-flowing water must have existed on Mars in the past. The many discoveries included Valles Marineris, a great canyon that stretched a quarter of the way around the planet.

In 1976, two American Viking orbiters reached Mars and each released a lander. These landers returned pictures from the surface of the planet showing a red, rocky landscape. Onboard laboratories tested the Martian soil and failed to find traces of bacterial life. In 1996, a meteorite from Mars recovered from the ice of Antarctica contained microscopic structures that some scientists interpreted as microfossils from a past era when Mars had been younger, wetter, and more congenial to life. The meteorite was designated ALH84001, being the first one recovered from the Allan Hills in Antarctica in 1984. Other scientists have disputed the evidence from the meteorite, but hold out hope that fossils will be found by some future expedition to the red planet.

In the 1990s, NASA started to build cheaper interplanetary probes with fewer capabilities and more limited scientific instruments, unlike their early more expensive spacecraft, which had more redundant systems aboard and a full range of instruments. The result was more failures of NASA spacecraft, especially some high-profile Mars probes in the 1990s. A success came with the Mars Pathfinder mission in 1997, which landed the robotic rover called *Sojourner*. In 2004, two further rovers were successfully landed to search for geological evidence of water and past life.

Mariner 10 flew by Venus on February 5, 1974, then went into orbit around the Sun. This orbit passed Mercury, the planet closest to the Sun, three times, on March 29, 1974, September 21, 1974, and March 16, 1975. The probe's instruments were specially hardened to withstand the fierce radiation of the Sun, and a sunshade, shaped like an umbrella, also protected the spacecraft. *Mariner 10* found that Mercury possessed a slight magnetic field and no atmosphere. The first flyby took the probe within 435 miles of the surface of the planet's dark side. The second flyby went past the sunlit side of the planet. The final encounter passed only 125 miles from the dark side. Pictures returned by the

The seventh space shuttle launch, Cape Canaveral, Florida, June 18, 1983 (Corbis)

probe showed a surface covered with craters and ancient lava flows. No other U.S. or Soviet spacecraft have visited Mercury.

The Americans also launched probes to investigate the outer planets. *Pioneer 10* reached Jupiter in 1973, taking over three hundred pictures. *Pioneer 11* reached Jupiter a year later and used the planet's gravity to propel it toward an encounter with Saturn. Both of the Pioneer craft eventually left the solar system, and each carried a plaque of greeting in case an alien intelligence should encounter the craft at some unknown time in the future. *Voyager 1* and *2* also toured the gas giants before leaving the solar system, each carrying a 12-inch, gold-plated copper disk with 115 images from Earth on it and a selection of sounds, including natural sounds and voice greetings in fifty-five languages.

The astronomer Carl Sagan coordinated the selection of the material.

The *Galileo* spacecraft arrived around Jupiter in December 1995. On its way to the gas giant, it flew by two different asteroids, and discovered that the second asteroid, named Ida, had its own moon. Ida was only 34 miles (55 kilometers) wide at its longest axis, and the moon was under a mile (1.5 kilometers) in diameter. *Galileo* also took pictures of the Shoemaker-Levy comet impact on Jupiter in July 1994. *Galileo* released a descent probe into Jupiter that used the atmosphere to break its fall before discarding its heat shield and deploying a parachute. The probe radioed back data until pressure within the atmosphere of Jupiter crushed it. *Galileo* itself orbited around Jupiter, making repeated flybys of the gas giant's moon. Evidence for liquid water under a surface of frozen ice on Europa and other large moons was discovered, and scientists hold out hope that primitive life-forms have evolved there. Running low on maneuvering fuel, *Galileo* was plunged into the atmosphere of Jupiter in 2003, sending back data as it succumbed to radiation and pressure. *Galileo* was not sterilized before launch, and scientists were concerned that if they allowed the spacecraft to continue to orbit, without any way of maneuvering, it might crash into Europa and contaminate the ocean with Earth microbes.

Other spacecraft from the United States, the Soviet Union (now Russia), and the European Space Agency (ESA) have examined comets, asteroids, and the Sun. NASA, the ESA, and the Italian Space Agency launched the Cassini mission in 1997. In 2004, *Cassini* is scheduled to go into orbit around Saturn's moon, Titan, and release the Huygens probe, which will descend by parachute to Titan's surface.

See also Apollo Project; Big Science; Cold War; Computers; European Space Agency; International Geophysical Year; National Aeronautics and Space Administration; Plate Tectonics; Sagan, Carl; Satellites; Search for Extraterrestrial Life; Shoemaker-Levy Comet Impact; Soviet Union; Voyager Spacecraft

References
Beatty, J. Kelly, Carolyn Collins Peterson, and Andrew Chaikin, editors. *The New Solar System.* Fourth edition. New York: Cambridge University Press, 1999.
Burrows, William E. *Exploring Space: Voyages in the Solar System and Beyond.* New York: Random House, 1990.
Fischer, Daniel. *Mission Jupiter: The Spectacular Journey of the Galileo Spacecraft.* New York: Copernicus, 2001.
Gatland, Kenneth. *Illustrated Encyclopedia of Space Technology: A Comprehensive History of Space Exploration.* New York: Harmony, 1981.
Murray, Bruce. *Journey into Space: The First Three Decades of Space Exploration.* New York: Norton, 1989.
Sheehan, William. *The Planet Mars: A History of Observation and Discovery.* University of Arizona Press, 1996.
Walter, Malcolm. *The Search for Life on Mars.* Cambridge, MA: Perseus, 1999.

Steptoe, Patrick (1913–1988)

Patrick Steptoe and Robert Edwards (1925–) perfected the process of in vitro fertilization (IVF) and created the first human test-tube baby. Patrick Christopher Steptoe was born in Witney, Oxfordshire, England, where his father was a church organist and his mother was a social worker. His mother taught him perseverance in the face of disappointment, a trait that served him well later. In 1939, Steptoe graduated from the University of London's Saint George's Hospital Medical School, specializing in obstetrics and gynecology. With the coming of World War II, Steptoe joined the Royal Naval Reserve as a surgeon. When his ship was sunk in the Battle of Crete in 1941, Steptoe was taken prisoner by the Italians. He was exchanged later in the war.

After the war, he completed further postgraduate study in his specialty and set up a practice in Oldham General and District Hospital in Oldham, near Manchester. He worked for five years to perfect his ability with the laparoscope, a narrow tube with a fiber-optic light at the end that allowed a surgeon to peer into the abdominal cavity. Steptoe became the first surgeon in England to learn to use this new instrument and perfected its use despite much criticism from his colleagues. This skill helped Steptoe determine the causes of infertility and perform difficult operations. Steptoe published *Laparoscopy in Gynaecology* in 1967.

In 1968, Steptoe and Robert Edwards, a physiologist at Cambridge University, began to work together. Edwards had perfected a technique for in vitro fertilization, that is, fertilizing human eggs with donated sperm in the laboratory in a petri dish. A major cause of infertility in women is when blockage of the fallopian tubes or oviducts prevents eggs from reaching the uterus. Steptoe used his skill to extract mature eggs from the ovaries of volunteers, and Edwards fertilized them in the laboratory. In 1972, Steptoe implanted a fertilized egg into a uterus for the first time. None of the approximately thirty women in their test group who became pregnant carried the babies for more than a trimester. Funding for their experiments was cut amid a general outcry from Members of Parliament, religious leaders, and other scientists. Among the objections was fear that a successful technique would not stop at helping infertile couples, but would lead to genetically engineered babies and the use of surrogate mothers. Steptoe financed the continuation of their research from his practice, which included legal abortions, and the two scientists moved their work to a smaller hospital in Oldham, Dr. Kershaw's Cottage Hospital.

Continued criticism from their colleagues caused Steptoe and Edwards to suspend reporting on their research. They persisted amid many failures until late 1977, when Lesley Brown was impregnated with her own egg, fertilized by her husband's sperm. Steptoe and Edwards decided to implant the egg after it had divided only three times and was only eight cells large, instead of waiting for it to get larger as they had done up until then. On July 25, 1978, a daughter named Louise Joy was delivered by Caesarian section. The world press proclaimed the world's first test-tube baby.

Steptoe and Edwards continued their collaboration, and their clinic succeeded in producing over a thousand babies for infertile couples. Their technique of in vitro fertilization and successful implantation of the egg into a uterus is now used throughout the industrialized world. Surrogate mothers have also carried to term babies created with eggs that are not their own, leading to legal issues.

In his personal life, Steptoe's marriage led to a son and a daughter. Steptoe was elected a Fellow of the Royal Society in 1987 a year before his death.

> *See also* Biotechnology; Edwards, Robert; In
>   Vitro Fertilization; Medicine
> **References**
> Edwards, Robert G. "Patrick Christopher Steptoe,
>   C.B.E., 9 June 1913—22 March 1988:
>   Elected F.R.S. 1987." *Biographical Memoirs of the*
>   *Fellows of the Royal Society* 33 (1987): 213–233.

Symbiosis

Symbiosis, from the Greek word for "living together," is defined as the intimate association of two different types of organisms. The German botanist Heinrich Anton de Bary (1831–1888) first used this term at a meeting at the University of Strasbourg in 1878. He described lichens as an example of symbiosis: lichens consist of fungus and algae mutually existing together, with the algae providing food through photosynthesis and the fungi providing shelter. Although the term *symbiosis* is often used as though it were simply synonymous with the term *mutualism,* which describes a situation in which both symbionts benefit from their relationships, symbiosis also includes commensalism, in which two species coexist but are not as intimately associated with each other as in mutualism, and parasitism, in which one of the symbionts benefits at the expense of the other symbiont.

In his 1927 book, *Symbioticism and the Origin of Species,* Ivan E. Wallin (1883–1969) proposed the term *symbiogenesis,* to describe a situation where new species are created when older species combine on the microbial level. Other scientists condemned symbiogenesis as Lamarckianism, the discredited theory of the inheritance of acquired characteristics, as opposed to the Darwinian assertion that traits are randomly acquired by mutation and then succeed or fail through natural selection.

Interest in symbiosis was revitalized in the late 1960s by the microbiologist Lynn Margulis (1938–). Margulis published her serial endosymbiosis theory (SET) in 1967, and followed with a book, *Origin of Eukaryotic Cells,* in 1970. Margulis proposed that eukaryotic cells, which contain nuclei, evolved when non-nucleated bacteria fused together billions of years ago. She also argued that chloroplasts, which convert sunlight into energy via photosynthesis, were originally free-living microbes that symbiotically merged with other cells. As evidence accumulated in the 1970s, SET came to be accepted, though Margulis's contention that most evolutionary

Red polypores fungus grows on a tree. (Hal Horwitz / Corbis)

Queen hypothesis, first proposed in the 1970s and based on an allusion to the Red Queen in the novel *Alice in Wonderland,* hosts and parasites must continually evolve in order to survive. If it fails to evolve to meet new challenges from parasites or other species, a species will die off.

The founding of the International Symbiosis Society in 1997, along with growth in professional journals and conferences devoted to symbiosis, has laid the institutional framework for interdisciplinary work. Such work, by emphasizing the role of symbiosis in coevolution, has begun to change theories of evolution from a strictly Darwinist approach, which sees species as distinct and unique, to a view that sees microbes as often exchanging genes and microbial species boundaries as more fluid. Now scientists accept that microbes can transmit induced characteristics, and that organelles are really former bacteria that have fused with other cells.

See also Evolution; Genetics; Margulis, Lynn; Zoology

References

Margulis, Lynn. *Symbiotic Planet: A New Look at Evolution.* New York: Basic, 1998.

Margulis, Lynn, and Dorion Sagan. *Acquiring Genomes: A Theory of the Origins of Species.* New York: Basic, 2002.

———. *Microcosmos: Four Billion Years of Evolution from Our Microbial Ancestors.* New York: Summit, 1986.

Paracer, Surindar, and Vernon Ahmadjian. *Symbiosis: An Introduction to Biological Associations.* Second edition. New York: Oxford University Press, 2000.

Sapp, Jan. *Evolution by Association: A History of Symbiosis.* New York: Oxford University Press, 1994.

Wakeford, Tom. *Liaisons of Life: From Hornworts to Hippos, How the Unassuming Microbe Has Driven Evolution.* New York: Wiley, 2001.

change is the result of symbiogenesis is not widely accepted. Margulis expanded on symbiosis by proposing in her 1981 *Symbiosis in Cell Evolution* that cilia and other locomotion appendages of microbes originally came about when spirochetes merged with archaebacteria. This hypothesis is not widely accepted. If symbiosis is important in evolution, what is the role of sexual reproduction? Sexual reproduction mixes and combines the genes of parents to produce unique children, accelerating the pace of evolution.

In the continuous struggle between hosts and parasites, evolution keeps changing the nature of the host and parasites. In the Red

T

Telescopes

Ever since their invention in 1608, telescopes have extended the human senses into the cosmos, and remarkable technical advances since 1950 have unveiled new wonders. In 1948, the world's largest optical telescope was completed at Mount Palomar in California, with a reflecting mirror 5 meters in diameter. In 1976, the Soviets built a problem-plagued 6-meter reflector at Selencukskaja in the Caucasus Mountains. Creating even larger mirrors that would not collapse under their own weight proved difficult, and in the 1970s other technologies were developed for optical telescopes. Instead of using a single reflecting mirror, multiple mirrors were carefully configured to create the effect of a much larger telescope. The first effort, the Multiple Mirror Telescope on Mount Hopkins in Arizona, was finished in 1979 and used six 1.8-meter mirrors. The equivalent single collector mirror would have had to be 4.5 meters in diameter. The Keck telescope on Mauna Kea in Hawaii used thirty-six 1-meter mirrors to create the equivalent of a 10-meter telescope.

Another innovative technology for optical telescopes is active optics, in which thin mirrors are continually adjusted. An example of this is the 3.6-meter telescope at the European Southern Observatory (ESO) in La Silla, Chile, built in 1976. The ESO also introduced adaptive optics in 1989, which adjust the image with a computer-controlled mirror to remove the twinkle effect from light coming through the atmosphere. The proposed Very Large Telescope (VLT) of the ESO at Paranal Observatory in the Atacama desert in northern Chile will have four 8.2-meter mirrors, and will use active and adaptive optics to create the equivalent of a 16.4-meter telescope.

Computers have revolutionized astronomy by allowing tighter control of telescopes to help in analyzing and enhancing images. Charge-coupled devices (CCDs) have replaced film as a way to collect the images, leading to much more accurate images, since they can collect ten times the light that good film can record. Astronomers have also used optical telescopes to go beyond the range of the human eye to examine the cosmos in the infrared and ultraviolet ranges, though the atmosphere makes such observations difficult from the ground.

Astronomers recognized long before the space age that if they could get their telescopes beyond the atmosphere, they would see so much more. The premier satellite telescope is the Hubble Space Telescope (HST), named after the important American astronomer Edwin Powell Hubble (1889–1953); it was launched from a space shuttle in 1990. The HST is very much like the spy satellites developed by the United States and Soviet Union during the cold war, except that it is looking out instead of down at Earth. The engineers working on the HST received minimal help from their colleagues in the military because of the classified nature of the spy satellites. A major issue was overcoming problems with the extreme differences in temperature between the heat caused by direct sunlight and the cold caused by shadows in space. The technical marvel came with another problem: a flaw in the mirror had to be compensated for with computer processing until a repair mission by astronauts from another space shuttle added additional optics and electronics. The HST revolutionized astronomy, with clearer pictures than any possible from the ground. The HST greatly increased the number of estimated galaxies, increased the known size of the universe, and provided compelling photographs of the Shoemaker-Levy comet impact on Jupiter in 1994.

Another important extension of human perception came with the development of radio telescopes. Radio astronomy began in the 1930s when scientists recognized that radio signals, just another part of the electromagnetic spectrum, came from outer space. The skills acquired during World War II, especially in work with radar signals, proved useful when astronomers turned to radio astronomy after the war. In the 1950s and 1960s, ever larger steerable dishes were built to pick up signals from the sky. Steerable dishes allowed the radio telescopes to follow a radio source across the sky. The largest dish ever built was not steerable; it was 305 meters in diameter, and filled a small valley near Arecibo, Puerto Rico. The Cambridge University astronomer Martin Ryle (1918–1984) designed a way for com-

bining multiple smaller dishes into a configuration to create the equivalent of a much larger radio telescope. The largest of these observatories was built in New Mexico in the 1970s and consisted of twenty-seven steerable dish telescopes organized into three long arms. Radio astronomy led directly to the discovery of quasars and pulsars, and contributed to the confirmation of the existence of black holes.

X-ray telescopes in space have constituted another important development. X-rays are absorbed by Earth's atmosphere, so early efforts after World War II to detect x-rays from outer space relied on quick peeks with instruments mounted atop rockets. In 1970, *Uhuru,* the first satellite with an x-ray observatory, was launched, its observatory based on the use of an ionization chamber. The American space station *Skylab,* launched in 1973, contained an x-ray mirror telescope and was used to study x-rays from the Sun. American, European, and Japanese satellites have continued the effort since then. Satellite observatories have also examined the visible, infrared, and ultraviolet areas of the spectrum.

See also Astronomy; Black Holes; Pulsars; Quasars; Satellites; Shoemaker-Levy Comet Impact; Space Exploration and Space Science

References

Chaisson, Eric J. *The Hubble Wars: Astrophysics Meets Astropolitics in the Two-Billion-Dollar Struggle over the Hubble Space Telescope.* New York: HarperCollins, 1994.

Edge, David O., and Michael J. Mulkay. *Astronomy Transformed: The Emergence of Radio Astronomy in Britain.* New York: Wiley, 1976.

Preston, Richard. *First Light: The Search for the Edge of the Universe.* New York: Atlantic Monthly, 1987.

Smith, Robert W., and others. *The Space Telescope: A Study of NASA, Science, Technology, and Politics.* New York: Cambridge University Press, 1989.

Tucker, Wallace, and Karen Tucker. *The Cosmic Inquirers: Modern Telescopes and Their Makers.* Cambridge: Harvard University Press, 1986.

———. *Revealing the Universe: The Making of the Chandra X-Ray Observatory.* Cambridge: Harvard University Press, 2001.

Edward Teller, physicist and "father of the hydrogen bomb" (Library of Congress)

Teller, Edward (1908–2003)

The physicist Edward Teller championed the design and construction of a successor to the atomic bomb and became known as the "father of the hydrogen bomb." Ede Teller was born in Budapest, Hungary, then part of the Austro-Hungarian Empire, where his father was a lawyer. His mathematical ability was noticed early, and he graduated from gymnasium in 1925. He attended Karlsruhe Technische Hochschule for two years, studying chemical engineering to satisfy his father's desire that he learn a practical profession, while he also pursued his interests in physics and mathematics. With the development of quantum mechanics, his interest quickened, and his desire to drop chemical engineering prompted his father to talk to his professors. The professors convinced the elder Teller that his son would be able to find employment as a physicist. Teller began studies of physics at the University of Munich in 1928. When a street trolley severed his foot, he delayed his studies to recover and learn to use a prosthetic. He then moved to the University of Leipzig to study with the physicist Werner Heisenberg (1901–1976) and obtained his doctorate in theoretical physics in 1930 at the age of twenty-two. After another year at Leipzig, he worked as a research assistant at the University of Göttingen for two years.

When the Nazis came to power, Teller realized that his Jewish heritage put him at risk, and he moved to Denmark to study with the physicist Niels Bohr (1885–1962) for a year. Teller married in 1934 and fathered a son and a daughter. In Copenhagen Teller met the Russian-born George Gamow (1904–1968). Gamow invited Teller to come to George Washington University in Washington, D.C., as a professor of physics. In his six years at George Washington University, from 1935 to 1941, Teller turned from his research on applying quantum mechanics to physical chemistry to studying nuclear physics. Gamow and Teller collaborated to create the Gamow-Teller selection rules for beta decay.

Teller adopted American citizenship in 1941 as he became involved in the secret atomic research that grew into the Manhattan Project. He worked initially at the University of Chicago, then joined the theoretical study group led by J. Robert Oppenheimer (1904–1967) at the University of California, Berkeley. When Oppenheimer founded the Los Alamos Laboratory in New Mexico, Teller followed him there. In late 1941, conversations between Enrico Fermi (1901–1954) and Teller led to speculation that a more powerful type of bomb, a super-bomb based on fusion instead of fission, was possible. Teller advocated research into fusion bombs as well as fission bombs, but Oppenheimer chose to concentrate on the technically less challenging fission bomb.

Though disturbed by the destruction that the atomic bombs wrought on Hiroshima and Nagasaki, Teller did not question the morality of working on the bomb. Fascism was too great a threat. After three years at the University of Chicago, Teller returned to Los Alamos in 1949 as an assistant director. When the Soviets demonstrated their own atomic bomb in 1949, the American government decided to pursue the hydrogen bomb based on fusion. Teller led this effort. Flawed calculations by Teller initially delayed the super-bomb project, but eventually other scientists corrected the errors. In 1951, Stanislaw Ulam (1909–1984) and Teller solved the major technical problem by electing to use x-rays generated by an atomic bomb trigger to start the fusion reaction in the hydrogen bomb. The first hydrogen bomb was exploded in 1952 on the atoll of Eniwetok, and Teller became the father of the hydrogen bomb.

After a brief return to the University of Chicago, Teller convinced the Atomic Energy Commission (AEC), which set civilian and military nuclear policy for the nation, to establish the Lawrence Livermore Laboratory at the University of California at Berkeley. This laboratory combined important civilian work in nuclear physics with military contracts to design new generations of nuclear weapons. Teller was the associate director of the laboratory from 1954 to 1958 and from 1960 to 1975, with a brief two-year stint as director from 1958 to 1960.

A political conservative and ardent anti-communist, Teller encouraged the United States to build an ever larger nuclear arsenal to counter the Soviet Union. He testified against Oppenheimer in the infamous 1954 AEC hearing that resulted in Oppenheimer losing his security clearance and being forced from government service. This testimony caused a rift in the physics community, one group considering Teller something of a hero, the other a pariah. Teller also supported the civilian use of nuclear energy, including the use of atomic bombs to build canals and in other large excavation projects, though such plans were later abandoned.

In the 1980s Teller became a vocal advocate of the Reagan Administration's Strategic Defense Initiative (SDI), commonly referred to as "Star Wars." SDI was not popular with many physicists, an attitude that Teller blamed on fallout from the Oppenheimer decision rather than the technical difficulties inherent in the idea. In 1987 he published *Better a Shield Than a Sword: Perspectives on Defense and Technology* in defense of SDI. Teller received the 1962 Enrico Fermi award from the AEC and the National Medal of Science in 1982.

See also Big Science; Cold War; Ethics; Gamow, George; Hydrogen Bomb; Nuclear Physics; Oppenheimer, J. Robert; Sakharov, Andrei

References

Blumberg, Stanley A., and Louis G. Panos. *Edward Teller: Giant of the Golden Age of Physics.* New York: Charles Scribner's Sons, 1990.

Teller, Edward, with Judith Shoolery. *Memoirs: A Twentieth-Century Journey in Science and Politics.* Cambridge, MA: Perseus, 2001.

Third World Science

The term *Third World* came to be used during the cold war to describe the industrially underdeveloped nations not aligned with either the United States and its allies or the Soviet Union and its allies. In the 1990s it became more common to refer to the north-south divide, since most of the developed nations are located in the Northern Hemisphere and most of the underdeveloped nations are in the Southern Hemisphere. Of course there are exceptions, such as the countries of Central America, Australia and New Zealand, and the Middle East. Third World science can be defined as scientific activities that affect the Third World or the science practiced by scientists based in the Third World.

At the end of World War II, most Third World nations were colonies of Western European nations. In the next decade, most colonies achieved independence, sometimes as a result of peaceful action and sometimes after armed revolt. Most Third World nations struggled with burgeoning populations, underdeveloped economies, and underdeveloped educational systems. Students who sought higher education usually traveled to the universities in the United States or Western Europe. During the height of the cold war, scholarships to universities in the Soviet Union and China supported students from nations that the communists sought to influence. The United States and its allies also offered scholarships for similar purposes. For example, in 1985–1986, 42 percent of the physics graduate students in American universities came from foreign countries.

The career of one of the rare examples of a Third World scientist earning a scientific Nobel prize is instructive. Abdus Salam (1926–1996) grew up in Pakistan, where his father worked as an official in the Department of Education. Raised as a Muslim, he remained devout his entire life. At the age of fourteen he earned the highest scores ever recorded on entrance exams for the University of the Punjab, where he earned a master's degree in 1946. He traveled to England in order to attend Saint John's College at Cambridge University. Pakistan gained its independence from England while Salam was studying theoretical physics, earning a doctorate in 1951. Salam returned to Pakistan, intent on creating a school of research, but found that he needed to return to Cambridge to continue his own research. In 1957 he became a professor of theoretical physics at Imperial College, London. Salam made important contributions in the 1960s to combining the weak nuclear and electromagnetic forces into the electroweak theory of quantum electrodynamics. For this work he shared the 1979 Nobel Prize in Physics with Steven Weinberg (1933–) and Sheldon Lee Glashow (1932–).

Salam founded the International Centre for Theoretical Physics (ICTP) in Trieste, Italy, in 1964, later renamed the Abdus Salam International Centre for Theoretical Physics, as a place for Third World scientists to come during their summers so that they might engage in research and social networking with their peers before returning to their home universities for the school year. The Italian government funded the center, under the sponsorship of the United Nations Educational, Scientific and Cultural Organization (UNESCO) and the International Atomic Energy Agency (IAEA). Salam devoted his Nobel prize money to his quest to help Third World science.

Salam founded the Third World Academy of Sciences (TWAS) in 1983, with headquarters at the ICTP. Like other international and national academies, TWAS awarded prizes, offered small research grants, provided a forum for scientists to communicate, and published a regular newsletter, proceedings of its conferences, and monographs on Third World science. Somewhat more unusually for a scientific academy, TWAS also offered a program to supply spare parts for equipment and donations of scientific books and periodicals to Third World countries.

Besides basic scientific research, Third World scientists often concentrate on urgent issues within their countries, such as environmental degradation, health care problems, and agriculture. Although the scientists who launched the Green Revolution of the 1950s through 1970s used Third World locations, like the International Rice Research Institute in the Philippines at Los Banos, to conduct their research, the researchers were usually from First World nations.

See also Associations; Cold War; Glashow, Sheldon L.; Green Revolution; Nobel Prizes

References

"Abdus Salam—Biography." http://nobelprize. org/physics/laureates/index.html (accessed September 18, 2004).

Selin, Helaine. Encyclopaedia of the History of Science, Technology, and Medicine in Non-Western Cultures. Boston: Kluwer Academic, 1997.

Third World Academy of Sciences. http://www. ictp.trieste.it/~twas/index.html (accessed February 13, 2004).

Transistors
See Computers

Tsui, Daniel C. (1939–)

The Chinese-born experimental physicist Daniel C. Tsui and a colleague discovered the fractional quantum Hall effect. Daniel Chee Tsui was born in an impoverished village in Henan Province, China, where his illiterate parents hoped for something better than a life of drought, flood, and war. They managed to send him to Hong Kong in 1951, where he began his formal schooling. After a year and a half, he entered Pui Ching Middle School. He

was fortunate in that many of the teachers were educationally overqualified for their positions, a consequence of the shortage of university and research jobs after years of civil war in China and the additional disruptions of World War II. Tsui graduated in 1957, and though he had been accepted into a medical school at the National Taiwan University, he hesitated to leave Hong Kong before knowing how his parents were doing. The following year, his Lutheran pastor arranged for him to enter Augustana College in Rock Island, Illinois. The awarding of the 1957 Nobel Prize in Physics to Chen Ning Yang (1922–) and Tsung-Dao Lee (1926–) for their work in elementary particles inspired Tsui and many other Chinese students to realize what was possible. Tsui remained in the United States, intent on attending graduate school, and since Yang and Lee both had attended the University of Chicago, Tsui chose to attend the same university for his graduate studies. After graduating from Augustana College in 1961, Tsui moved to Chicago, earning his Ph.D. in physics in 1967. Tsui married an undergraduate at the University of Chicago in 1964 and fathered two children. He later became a U.S. citizen.

Tsui joined the research staff of the famed Bell Laboratories at Murray Hill, New Jersey, where he worked on solid-state physics and devoted himself to studying the Hall effect. Edwin H. Hall (1855–1938) discovered in 1879 that a transverse electric field is created when electrical current passes through a solid material. The power of the effect is directly related to the force of the magnetic field present. In 1980, the German physicist Klaus von Klitzing (1943–) created a Hall effect in silicon at temperatures near absolute zero and found that the electrical current changed in steps as the magnetic field was changed. This quantized Hall effect earned Klitzing the 1985 Nobel Prize in Physics.

The German-born physicist Horst L. Störmer (1949–) came to Bell Labs in 1977 and partnered with Tsui on his projects. They

Daniel C. Tsui, experimental physicist (Rick Maiman/Corbis Sygma)

started to examine the quantum Hall effect using a gallium-arsenide-based semiconductor and the powerful magnet at the Francis Bitter Magnet Laboratory at the Massachusetts Institute of Technology (MIT). In 1982, they discovered the fractional quantum Hall effect when they found that electrical current changes in even smaller steps than Klitzing had found, even fractional to the charge of an electron. The American theoretical physicist Robert B. Laughlin (1950–), also at Bell Labs, provided a theoretical explanation for the fractional quantum Hall effect. Laughlin speculated that at near-zero temperatures, and in a high magnetic field, electrons behave like a fluid rather than particles. Within a few months of their discovery, Tsui realized an ambition to teach by becoming a professor of electrical engineering at Princeton University in 1983 and remains active in teaching and research. Tsui shared the 1998 Nobel Prize in Physics with Störmer and Laughlin for their discovery.

See also Nobel Prizes; Particle Physics
References
"Daniel C. Tsui—Autobiography." http://www. nobelprize.org/physics/laureates/1998/ tsuiautobio.html (accessed September 18, 2004).
"Horst L. Störmer—Autobiography." http:// nobelprize.org/physics/ (accessed September 18, 2004).

Tsui, Lap-Chee (1950–)
The Chinese-born molecular geneticist Lap-Chee Tsui led one of the teams that identified the gene that causes cystic fibrosis (CF). Lap-Chee Tsui was born in Shanghai, China, and grew up in Hong Kong. He earned a B.Sc. in biology in 1972 and an M.Phil. in biology in 1974, both from the Chinese University of Hong Kong. Coming to the United States, he earned his Ph.D. in biological sciences in 1979 from the University of Pittsburgh. After a year as a postdoctoral investigator at the Oak Ridge National Laboratory, Tsui joined the Department of Genetics at the Hospital

for Sick Children in Toronto, Ontario, where he began to work on CF. In 1985, Tsui helped identify a genetic marker for CF in chromosome 7.

A joint group headed by physician-geneticist Francis Collins (1950–) at the University of Michigan and Tsui and John Riordan at the Hospital for Sick Children pursued the gene or genes that caused the disorder. Their efforts were funded by private and public sources: the Howard Hughes Medical Center, Cystic Fibrosis Foundation, Canadian Cystic Fibrosis Foundation, and National Institutes of Health. In 1989, the group found success and spent five months confirming their results (a single gene) before publishing their discovery in three articles in the same issue of *Science*. The discovery of this gene is the necessary first step to eventually overcoming the disease, and Tsui has remained active in organizing international research in this area. Later research has showed that the CF gene is part of the genetic instructions to create a protein called the cystic fibrosis transmembrane regulator (CFTR).

Tsui participated in the Human Genome Project, eventually serving as president of Human Genome Organization, the international collection of scientists involved in the Human Genome Project. Tsui received many honors, including an appointment as university professor in the Department of Medical Genetics and Microbiology at the University of Toronto, election as a Fellow of the Royal Society in London and of the Royal Society of Canada, and numerous honorary degrees. Tsui is married and is the father of two sons. Tsui later became active in scientific efforts in Hong Kong, in addition to his activities in Canada, becoming head of the Genome Research Centre at the University of Hong Kong. In 2002, Tsui became the vice-chancellor of the University of Hong Kong.

See also Biotechnology; Foundations; Genetics; Human Genome Project; Medicine; National Institutes of Health

References
"Biography of Lap-Chee Tsui." http://www.genet.sickkids.on.ca/~lapchee/bio.html (accessed September 18, 2004).
"Lap-Chee Tsui." http://www.genet.sickkids.on.ca/~lapchee/ (accessed September 18, 2004).

U

Undersea Exploration

Up until the twentieth century, much of what occurred under the waters of the seas of Earth remained a mystery. Then technological innovations allowed scientists to explore under the oceans. Initially deployed as instruments of war, submarines allowed exploration of the ocean depths. After World War II, engineers built specialized submarines called bathyscaphes to survive the intense pressures of the deep ocean. A pioneer in bathyscaphes was the Swiss-born Belgian physicist Auguste Piccard (1884–1962), who had set the world altitude record in a balloon in 1932. Having turned his attention to the sea, Piccard and his son, Jacques Piccard (1922–), built a bathyscaphe called the *Trieste* after World War II. The French bathyscaphe *FNRS-3* and *Trieste* competed in 1953 with deep dives into the Mediterranean. In dives below 10,000 feet (3,000 meters), the *FNRS-3* and *Trieste* discovered previously unknown species and unusual geological features, despite their limited ability to move laterally underwater. The U.S. Navy purchased the *Trieste,* and the younger Piccard accompanied the U.S. naval officer Don Walsh (1931–) on a record-setting dive to the deepest part of the ocean. On January 23, 1960, the *Trieste* reached a depth of 35,800 feet (10,912 meters) in the Challenger Deep of the Mariana Trench, located near Guam in the Pacific Ocean. The Challenger Deep had been discovered in 1951 by the British research ship *Challenger II,* and was not again visited by scientists until a Japanese robotic submersible explored the Deep in 1995.

The next step into undersea exploration was to move beyond the ungainly bathyscaphes to create deep-sea submersibles. The *Alvin,* the first of these craft, was built in 1964 by the Woods Hole Oceanographic Institution and the Office of Naval Research. In 1977, the *Alvin* discovered deep-sea hydrothermal vents at about 2,700 meters in the Galapagos Rift near the Galapagos Islands. The unique ecology surrounding the vents includes large tubeworms, giant clams, and other new species. Bacteria around the vents use chemosynthesis to create energy, converting the sulfides into organic carbon. The more complex life-forms form symbiotic relationships with the chemosynthesis bacteria and survive off them. Besides the small scientific submersibles that carried a few crew members, unmanned robotic submersibles called remotely operated vehicles (ROV) were developed in the 1970s and have proved a low-cost alternative to manned submersibles and are able to penetrate depths that the manned craft do not normally reach. In 1990, the Japanese finished the *Shinkai*

The deep-sea submersible Alvin *(Ralph White / Corbis)*

6500, the world's deepest-diving manned submersible. With the end of the cold war, American military and Russian military submersibles and resources began to be used by civilian scientists. Also, previously classified information, such as sea floor gravity data collected by the U.S. Navy, was released.

In 1943, the French naval officer Jacques Cousteau (1910–1997) successfully developed his "Aqua-Lung," based on the invention of an automatic gas-feeder valve by Émile Gagnon. This scuba (*s*elf-*c*ontained *u*nderwater *b*reathing *a*pparatus) gear allowed humans to freely venture into shallow waters. Diving before scuba was confined to using inverted bells for air supply, physical conditioning to hold one's breath during a free dive, or hardhat diving that obtained oxygen through hoses from the surface.

The development of saturation diving in the late 1950s by the U.S. Navy allowed divers to remain underwater for days at a time, decompressing only once at the end of the dive. Saturation diving required divers to have undersea habitats to live in. The 1960s and 1970s saw numerous habitats developed around the world to serve as observation stations, laboratories for oceanographers, and homes for divers. Among the more well-known habitats were *Conshelf* (France), *Chernomor* (Soviet Union), *Helgoland* (West Germany), and the U.S. *Sealab, Tektite, Hydrolab,* and *Aquarius* habitats. The success of these habitat programs encouraged some futurists to predict a future of undersea cities and vigorous exploitation of the ocean's resources by a new underwater civilization. The number of habitats faded until only three were in use by 2000.

The deep-sea drilling ship *Glomar Challenger* was launched in 1968, funded by the National Science Foundation (NSF). This ship traveled the world as part of the Deep Sea Drilling Project, drilling holes around the world and collecting core samples. These samples definitively proved the existence of continental drift, hypothesized by the theory of plate tectonics. In 1970, the *Challenger* crew used sonar and a reentry cone to successfully reenter a previously drilled hole, a technique that allowed replacing worn drill bits. The *Challenger* retired in 1983 after extracting over 19,000 core samples. In 1985, the Ocean Drilling Program (ODP) and Joint Oceanographic Institutions for Deep Earth Sampling (JOIDES) converted an oil exploration vessel into a drilling ship, the *JOIDES Resolution,* to continue the work of the *Challenger.*

Institutional support of undersea exploration has come from private institutions like the Scripps Institution in California, Woods Hole in Massachusetts, Bedford Institute in Nova Scotia, Alfred-Wegener Institute in Bremerhaven, Germany, and P.P. Shirshov Institute in Moscow. Government organizations, such as the National Science Foundation (NSF); the American National Oceanic and Atmospheric Administration (NOAA), founded in 1970; and their counterparts in Japan, Russia, and various European countries, have also supported undersea exploration.

Undersea exploration has been driven by more than scientific curiosity. The sea is a major source of protein to feed the planet's human population, and international bodies coordinate the management of fisheries. Offshore oil drilling is a major source of oil, with further efforts being made to drill into ever greater depths. Underwater archaeology, examining shipwrecks and drowned human habitation sites, received a tremendous boost from the invention of scuba gear. Since the 1960s, underwater archaeology has flourished. Our emerging understanding of global warming is greatly aided by increasing knowledge of the oceans, since the oceans are the main mechanism regulating the temperature of the planet.

See also Archaeology; Cold War; Cousteau, Jacques; Deep-Sea Hydrothermal Vents; Geology; International Geophysical Year; Meteorology; National Science Foundation; Oceanography; Plate Tectonics

References
Ballard, Robert D., with Will Hively. *The Eternal Darkness: A Personal History of Deep-Sea Exploration.* Princeton: Princeton University Press, 2000.
Bascom, Willard. *A Hole in the Bottom of the Sea: The Story of the Mohole Project.* New York: Doubleday, 1961.
Broad, William J. *The Universe Below: Discovering the Secrets of the Deep Sea.* New York: Simon and Schuster, 1997.
Kaharl, Victoria A. *The Water Baby: The Story of Alvin.* New York: Oxford University Press, 1990.
Lawrence, David M. *Upheaval from the Abyss: Ocean Floor Mapping and the Earth Science Revolution.* New Brunswick, NJ: Rutgers University Press, 2002.
Miller, James W., and Ian G. Koblick. *Living and Working in the Sea.* New York: Van Nostrand Reinhold, 1984.

Universities

Universities in the Western world have been the wellspring of scientific training and innovation ever since medieval times. Prior to the twentieth century, scientists in the United States usually funded their research out of their own pockets or from university funds. In the first half of the twentieth century, private foundations became important sources of funding, especially in the health sciences. During World War II, federal funding in the United States poured into university-based scientific research. After the war, the federal government chose to continue supporting basic and applied scientific research at universities, using many mechanisms to deliver funding, including the military, the National Science Foundation (founded in 1950), the National Aeronautics and Space Administration (founded in 1958), and the National Institutes of Health (founded in its current form in 1948).

In the postwar world, the number and size of universities grew, as did the amount of scientific research undertaken and the number of students being educated. Big science projects, especially particle accelerators and telescopes, were often put under the sponsorship of a university or a consortium of universities. Most scientific research funding in the United States now comes from sources outside the university itself, from commercial companies, government organizations, and the military, even when that research is being carried on at the universities. At the turn of the millennium, approximately half of all basic research in the United States was being conducted in universities, with the rest being conducted in government institutes, in commercial companies, or privately. Much of the research done at universities now follows the priorities of funding sources rather than the personal interests of researchers.

The story of Stanford University is a good example of the changes in the U.S. university. Founded in 1885 by a railroad baron and his wife as a memorial to their son, the private university included scientific research among its founding goals. Stanford became a leading West Coast university, and postwar federal government funding transformed the university. The federally funded Stanford linear accelerator (SLAC), completed in 1966, became the largest linear accelerator in the world. Researchers from Stanford helped invent integrated circuits and create the microelectronics, computer, and software industries in the surrounding "Silicon Valley." The Stanford Research Institute (SRI) became a leading research think tank for the American military. Stanford in the form of SRI also became one of the original four nodes of ARPAnet, the precursor to the Internet.

For centuries, the premier universities in England were at Oxford and Cambridge, though in the postwar period, regional universities gradually grew in stature and now support leading researchers. The United States developed and continues to support a decentralized university system, with major funding coming from state governments, the federal government, university alumni, private donations, and commercial companies. Scientific institutes in France and Germany are important centers of basic scientific research in addition to the universities. In the Soviet Union, the Soviet Academy of Sciences conducted most basic research in its own academy institutes, often in association with universities.

Much of the research in Japan occurs in corporations rather than universities. Some critics, both inside and outside the country, have argued that science that breaks boundaries and pushes the edges is inhibited by the Japanese cultural predilection toward conformity and respect for authority. The elite universities of the Japanese university system also strongly tend to hire from among their own graduates rather than on merit. Laboratory space and research funding are usually allocated equally among Japanese university professors, rather than through any sort of competitive process.

See also Associations; Big Science; Foundations; Integrated Circuits; Internet; National Institutes of Health; National Science Foundation; Particle Accelerators; Soviet Academy of Sciences; Telescopes

References
Coleman, Samuel. *Japanese Science: From the Inside.* New York: Routledge, 1999.
Crease, Robert P. *Making Physics: A Biography of Brookhaven National Laboratory, 1946–1972.* Chicago: University of Chicago Press, 1999.
Geiger, Roger L. *Research and Relevant Knowledge: American Research Universities since World War II.* New York: Oxford University Press, 1993.
Leslie, Stuart W. *The Cold War and American Science: The Military-Industrial-Academic Complex at MIT and Stanford.* New York: Columbia University Press, 1994.
Lowen, Rebecca S. *Creating the Cold War University: The Transformation of Stanford.* Berkeley: University of California Press, 1997.

V

Vitalism

See Biology and the Life Sciences

Volcanoes

Geologists watched with fascination as a volcano named Parícutin was born in a farmer's field in Mexico on February 20, 1943. The volcano continued to grow, closely monitored by geologists, until March 4, 1952. Scientists from countries with active volcanoes—such as Iceland, the United States, Japan, and Italy—have often led the world in volcanology. The Icelander Sigurdur Thorarinsson (1912–1983) pioneered work in tephrochronology, the study of the strata laid down by volcanic debris to determine the timing of past eruptions. Takeshi Minakami (1909–1985) created a method to classify earthquakes caused by volcanic activity.

Geologists knew that volcanoes were more frequent in some areas, such as along the Pacific coasts of America, Japan, and Southeast Asia, but they did not understand why until the 1960s. During that decade, geology experienced a revolution, as geologists became convinced that continents really did drift and a 1970 article first used the term *plate tectonics*. The theory of plate tectonics provided an explanation for continental drift; indeed, it proved to be such a powerful explanatory mechanism that it transformed geology from a science of particulars into a mature science with a body of integrated knowledge. Plate tectonics explains that, as plates grid together, they create mountains from uplift, and where plates pull apart, they create rift valleys and opportunities for magma to well up from deeper in the Earth. Volcanoes and greater earthquake activity exist on the edges of continental plates. The rim of the Pacific Ocean is called the Ring of Fire because of the numerous active volcanoes.

Volcano observatories around the world watch some 150 current active volcanoes in Iceland, Hawaii, Japan, Italy, and other places, collecting measurements for sustained studies of volcanoes in action. Lack of funding prevents many possibly active volcanoes from being regularly watched. An opportunity to closely monitor an erupting volcano came with the eruption of Mount Saint Helens in Washington State. Small eruptions began on March 27, 1980, prompting an evacuation of the area; scientists set up monitoring equipment. The May 18 eruption killed only fifty-seven people, but it created a moonlike wasteland. Major eruptions occurred later that year before the volcano settled back into dormancy.

Voluminous plumes of volcanic ash and rock blast from the side of Mount Saint Helens in southwestern Washington on July 22, 1980. (Gary Braasch/Corbis)

Other volcanoes have provided further opportunities for study. The ash and gases from El Chichón in Mexico in 1982 reached as high as twenty miles and prompted a greater appreciation for the effect of volcanic eruptions in creating climatic changes. After a year of rumbling warnings, Nevado del Ruiz in Colombia erupted on November 13, 1985. Although the eruption was not large, melting ice on the summit created a mud flow that killed about 23,000 people. In 1991, Mount Pinatubo erupted in the Philippines. During the twentieth century, only the 1912 eruption of Katmai in Alaska threw more material in the atmosphere. Seismic activity, surface temperatures, emission of gases, and landscape changes helped geologists predict with good probability what Pinatubo would do. Given the magnitude of the eruption, it was considered a good result that only 1,202 people died, a testament to evacuation lessons learned during earlier eruptions and the in-creasing ability of scientists to make effective predictions. For volcanologists concerned about their science being used to save lives, Pinatubo became their example of success.

Dangers from volcanoes can come from other sources. On August 21, 1986, a cloud of carbon dioxide bubbled out of Lake Nyos in Cameroon, suffocating 1,742 nearby villagers and animal life. Teams of scientists from various countries examined this curious incident. Some scientists argued that the carbon dioxide had come from lower waters in the lake that had been disturbed somehow. Others argued that a volcanic eruption under the lake had emitted the gas. The dispute is still not settled.

Volcanoes have been detected on Jupiter's moon Io, and the extinct volcano of Olympus Mons on Mars is higher than any mountain on Earth. Study of the geology of other planets and moons may lead to greater understanding of the geology, plate tectonics, and volcanoes of Earth.

See also Earthquakes; El Niño; Geology; Kuno, Hisashi; Plate Tectonics; Space Exploration and Space Science

References

Boer, Jelle Zeilinga de, and Donald Theodore Sanders. *Volcanoes in Human History: The Far-Reaching Effects of Major Eruptions.* Princeton: Princeton University Press, 2002.

Fisher, Richard V., Grant Heiken, and Jeffrey B. Hulen. *Volcanoes: Crucibles of Change.* Princeton: Princeton University Press, 1997.

Frankel, Charles. *Volcanoes of the Solar System.* Cambridge: Cambridge University Press, 1996.

Scarth, Alwyn. *Vulcan's Fury: Man against the Volcano.* New Haven: Yale University Press, 1999.

Thompson, Dick. *Volcano Cowboys: The Rocky Evolution of a Dangerous Science.* New York: St. Martin's, 2000.

Voyager Spacecraft

Both *Voyager 1* and *Voyager 2* were launched in 1977 to take advantage of an alignment of the outer planets that only occurs approximately every 175 years. The Jet Propulsion Laboratory (JPL) designed and managed the National Aeronautics and Space Administration (NASA) mission. As launched, the two probes were designed to last only five years and to visit only Jupiter and Saturn, but *Voyager 2* was remotely reprogrammed during transit to continue its mission and visit Uranus and Neptune. *Pioneer 10* had already reached Jupiter in 1973, taking over three hundred pictures. *Pioneer 11* reached Jupiter a year later and used the planet's gravity to propel it on to an encounter with Saturn. Both of the Pioneer craft eventually left the solar system.

Because the outer planets are so far away from the Sun, solar panels cannot provide power for the spacecraft, and the Voyager spacecraft relied on radioisotope thermoelectric generators (RTGs), which derived electricity from the heat generated by decaying plutonium. *Pioneer 10* and *11* had each carried a plaque of greeting in case an alien intelligence encountered the craft at some unknown time in the future; each of the Voyager spacecraft carried a 12-inch, gold-plated copper disk with 115 images from Earth on it and a selection of sounds, including natural sounds and voice greetings in fifty-five languages. The astronomer Carl Sagan coordinated the selection of the material. Each Voyager spacecraft carried ten experiments, including television cameras, magnetometers, and sensors, to take infrared and ultraviolet pictures, as well as detect plasma, cosmic rays, and charged particles.

Both Voyagers reached Jupiter in 1979. The moons of the planet surprised scientists. Gravitational stress from Jupiter and nearby moons keeps the moon Io geologically active, with numerous volcanoes that regularly erupt. Pictures showed sulfur plumes from volcanoes reaching 190 miles (300 kilometers) above the moon's surface. The other three large moons of Jupiter—Europa, Ganymede, and Callisto—all displayed surfaces with considerable ice. Scientists surmised from the evidence that Europa has a thick crust of ice floating atop a deep liquid ocean, perhaps a place that life-forms might be found in the future. Faint rings around Jupiter were also discovered, showing that all the gas giants possessed rings, with the most spectacular rings girdling Saturn.

Both Voyagers reached Saturn in 1981. Scientists were intrigued to find from Voyager pictures that the activity of two shepherd moons is part of the reason that Saturn's rings remain stable. Saturn's moon Titan was found to have a thick nitrogen-methane atmosphere, with a surface temperature of –292 degrees Fahrenheit (94 degrees kelvin) and a surface pressure 1.5 times the pressure of Earth's atmosphere at sea level. As a result of atmospheric chemistry, hydrocarbons and other organic molecules might carpet the surface. Photographs revealed on Saturn's moon Mimas a massive crater from an impact that must have come close to shattering the moon.

Voyager 2 visited Uranus in 1986, where its cameras found shepherd moons straddling that gas giant's rings. Titania, the largest moon of Uranus, was characterized by rift canyons, probably formed by water flowing during past volcanic activity. *Voyager 2* continued on to the farthest gas giant, reaching Neptune in 1989. Pictures of Neptune's largest moon, Triton, showed active geysers spraying nitrogen and dust several kilometers into the thin atmosphere.

Both Voyager spacecraft eventually left the solar system. Their RTGs should provide enough power to work until 2020. Scientists hope to detect the end of the heliosphere, called the heliopause, where the solar wind from the Sun peters out and true interstellar space begins.

The total cost of both the Voyager missions came to $865 million. The Voyager craft discovered three new moons around Jupiter, seven new moons around Saturn, ten new moons around Uranus, and six new moons around Neptune. NASA followed up the Voyager discoveries by dispatching the Galileo mission to Jupiter, consisting of an orbiter and descent probe, which arrived at Jupiter in 1995. NASA, the European Space Agency, and the Italian Space Agency launched the Cassini mission in 1997. In 2004, Cassini is scheduled to go into orbit around Saturn's moon, Titan, and release the Huygens probe, which will descend by parachute to Titan's surface.

See also Big Science; European Space Agency; Geology; Jet Propulsion Laboratory; National Aeronautics and Space Administration; Plate Tectonics; Sagan, Carl; Search for Extraterrestrial Life; Shoemaker-Levy Comet Impact; Space Exploration and Space Science

References

Burgess, Eric. *Far Encounter: The Neptune System.* New York: Columbia University Press, 1991.

Cruikshank, Dale P., editor. *Neptune and Triton.* Tucson: University of Arizona Press, 1995.

Littmann, Mark. *Planets Beyond: Discovering the Outer Solar System.* Updated edition. New York: Wiley, 1990.

Miner, Ellis D. *Uranus: The Planet, Rings, and Satellites.* Second edition. New York: Wiley, 1998.

National Aeronautics and Space Administration. *Voyager, the Grandest Tour: The Mission to the Outer Planets.* Washington, DC: United States Government Printing Office, 1991.

Sagan, Carl. *Murmurs of Earth: The Voyager Interstellar Record.* New York: Random House, 1978.

"Voyager: Celebrating 25 Years of Discovery." http://voyager.jpl.nasa.gov/ (accessed February 13, 2004).

W

War

Nations have always sought advantages over their foes when going to war. Weapons technologies can mean the difference between victory and defeat, and governments have supported the development and production of new weapons. Although most technological development before the 1860s was driven by craftsmen and engineers, by the twentieth century, science had become the primary driver of technological development. World War II (1939–1945) proved to be a war of scientists as much as a war fought by soldiers, sailors, and airmen. Radar, advanced sonar, synthetic gasoline from coal, penicillin, computers, jet airplanes, short-range missiles, and the atomic bomb—all these changed how nations fought wars. Spending on science and technology during the cold war by the governments of the United States, the Soviet Union, and their allies remained at near-wartime levels. In what may be characterized as the militarization of science, significant advances in electronic computers, integrated circuits, rocket technology, and satellites all occurred because of military funding. Although many technological advances included possible civilian applications, other advances were purely military in nature, such as stealth technology to defeat radar detection of aircraft.

Advanced radio communications and satellite technologies have allowed the militaries of the United States and other advanced nations to more effectively control their armed forces. The need for the U.S. military to always be able to precisely locate their personnel and equipment anywhere on Earth's surface led to the U.S. Global Positioning System (GPS). The GPS satellites have many important civilian and scientific uses.

Biological weapons have only rarely been used in wars, but their possible impact as weapons of mass destruction has encouraged governments to fund scientific research into bacteria and viruses as weapons and into ways to counter such weapons. Research into offensive biological measures is very similar to research into defensive biological measures, making it difficult to enforce a ban on offensive-minded research as opposed to legitimate defensive research. Though it was already banned by a 1925 Geneva treaty, such was the danger of this kind of weapon backfiring and killing one's own population that the Soviet Union, the United States, and over seventy other nations signed a 1972 Biological and Toxin Weapons Convention. The treaty banned the continued manufacture and development of biological weapons and required the destruction of existing stocks. It is now known that the Soviet Union did not adhere to the treaty and continued a large biological weapons program well into the 1990s.

Since scientists form an international community and body of knowledge independent of political ideology, some have asked why scientists do not put the greater good of the world ahead of nationalism. Indeed, some scientists have been famous pacifists, such as Albert Einstein (1879–1955) and Linus Pauling (1901–1994). Pauling worked with the Office of Scientific Research and Development during World War II developing explosives and rocket propellants, but declined to work on the Manhattan Project to develop the atomic bomb. After becoming a Nobel laureate in chemistry in 1954, Pauling became more aggressive in his antinuclear activities. Concerned about the accumulation of radiation in the atmosphere from open-air nuclear weapons testing, Pauling and other scientists campaigned for the United States and Soviet Union to at least stop open-air testing in the interest of public health. His efforts helped lead to the 1963 Nuclear Test Ban Treaty, and he received the 1962 Nobel Peace Prize. Nevertheless, it is apparent that enough scientists and engineers are motivated by nationalism or monetary concerns to fund whatever weapons research a nation wants to undertake.

Some scientists and engineers who have developed powerful new weapons, such as the machine gun, chemical weapons, and atomic weapons, have thought that their inventions would actually end war because they would be too horrible to use. Only in the case of atomic bombs has this possibly been true. The nuclear standoff between the Soviet Union and United States during the cold war restrained that ideological conflict from igniting into a third world war.

The threat of war has led to the creation of large military bases and fortified zones along tense national borders. Ironically, these bases and border zones, where people are not allowed to settle, have actually promoted the preservation of nature in those areas. On the other hand, military bases are also some of the most contaminated places on Earth, especially bases involved in nuclear and chemical weapons production or storage.

Without a doubt, military spending has set priorities for scientific investigation and technological development, and has meant that more has been done than would have been the case without military involvement. Absent military involvement, investment by governments in scientific inquiry would probably have been much lower. Although scientists have benefited from military spending, they have also often been constrained from quickly publishing their results because of concerns over classified information sources and possible national security concerns. This restriction strikes at the heart of the scientific enterprise, which has always thrived best in an open environment of free publication, free inquiry, and unrestrained criticism.

See also Cold War; Computers; Ethics; Hydrogen Bomb; Integrated Circuits; Satellites; Soviet Union

References

Leslie, Stuart W. *The Cold War and American Science: The Military-Industrial-Academic Complex at MIT and Stanford.* New York: Columbia University Press, 1994.

Macksey, Kenneth. *Technology in War: The Impact of Science on Weapon Development and Modern Battle.* New York: Prentice Hall, 1986.

Volkman, Ernest. *Science Goes to War: The Search for the Ultimate Weapon—From Greek Fire to Star Wars.* New York: Wiley, 2002.

Watson, James D. (1928–)

James D. Watson and Francis Crick (1916–2004) discovered the double-helix shape of deoxyribonucleic acid (DNA) that is the basis of modern genetics and molecular biology. James Dewey Watson was born in Chicago, Illinois, where his father was a businessman and his mother worked as an admissions officer at the University of Chicago. His father imparted to Watson a love of birdwatching, and the young man wanted to become an ornithologist. Watson entered the University of Chicago at the age of fifteen in an experimental program. In 1947, he graduated with a B.S. in zoology, and moved to Indiana University at Bloomington, earning a Ph.D. in zoology in 1950. He then spent a year studying in Copenhagen, where he became interested in the problem of the molecular structure of DNA.

As a twenty-three-year-old wunderkind, Watson traveled to Cambridge University in 1951 to do further postdoctoral study at the Cavendish Laboratories. He met Francis Crick (1916–2004), who was also working on the structure of DNA. Watson and Crick developed a close relationship, and together they drew on the x-ray diffraction pictures of DNA taken by Rosiland Franklin (1920–1958) of the University of London, an associate of Crick's friend, the physicist Maurice H. F. Wilkins (1916–). In February 1953, Watson realized that a double helix would match the shape that they sought. They created a model of their double helix out of beads, wire, and cardboard, and published their discovery in a letter to the journal *Nature* in April. In their landmark letter, they also observed that the double helix, with its zipper-like nature, offered a mechanism for gene replication.

Watson spent 1953–1955 at the California Institute of Technology as a Senior Research Fellow in Biology, then spent another year at Cambridge with Crick, before joining the faculty of Harvard University in 1956. He continued to conduct research and became a full professor in 1961. Watson, Crick, and Wilkins received the 1962 Nobel Prize in Physiology or Medicine. (Franklin had died by that time, and the Nobel is not awarded posthumously.) Watson's controversial 1968 memoir, *The Double Helix: A Personal Account of the Discovery of the Structure of DNA*, chronicled his version of this story. Throughout his life, Watson has forthrightly expressed himself, viewing this trait in himself as absolute honesty, while some others have

James D. Watson, zoologist, 1962 (Bettmann/Corbis)

regarded the trait as a lack of tact or even as an inability to distinguish between his own opinions and facts.

Watson married in 1968 and later fathered two sons. Also in 1968, he assumed the position of director of the Cold Spring Harbor Laboratory on Long Island, New York. Watson resigned from Harvard in 1976 to devote himself to the directorship, building the laboratory from an underfunded declining institution into a well-funded center of biological research, emphasizing research into the complex problem of cancer. In 1988, the National Academy of Sciences chose Watson to be the first director of the Human Genome Project. He served until 1992, when he resigned after a dispute over intellectual property rights. Watson returned to Cold Spring Harbor Laboratory, became its first president in 1994, and became chancellor ten years later.

See also Biotechnology; Crick, Francis; Franklin, Rosiland; Genetics; Human Genome Project
References
McElheny, Victor K. *Watson and DNA: Making a Scientific Revolution.* Cambridge, MA: Perseus, 2003.
Watson, James D. *Genes, Girls, and Gamow: After the Double Helix.* London: Oxford University Press, 2001.

Wheeler, John A. (1911–)

The physicist John A. Wheeler made important contributions to the development of nuclear weapons and nuclear reactors, as well as the theories of gravitation, relativity, and quantum mechanics. John Archibald Wheeler was born in Jacksonville, Florida, where his father was a librarian and his mother had been a reference librarian before marriage. The family followed his father's librarian jobs to Los Angeles, then Youngstown, Ohio, and finally Baltimore, Maryland. In 1926, Wheeler entered the Baltimore City College, intent on following his interest in science. A year later, a scholarship took him to Johns Hopkins University, where a combined undergraduate-graduate program accelerated him on the path to a Ph.D. in physics in 1933. A fellowship allowed him to travel to Copenhagen and study with the celebrated Niels Bohr (1885–1962).

In 1935, Wheeler became an assistant professor of physics at the University of North Carolina. He conducted research in nuclear structure and in 1937 worked with Edward Teller (1908–2003) on nuclear rotation. Wheeler joined the Princeton University faculty at the Institute for Advanced Study in 1938. In 1939, Bohr and Wheeler published "The Mechanism of Nuclear Fission" in the journal *Physical Review.* This article and two follow-up articles described the possible fission reactions inherent in uranium-235 and plutonium-239. This insight was to lead to making a nuclear bomb and power-generating nuclear reactors. In 1939, Wheeler began to mentor the young Richard P. Feynman

(1918–1988), who later became perhaps the most brilliant American physicist of the twentieth century and refined quantum electrodynamics (QED) into its modern form.

During World War II, Wheeler worked on the secret Manhattan Project. He initially worked at the University of Chicago before moving to Hanford, Washington, to advise on the development of nuclear reactors. He solved several key technical problems. The Hanford reactors produced the plutonium-239 used in the atomic bomb dropped on Nagasaki. After the war, Wheeler returned to Princeton, but in 1949 joined Teller in the project to create the hydrogen bomb. From 1951 to 1953, Wheeler directed the top-secret Project Matterhorn at Princeton, designing thermonuclear weapons. Though he remained involved with the nuclear weapons establishment for the next several decades, serving on boards and committees, Wheeler returned to academic research in 1953.

In his pure scientific work, Wheeler characterized his early theorizing as "everything is particles." In 1952, he became intellectually engaged with relativity and gravitation and entered his "everything is fields" period. Late in life, he changed into his self-described "everything is information" period. Among his more important innovations was the description in 1955 of the geon, a collection of electromagnetic radiation kept together by its own gravity. He later showed that geons would be too unstable to exist in nature, but the idea helped move forward work on gravitation and relativity. In 1968, Wheeler was the first to use the term *black hole* in a published paper. He had initially opposed the idea of singularities, points in space of infinite density and infinitesimal volume, a possible consequence of the theories of relativity and quantum mechanics, but had come to advocate the idea that massive stars would collapse on their death into black holes. Wheeler later worked to develop a grand unified theory of physics and made incremental contributions to the search for such a theory.

Wheeler married a medical social worker in 1935 and they had two daughters and a son. In 1976, Wheeler accepted an invitation to join the University of Texas at Austin. He left Princeton, but in 1986, when he finally retired, Wheeler and his wife moved back to the Princeton area to be near their children and grandchildren, and for years Wheeler maintained an office at his old institution. Wheeler has received numerous awards and honorary degrees for his work, including the Niels Bohr International Gold Medal in 1982. A lifelong Unitarian, Wheeler has always been active in church and campus affairs.

See also Feynman, Richard; Grand Unified
    Theory; Hydrogen Bomb; Nuclear Physics;
    Particle Physics; Physics; Teller, Edward
References
Wheeler, John Archibald. *At Home in the Universe.*
    Woodbury, NY: American Institute of Physics,
    1994.
Wheeler, John Archibald, with Kenneth Ford.
    *Geons, Black Holes and Quantum Foam: A Life of
    Physics.* New York: Norton, 1998.

Wiles, Andrew (1953–)

The mathematician Andrew Wiles proved Fermat's last theorem, long thought impossible to prove. Andrew John Wiles was born in Cambridge, England, where his father was a theologian at Oxford University. He became fascinated with Fermat's last theorem as a child and spent his teenage years trying to prove it. The French mathematician Pierre de Fermat (1601–1665) had written in the margin of a book in 1637 that he had proved the ancient problem posed by Diophantus, that for the equation $xn + yn = zn$, there exist no solutions where *n* is a positive whole number greater than 2. Many eminent mathematicians had worked on the problem and proved the equation for specific numbers, but the problem was thought unprovable, so only amateurs usually tried to solve it. After entering Merton College at Oxford University, Wiles put away the problem. He graduated in

1974 with a B.A. and moved to Clare College at Cambridge University to obtain his Ph.D. in 1980. After two years at Harvard University, Wiles settled into a faculty appointment at Princeton University in 1982. Wiles married and fathered two daughters.

In the summer of 1986, Wiles informed his family and one colleague that he was going to work on Fermat's last theorem. For the next seven years he worked extensively on the problem in private, producing a thousand-page manuscript. Even though many mathematicians thought the theorem unprovable, a computer simulation in 1987 run by other mathematicians had shown that there were no solutions for the first 150,000 positive whole numbers. By 1993 another computer simulation had shown that there were no solutions for the first four million positive whole numbers. In proving Fermat's last theorem, Wiles also solved the Taniyama-Shimura Conjecture, a problem posed by the Japanese mathematicians Goro Shimura (1930–) and Yutaka Taniyama (1927–1958) in the 1950s.

In a trio of lectures at Cambridge University in 1993, Wiles announced his proof. When other mathematicians reviewed his work, they found a few minor problems that took Wiles two years to solve. One last problem required the help of a former student, and in 1995 the proof was published in the *Annals of Mathematics.* Wiles shared the 1995 Wolf Prize and received much popular acclaim. Most mathematicians make their great breakthroughs at a younger age than Wiles, making his achievement even more impressive.

See also Mathematics
References
Aczel, Amir D. *Fermat's Last Theorem: Unlocking the
    Secret of an Ancient Mathematical Problem.* New
    York: Four Walls Eight Windows, 1996.
Mozzochi, C. J. *The Fermat Diary.* Providence, RI:
    American Mathematical Society, 2000.
Singh, Simon, and Kenneth A. Ribet. "Fermat's
    Last Stand." *Scientific American* 277, no. 5
    (November 1997): 68–73.

Wilson, Edward O. (1929–)

The biologist Edward O. Wilson became the world's recognized authority on ants and created the theories of sociobiology and consilience. Edward Osborne Wilson was born in Birmingham, Alabama, where his father was an accountant for the federal government. An only child, Wilson throve in the outdoors and profited from his family's frequent moves, including two years in Washington, D.C., where the young boy regularly visited the Smithsonian museums and the National Zoo. Wilson decided that he wanted to be an entomologist and chose to focus on ants while still in high school. He attended the University of Alabama, majoring in biology, earning a B.S. in 1949 and an M.S. in 1950. While in college, he worked for the Alabama State Department of Conservation as an entomologist and conducted the first thorough study of the fire ant. After a year of graduate study at the University of Tennessee, Wilson entered Harvard University in 1951. He earned his Ph.D. in biology in 1955 and became a faculty member the following year. Wilson became a full professor of zoology in 1964, the curator of entomology at Harvard's Museum of Comparative Zoology in 1973, and Frank B. Baird Jr. Professor of Science in 1976. Wilson married in 1955 and fathered a single daughter.

Wilson established his reputation with his studies of ants, including fieldwork in the United States, Central America, Australia, New Guinea, and the South Pacific. In collaboration with the entomologist William L. Brown Jr. (d. 1997), while working on the ant genus *Lasius,* Wilson made important contributions to taxonomy and understanding natural selection by developing the concept of character displacement. Sometimes, when two closely related species come into close proximity to each other, natural selection causes hybridization to merge the two gene pools. Character displacement, however, is what happens when natural selection increases the differences between the two species by selecting for traits that make each species more unlike the other. Research in the South Pacific led Wilson to develop the theory of the taxon cycle to explain how new species enter new territories and push out old species as part of a cycle of repeated conquest by still more recent species.

Wilson combined fieldwork with laboratory work, and in a series of important experiments in the late 1950s, he showed that ants communicate via pheromones. With a colleague, he later created a theory of pheromone transmission and communication, showing how seemingly complex behaviors could be induced by communication via a limited number of chemical signals. He also developed important insights into the social order and caste system of ants. His book *The Insect Societies* (1971) presented a definitive synthesis of his own work and the work of other scientists on social insects, such as ants and wasps.

Although some scientists sit back on their laurels after achieving a major research success, seemingly having accomplished their goals in life, others, like Wilson, use their success to create new opportunities for research. Wilson also combined the abilities of a scientist of the first rank with skilled writing. In *Sociobiology: The New Synthesis* (1975), he caused a storm of controversy with his assertion that what biologists had learned about innate behavior in insect and animal societies, along with the insights provided by the theory of natural selection, should be applied to understanding human behavior and human society. His critics worried that this approach could lead to theories reminiscent of past pseudoscientific justifications that condoned racism, viewing humans as genetically determined rather than individuals more subject to nurture than to nature. The paleontologist Stephen Jay Gould (1941–2002), a colleague at Harvard, was among many who vehemently rejected sociobiology. The emphasis on genes as the focus of evolution was also promoted by Richard Dawkins (1941–) in his push for a science of evolutionary psychology.

Edward O. Wilson, biologist and science writer (Rick Friedman / Corbis)

Wilson's next book, *On Human Nature* (1978), defended his ideas against his critics and won a Pulitzer Prize. He acknowledged that most human behavior was learned, not inborn, but also argued that the urge toward religious belief seemed innate. In his own life, he believed in scientific materialism, rather than any conventional religion. He also argued that altruistic behavior, which seems contrary to natural selection, is consistent with natural selection if an individual's altruism is directed toward close relatives. This perspective makes the gene, rather than the individual, the mechanism of natural selection.

Wilson also believed that all humans, like himself, possessed a strong innate attraction toward other forms of life. His *Biophilia: The Human Bond to Other Species* (1984) emphasized this point. Wilson's strong interest in environmental issues also led to later books:

The Diversity of Life (1992) and *The Future of Life* (2002). Wilson and his longtime research associate, the German-born entomologist Bert Hölldobler (1936–), published the definitive study *The Ants* in 1990. The book earned Wilson his second Pulitzer Prize.

In his ambitious *Consilience: The Unity of Knowledge* (1998), Wilson argued that scientists and scholars will eventually discover a fundamental unity underlying all forms of knowledge. The definition of consilience is "a jumping together" and Wilson argues that the basis of all such knowledge must be science and scientific materialism. Even moral reasoning must use the scientific process to arrive at its conclusions.

Wilson shared the 1990 Crafoord Prize with Paul R. Ehrlich (1932–). The Crafoord prize is awarded by the Royal Swedish Academy of Sciences in those areas not covered by the Nobel prizes. Wilson also received a National Medal of Science in 1977 and awards from the Ecological Society of America and the Academy of Natural Sciences. His prolific output includes numerous other books, over two hundred articles, and an autobiography, *Naturalist* (1994). Wilson later became the Pellegrino University Research Professor and Honorary Curator in Entomology at Harvard University and remains active as a researcher and writer.

> *See also* Biology and the Life Sciences; Dawkins, Richard; Ehrlich, Paul R.; Environmental Movement; Ethics; Evolution; Gould, Stephen Jay; Media and Popular Culture; Sociobiology and Evolutionary Psychology
>
> *References*
> Wilson, Edward O. *Naturalist.* Washington, DC: Island Press, 1994.

Wilson, John Tuzo (1908–1993)

The prominent geophysicist John Tuzo Wilson championed the idea of plate tectonics in the early 1960s and proposed a third method of plate movement. Wilson was born in Ottawa, Ontario, Canada, where his father, John Armistead Wilson, a Scottish immigrant trained as an engineer, eventually became director of Air Services for the Canadian government. His mother, the former Henrietta Loetitia Tuzo, was active in political and charitable organizations. Both parents were active mountain climbers, and Mount Tuzo near Banff, Alberta, was named after his mother because she was the first to climb it. Wilson was a vigorous man his entire life, known for his endurance.

Wilson attended Trinity College at the University of Ontario, where he changed his major from chemistry to a combination of physics and geology because he preferred to work in the field instead of the laboratory. After receiving his B.A. in 1930, the first student to graduate in geophysics from a Canadian university, he moved to England to attend Saint John's College at Cambridge University. He received a B.A. from Cambridge in 1932, then moved to Princeton University in America, where he obtained his Ph.D. in geology in 1936. From 1936 to 1939, Wilson worked for the Geological Survey of Canada. With the coming of World War II in 1939, he served in the Royal Canadian Engineers, rising to the rank of colonel before retiring in 1946. He became involved in Arctic science and exploration during this time.

Wilson became a professor of geophysics at the University of Toronto in 1946 and began a vigorous career of publication and leadership in geophysics. As president of the International Union of Geodesy and Geophysics from 1957 to 1960, he provided important leadership during the International Geophysical Year (IGY) of 1957–1958. During the IGY, he became acquainted with scientists isolated within communist-controlled China and wrote two books, *One Chinese Moon* (1959) and *Unglazed China* (1973), on Chinese science. He also wrote a book on the IGY effort, *IGY: Year of the New Moons* (1961).

Wilson became convinced of plate tectonics in the early 1960s and became a leading advocate for the theory. His 1965 paper, "A New Class of Faults and Their Bearing on Continental Drift," introduced his innovation, the mechanism of transform faults. Earlier theorists had found that continental plates pull away from each other or push toward each other. Transform faults move laterally, rubbing along each other. Wilson also developed the concept of hot spots, places in the center of plates where volcanic activity pushes upward. Hawaii and Yellowstone National Park are examples of such hot spots, leaving a trail of seamounts in the Pacific and lava fields in the American Northwest as the locations of Hawaii and Yellowstone on their continental plates have drifted across an upwelling of magma.

Though he claimed that he disliked administration, Wilson obviously excelled at organizing efforts in professional organizations and at academic institutions. After his retirement from the University of Toronto in 1974,

Wilson became the director general of the Ontario Science Centre in Toronto. He served as chancellor of York University from 1983 to 1986. Wilson married in 1938, fathered two daughters, and received many honors during his lifetime, including the Order of the British Empire.

See also Geology; International Geophysical Year; Oceanography; Plate Tectonics

References

Bingham, Roger. "Tuzo Wilson: Earthquakes, Volcanoes, and Visions." Pp. 189–196 in *A Passion to Know: 20 Profiles in Science*. Edited by Allen L. Hammond. New York: Charles Scribner's Sons, 1984.

World Health Organization

The founding of the United Nations (UN) during the course of World War II prompted the creation of a new international health organization. The World Health Organization (WHO), or Organisation Mondiale de la Santé, came into existence in 1948 with a structure modeled on the UN: a World Health Assembly composed of delegates from each of the member nations, meeting once a year to set policy; an Executive Board that meets twice a year to act as the executive organ of WHO; and a Secretariat, which is headed by a director-general and manages the day-to-day technical and administrative functions of the organization. Headquartered in Geneva, Switzerland, WHO also maintains regional offices in Washington, D.C., Copenhagen, Cairo, Brazzaville, New Delhi, and Manila.

WHO absorbed earlier organizations, including the Health Organization of the League of Nations and various international sanitation organizations. WHO began with an annual budget of $5 million in 1948, peaked at a budget of $310 million in the early 1980s, then declined to $264 million in the mid-1990s. In 1951, WHO issued its first set of updated international sanitary regulations, guidelines for member states to follow to alleviate health problems created by poor water, inadequate sewage disposal, and other environmental problems. WHO works

Headquarters of the World Health Organization in Geneva, Switzerland (World Health Organization / P. Virot)

closely with other UN programs, such as the United Nations Children's Fund (UNICEF) and the United Nations Educational, Scientific and Cultural Organization (UNESCO); and WHO also works with regional and national health care organizations, such as the prominent U.S. Centers for Disease Control and Prevention (CDC) in Atlanta, Georgia.

WHO successfully directed the international campaign (1958–1979) to eradicate the deadly disease of smallpox. Smallpox proved to be the ideal candidate for absolute eradication because a single vaccination offered protection from all strains of the variola virus and the virus only lived in human beings. After its success with smallpox, WHO turned its attention to polio, coming close to eradicating it by the end of the twentieth century. Other candidates proposed for future eradication are the guinea worm, measles, rubella, and hepatitis B.

In 1955, WHO began to plan for the eradication of malaria, a major killer and economic drain in tropical nations. WHO used dichloro-diphenyl-trichloro-ethane (DDT) and other insecticides to destroy anopheles mosquitoes, the carrier of the disease, and promoted the use of drug regimes. The result was the evolution of mosquitoes resistant to DDT and the evolution of strains of the malaria protozoa able to resist the drugs. Thus, although the disease declined for a time, it began to resurge in the 1970s. In 1977, India experienced sixty times more malaria cases than in 1960. By the 1990s, 2 million people a year died from malaria, half of them in Africa, while hundreds of millions of other people suffered from the disease. WHO gave up on its antimalarial campaign in 1992.

The initial concentration on disease-specific programs by WHO began to shift in the 1960s toward emphasizing access to primary health care. WHO now sponsors programs in public health throughout the world and offers technical help to member nations when requested. Among their current ongoing programs are a family-planning effort initiated in 1970 to reduce population growth, a Safe Motherhood Initiative launched in 1987 to reduce maternal morbidity, various programs to vaccinate more children against diseases, a tuberculosis control program initiated in 1948, and a large anti-HIV/AIDS program launched in 1987. WHO has been criticized as being too bureaucratic and even a hindrance in some public health emergencies and international programs. WHO's efforts to provide health standards and technical assistance have not been criticized, but efforts to encourage health policy changes among member nations are at times criticized as being too political. WHO also functions as a clearinghouse on medical research and public health issues, and issues statistical reports on world health, detailed by nation, age, and causes of health problems or mortality.

See also Centers for Disease Control and Prevention; Population Studies; Smallpox Vaccination Campaign

References
Lee, Kelley. *Historical Dictionary of the World Health Organization.* Lanham, MD: Scarecrow, 1998.
"WHO at Fifty: Highlights of the Early Years until 1960." *World Health Forum* 19, no. 1 (1998): 21–37.
"WHO at Fifty: 2. Highlights of Activities from 1961 to 1973." *World Health Forum* 19, no. 2 (1998): 140–155.
"WHO at Fifty: 3. Highlights of Activities from 1974 to 1988." *World Health Forum* 19, no. 3 (1998): 219–233.
"WHO at Fifty: 4. Highlights of Activities from 1989 to 1998." *World Health Forum* 19, no. 4 (1998): 441–455.
The World Health Report 1998: Life in the 21st Century. A Vision for All. Geneva: World Health Organization, 1998.

World Wide Web

See Berners-Lee, Timothy (1955–)

Wu, Chien-Shiung (1912–1997)

The experimental physicist Chien-Shiung Wu performed important experiments in beta decay, showing that parity is not conserved during weak nuclear reactions. Chien-Shiung

Chien-Shiung Wu, experimental physicist, 1958 (Bettmann / Corbis)

Wu was born in Liuho, near Shanghai, China, where her progressive father opened an elementary school for girls, a rare opportunity for Chinese girls of that time. After she had graduated from her father's school, her family sent Wu to a Westernized boarding school from the ages of ten to seventeen. She entered the National Central University in Nanjing, having studied on her own to learn enough mathematics and science to qualify as a student of mathematics and physics. As a student in her boarding school and at the university, Wu became a leader of student protesters, even once confronting Chiang Kaishek, the president of China.

Wu graduated in 1934 and taught for a year while conducting research in x-ray crystallography. A female physics professor, who

had graduated from the University of Michigan with a Ph.D., urged Wu to do the same. A wealthy uncle paid for her to leave China in 1936. While in San Francisco, Wu decided to attend the University of California at Berkeley, then becoming a major center of physics in the United States. She also met Luke Yuan, a fellow graduate student in physics, whom she later married. A petite woman who charmed her professors and impressed them with her intellectual intensity, Wu earned her Ph.D. in physics in 1940, becoming a well-known expert on nuclear fission. After she had spent some time teaching at Smith College in Massachusetts, the shortage of physicists during the war allowed Wu to overcome the strong barriers to women in university positions and obtain a position teaching at

Princeton University. Despite her obvious qualifications, only in 1944 was she asked to come to Columbia University in New York City to join the Manhattan Project.

After Japan invaded China in 1937, Wu was unable to contact her family until the end of World War II in 1945. After the war, she found her family was well, and her father had played an important part in building the Burma Road, a remarkable road cut through the Himalayas to carry military supplies from Burma to China. After the war, Columbia asked Wu to remain as a research associate. Her husband obtained a position at Brookhaven National Laboratory on Long Island, and they lived together on weekends. Wu bore a son in 1947, hired help to care for him, and was proud when he later became a physicist. After the Chinese communists won the civil war, Wu and her husband elected to remain in America, becoming citizens in 1954, and Wu did not see her remaining family members until 1973.

At Columbia, Wu concentrated on beta decay as her problem of choice. The Italian-born physicist Enrico Fermi (1901–1954) had proposed in 1933 how the nucleus might behave during beta decay, a form of radiation. No one had been able to confirm his theory until Wu demonstrated that earlier attempts had been confused by using uneven materials. This established her reputation as someone who had the ability to perform complex, delicate experiments and as a demanding taskmaster of her students and staff. Wu became an associate professor at Columbia in 1952, a professor in 1958, and Michael I. Pupin Professor of Physics from 1971 until her retirement in 1981.

When Tsung-Dao Lee (1926–) of Columbia and Chen Ning Yang (1922–) of the Institute of Advanced Study at Princeton University proposed that perhaps parity was not always conserved during weak nuclear reactions, Wu set out to construct an experiment, even though she, like others, thought that Lee and Yang's idea was not true. She arranged to use the facilities at the National Bureau of Standards in Washington, D.C., which could cool materials to close to absolute zero. The experiment took six months for her team and her to construct, and in January 1957, she showed that Lee and Yang were right. This result so surprised the physics community that Lee and Yang shared the 1957 Nobel Prize in Physics that very year. Wu was not included in the Nobel prize, since the idea was not hers, just the proof, though she did receive the 1963 Comstock Prize from the National Academy of Sciences, a National Medal of Science in 1975, and the first Wolf Prize in Physics in 1978, and in 1975 she became the first woman to serve as president of the American Physical Society.

See also Nuclear Physics; Particle Physics; Physics
References

McGrayne, Sharon Bertsch. *Nobel Prize Women in Science.* Second edition. Secaucus, NJ: Carol, 1998.
"Obituaries: Chien-Shiung Wu." *Physics Today,* October 1997, 120–121.

Y

Yalow, Rosalyn (1921–)

The medical physicist Rosalyn Yalow developed radioimmunoassay (RIA) to measure minute amounts of biological materials. She was born Rosalyn Sussman in New York City to Jewish immigrant parents. She attended the women-only Hunter College, where her professors urged her to pursue her interest in physics, though her family thought an education as an elementary school teacher held more promise. She graduated with a B.A. in physics and chemistry in 1941. After training as a secretary for a short time, she felt fortunate to be offered a teaching assistantship in physics and entry into graduate school at the University of Illinois. She found that she was the only woman student in the entire College of Engineering and met her future husband, Aaron Yalow, on the first day of school. She married in 1943, became skilled in working with radioactive materials, and received a Ph.D. in nuclear physics in 1945. Yalow had two children and relied on hired help to raise her children when they were younger.

After working briefly as an engineer, Yalow returned to Hunter College to teach physics to men in a veterans pre-engineering program. Yalow began to work as a part-time consultant at the Bronx Veterans Administration (VA) Hospital, and in 1950 she left teaching to join the VA full-time. She met a medical resident that same year, Solomon A. Berson (1919–1972), and began to work with him. After completing several projects together and essentially teaching each other how to conduct research, the pair turned to the problem of diabetes. Type II diabetes was treated with insulin from animals, and human patients developed a resistance to that insulin that meant ever greater doses were required. Yalow and Berson developed a theory that antibodies were clinging to the insulin and preventing it from entering the human cells. As more insulin was used, its foreign nature stimulated the immune system to produce ever more antibodies. Although their theory was not widely accepted at first, they combined their skills in immunology and radioisotope tracing to develop the RIA technique in 1959 to measure minute amounts of antibodies. This technique was also applicable to other substances that reacted biologically. RIA is now used to screen blood at blood banks and in numerous other applications.

After Berson died in 1972, Yalow requested that their laboratory be renamed the Solomon A. Berson Research Laboratory. In 1968, Yalow became a research professor at Mount Sinai School of Medicine, and after various academic promotions, became the Solomon A.

Rosalyn Yalow, medical physicist, 1977 (Bettmann / Corbis)

Berson Distinguished Professor at Large at Mount Sinai in 1986. Yalow became the first woman to receive the Albert Lasker Basic Medical Research Award in 1976. Yalow shared the 1977 Nobel Prize in Physiology or Medicine with Polish-born Andrew V. Schally (1926–) and French-born Roger Guillemin (1924–). Yalow received half of the prize, and Schally and Guillemin shared the other half for their discoveries of peptide hormone production in the brain. Yalow thought that if Berson had not already died, he would have shared in the prize. She also received the 1988 National Medal of Science.

See also Medicine; Nobel Prizes; Nuclear Physics
References
"Rosalyn Yalow—Autobiography." http://
    nobelprize.org/medicine/laureates/index.html
    (accessed September 19, 2004).
Straus, Eugene. *Rosalyn Yalow, Nobel Laureate: Her
    Life and Work in Medicine. A Biographical Memoir.*
    New York: Plenum, 1998.

Chronology

1950   Total world population stands at 2.53 billion humans; the National Science Foundation (NSF) is founded in the United States; John F. Nash publishes his work on noncooperative games, extending game theory and setting the groundwork for important innovations in economic theory; Hans Selye publishes *Stress,* demonstrating how physiological stress affects the human body.

1951   First commercial digital electronic computer, the UNIVAC; U.S. Communicable Disease Center (CDC) creates the Epidemic Intelligence Service to respond to epidemic outbreaks around the world; a small experimental breeder reactor in Idaho generates its first electricity in 1951, though just enough power to light the reactor building; Carl R. Rogers publishes *Client-Centered Therapy,* a cornerstone of humanistic psychology.

1952   The United States explodes the first hydrogen bomb; London fog traps enough pollution from coal furnaces to kill 4,000 people; philanthropist John D. Rockefeller III founds The Population Council; bacteriophages are shown to contain deoxyribonucleic acid (DNA) surrounded by a protein shell, and the conclusion is reached that DNA must be where the elusive genes of heredity are located; first maps of ocean floor topography unexpectedly show a rift valley in the middle of the Mid-Atlantic Ridge.

1953   The Soviet Union explodes its first hydrogen bomb; the Piltdown Man is shown to be a hoax; a heart-lung machine is first used; James D. Watson and Francis Crick discover the molecular structure of the DNA molecule; the complete chemical sequence of insulin is described; Alfred C. Kinsey publishes *Sexual Behavior in the Human Female,* a follow-up to his 1948 book, *Sexual Behavior in the Human Male;* the Miller-Urey experiment shows that the early Earth's atmosphere could have spontaneously produced amino acids.

1954   The first successful organ transplant occurs in Boston when a patient receives a kidney from his identical twin brother; the first maser is built; CERN (Conseil Européen pour la Recherche Nucléaire) is founded; nerve growth factor (NGF), the agent that activates nerve growth, is discovered; microscopic photographs of Precambrian bacteria and algae are published, creating the field of microfossils and pushing back the known age of life on Earth.

1955   Antiprotons are created in the Bevatron particle accelerator at Berkeley; the Salk killed-virus vaccine solves the polio problem.

1956   Mexico's wheat harvest becomes self-sufficient for that nation in 1956, the first fruits of the Green Revolution; neutrinos are detected for the first time in an experiment outside a nuclear reactor in South Carolina; the term *artificial intelligence* (AI) is first used in a summer workshop at Dartmouth College.

1957   The International Geophysical Year of 1957–1958 begins; the first artificial satellite, *Sputnik I,* is launched; the theoretical foundation for superconductivity is published; Dorothy Crowfoot Hodgkin uses x-ray crystallography to determine the chemical structure of vitamin B_{12}; an experiment shows that parity is not always conserved during weak nuclear reactions; Noam Chomsky publishes *Syntactic Structures.*

1958   The World Health Organization launches the Smallpox Eradication Program; the National Aeronautics and Space Administration (NASA) is established; the Van Allen Radiation Belts are discovered by the first American satellite; Harry F. Harlow shows through primate research that the need for love and affection is a primary drive in primates and humans.

1959   The English molecular physicist and novelist C. P. Snow publishes *The Two Cultures and the Scientific Revolution;* the first photograph of the far side of the Moon is taken, by the Soviet *Luna 3* spacecraft; Rosalyn Yalow and Solomon Berson develop the radioimmunoassay (RIA) technique to measure minute amounts of biological materials.

1960   The first birth control pill is approved for sale in the United States; the first laser is built; *Tiros 1,* the first meteorological satellite, is placed in orbit; the first attempt in the search for extraterrestrial intelligence (SETI) involves listening to two nearby stars, Tau Ceti and Epsilon Eridani; the bathyscaphe *Trieste* dives to the deepest part of the ocean in the Mariana Trench; electroweak theory, combining the weak nuclear force and the electromagnetic force, is developed; Jane Goodall begins her field research on chimpanzee behavior.

1961   The Soviet Union puts the first human into orbit; the Apollo project is begun; the publication of *The Genesis Flood* revives biblical creationism; Fairchild Semiconductor brings the first integrated circuit to market.

1962   The historian Thomas S. Kuhn publishes *The Structure of Scientific Revolutions;* Rachel Carson, a biologist and environmental activist, publishes her prophetic classic, *Silent Spring;* neutrinos associated with the formation of a muon particle are first discovered; *Telstar 1,* the first commercial communications satellite, is placed in orbit; *Mariner 2* flies past Venus in the first successful interplanetary mission; Harry Hess proposes that magma is the lubricant that allows continental plates to move.

1963   The Nuclear Test Ban Treaty is signed, banning aboveground testing; the first successful human liver transplant is conducted; quasars are discovered; meteorologist Edward N. Lorenz begins the study of complex physical systems that leads to chaos theory.

1964   The International Biological Programme (1964–1974) begins; the Soviet geneticist Trofim Lysenko is finally discredited and exiled; radio astronomers at Bell Labs in New Jersey detect background radiation in the universe, a discovery that establishes the Big Bang theory of the origin of the universe; Murray Gell-Mann proposes the existence of quarks.

1965   *Mariner 4* flies past Mars, revealing a cratered surface; the scanning electron microscope, invented in 1942, is finally commercialized; John Tuzo Wilson suggests that transform faults occur when continental plates rub laterally along each other.

1966   The Stanford linear accelerator (SLAC) is completed, the largest linear accelerator in the world; the U.S. Army Cold Regions Research and Engineering Laboratory project drills all the way from the surface of Greenland's ice cap to bedrock; the repressor molecule used to stop proteins from being produced is discovered; William H. Masters and Virginia Johnson publish *Human Sexual Response.*

1967   The first successful cardiac transplant is carried out; pulsars are discovered; Lynn Margulis publishes her serial endosymbiosis theory (SET); James Lovelock first presents his Gaia hypothesis that the chemical composition of the Earth's atmosphere is actively controlled by life in order to keep our planet habitable.

1968   The entomologist Paul R. Ehrlich publishes *The Population Bomb;* the term *black hole* is first used to describe a gravitational singularity; Doug Engelbart demonstrates on-screen video conferencing, an early form of hypertext, the use of windows on the screen, mixed graphics-text files, structured document files, and the first mouse.

1969   The *Apollo 11* astronauts make the first manned landing on the Moon; the first successful bone-marrow transplant is carried out; ARPAnet is created, the precursor to the Internet; the structure of the insulin molecule, containing 777 atoms, is described; Elisabeth Kübler-Ross publishes *On Death and Dying,* describing the stages of grief.

1970   The International Decade of Ocean Exploration (1970–1980) begins; the United States Environmental Protection Agency is established; astrophysicist Brandon Carter coins the term *anthropic principle;* Roger Penrose and Stephen Hawking publish a theory that argues that black holes may be detected by relativistic radiation from near the event horizon of black holes; plate tectonics is finally accepted as the main theory explaining the phenomena observed by geology; Yoichiro Nambu founds string theory.

1971   An orbiting x-ray observatory discovers Cygnus X-1, a possible black hole in orbit around a blue supergiant some 6,000 light-years away; victims of the rare Creutzfeldt-Jakob disease (CJD) are shown to display symptoms similar to victims of kuru and scrapie, eventually leading to the discovery of prions; the first microprocessor, Intel 4004, is brought to market.

1972   The Tuskegee syphilis study is revealed; the United Nations Environmental Programme is established; the Soviet Union, the United States, and many other nations sign the Biological and Toxic Weapons Convention; the evolutionary theory of punctuated equilibria is published; Herbert Boyer and Stanley Cohen meet at a conference and combine their techniques to produce genetic engineering.

1973   *Skylab,* the first space station, is launched by the United States; *Pioneer 10* reaches Jupiter; positron emission tomography (PET) is invented.

1974   Mario Molina and F. Sherwood Rowland discover that chlorofluorocarbons (CFCs) are eroding the ozone layer; Donald Johanson finds "Lucy," the first nearly complete skeleton of *Australopithecus afarensis;* the royal tomb of the first emperor of China, Qin Shi Huang, is unearthed.

1975   During the Apollo-Soyuz mission of 1975, U.S. astronauts and Soviet cosmonauts rendezvous in orbit; Edward O. Wilson, a prominent biologist, publishes *Sociobiology: The New Synthesis;* the European Space Agency is created.

1976   Richard Dawkins, a zoologist, publishes *The Selfish Gene;* the Centers for Disease Control and Prevention (CDC) discover the cause of an outbreak of Legionnaire's Disease in Philadelphia; Viking landers on Mars fail to find life; a formal proof of the four-color theorem is constructed, which marks the first time that a computer has been used to construct a formal mathematical proof; asymmetric encryption, also called public key encryption, is developed.

1977   Deep-sea submersible *Alvin* finds deep-sea hydrothermal vents in the Galapagos Rift; the search for extraterrestrial intelligence (SETI) effort at Ohio State University receives an intense signal, dubbed the "wow" signal, that is never repeated; the first case of tuberculosis that antibiotics cannot cure appears in South Africa; geneticists successfully sequence every base sequence used in a protein.

1978   The first successful test-tube baby is created via in vitro fertilization; Mary Leakey finds at Laetoli the fossilized footprints of three hominids dated 3.6 million years old.

1979   Smallpox is officially eradicated; a minor nuclear accident occurs at Three Mile Island in Pennsylvania; Voyager spacecraft reach Jupiter and discover wonders such as the exploding volcanoes on the moon Io; Benoit Mandelbrot develops his Mandelbrot Set to support his theory of fractals; the scanning tunneling microscope is invented; the first complete genome sequence of an organism is described, the 10 bases and 5,000 base pairs of the bacteriophage phi-X174.

1980   The Mount Saint Helens volcano erupts in the United States; a decision by the U.S. Supreme Court in the case of *Diamond v. Chakrabarty* permits life created in a laboratory to be patented; the German physicist Klaus von Klitzing discovers the quantized Hall effect, finding that an electrical current passing through solid matter changes in steps as the magnetic field is changed; high concentrations of iridium in a 65-million-year-old layer of clay lead to the conclusion that the dinosaurs were driven to extinction by the impact of an asteroid or comet.

1981   First space shuttle is launched; Voyager spacecraft reach Saturn.

1982   The El Niño Southern Oscillation of 1982–1983 attracts the attention of scientists; prions are identified as a protein that causes Creutzfeldt-Jakob disease (CJD); experimental physicist Daniel C. Tsui and a colleague discover the fractional quantum Hall effect, showing that at near-zero temperatures, and in a high magnetic field, electrons behave like a fluid rather than like particles.

1983   Abdus Salam founds the Third World Academy of Sciences; the first artificial heart transplant is carried out, and the patient lives 112 days; bosons are discovered in a particle accelerator, as predicted by theory.

1984   Human immunodeficiency virus (HIV), the cause of acquired immunodeficiency syndrome (AIDS), is identified; Baby Fae, a twelve-day-old girl, receives a baboon heart transplant and dies; the Lindow Man is excavated from an English bog; forensic analysis using DNA fingerprinting is proposed.

1985   Nevado del Ruiz volcano erupts in Colombia; a metal that remains superconductive at a much higher temperature is discovered, setting off a race to find more such materials.

1986   Most whaling is banned worldwide; a major accident occurs at the Soviet nuclear plant near Chernobyl in the Ukraine; *Voyager 2* reaches Uranus; atomic

1986
(cont.)
force microscope is developed; K. Eric Drexler's *Engines of Creation: The Coming Era of Nanotechnology* is published, predicting a coming age of microscopic machines; Prozac is introduced, an antidepressant and the first of the selective serotonin reuptake inhibitors.

1987
The first of the Montreal Protocol international agreements of 1987, 1990, and 1992 is signed, agreeing to phase out the industrial and commercial use of chlorofluorocarbons.

1988
The United Nations establishes the Intergovernmental Panel on Climate Change to study global warming.

1989
The Berlin Wall falls and the cold war ends; the discovery of cold fusion is announced and then discredited; CERN (Conseil Européen pour la Recherche Nucléaire) builds the Large Electron-Positron Collider (LEP), the largest scientific instrument ever constructed; the U.S. Global Positioning System (GPS) becomes operational, allowing users to pinpoint their location on Earth; *Voyager 2* reaches Neptune; the gene that causes cystic fibrosis is identified.

1990
The Human Genome Project begins; the Hubble Space Telescope is deployed into orbit; European-U.S. spacecraft *Ulysses* is launched to fly over the poles of the Sun and explore sections of space outside the ecliptic plane.

1991
The Soviet Union collapses; the Mount Pinatubo volcano erupts in the Philippines; the World Wide Web is created; a receding glacier in the Alps reveals Similuan Man (popularly called the Iceman), the body of a 5,000-year-old man preserved by the ice.

1992
The United Nations Conference on Environment and Development, also known as the Earth Summit, is held in Rio de Janeiro; scientists at Cambridge University create a breed of transgenic pigs by adding a human gene to the pig's genome; the World Health Organization gives up on its antimalarial campaign.

1993
Greenland Ice Core Project 2 recovers the deepest ice core drilled in Greenland; the Superconducting Supercollider (SSC) in Texas is canceled.

1994
Fragments of the Shoemaker-Levy 9 comet impact Jupiter; Deep Green, the Green Plant Phylogeny Research Coordination Group, examines the phylogenetic makeup of numerous plants in an attempt to reconstruct the evolutionary relationships among all green plants.

1995
First extrasolar planet found orbiting 51 Pegasi; Andrew Wiles proves Fermat's last theorem; the Ice Maiden, a child sacrifice, is found in the Andes mountains, preserved by the high-altitude air.

1996
Dolly the sheep, the first clone of a large mammal from an adult somatic cell, is born.

1997   The United Nations Conference on Climate Change is held in Kyoto, Japan, leading to the Kyoto Protocol; the El Niño Southern Oscillation of 1997–1998 is the largest and warmest in the twentieth century; Deep Blue, an IBM supercomputer, defeats the world chess champion in a chess tournament.

1998   A New Zealander receives the first hand transplant, which his body eventually rejects.

1999   World population passes 6 billion humans.

2000   The Bill and Melinda Gates Foundation is established; the Shuttle Radar Topography Mission confirms the impact crater in Chicxulub, Yucatán, Mexico, as the source of the K-T boundary iridium that probably killed the dinosaurs.

2001   The U.S. government permits research into embryonic stem cells, while strictly limiting which cell lines are available for study; the terrorist attacks on the World Trade Center and Pentagon on September 11, together with the sending of anthrax spores through the U.S. mail, heighten concern about science-based terrorist threats involving the use of nuclear and biological weapons.

2002   Scientists synthesize polio using its commonly known genetic sequence, downloaded from the Internet, increasing fears of bioterrorism; the Larsen Ice Shelf in Antarctica splinters into icebergs following the warmest summer ever recorded, lending further support to the theory of global warming.

2003   The Human Genome Project is completed; the U.S. space shuttle *Columbia* disintegrates on atmospheric reentry, killing seven crew members and calling the future of the shuttle program into question; the Galileo probe completes its mission and falls into Jupiter; a severe acute respiratory syndrome (SARS) outbreak in China reinforces fears that air travel has made the entire world a single microbiological environment, though the outbreak is successfully fought and followed by rapid identification of the virus responsible and its gene sequence.

2004   Prominent scientists criticize the U.S. presidential administration of George W. Bush, alleging an unprecedented politicization of science; a new planetoid smaller than Pluto and currently about three times as far from the Sun as Pluto is discovered and named Sedna.

Selected Bibliography

This bibliography contains only selected books, comprehensively covering the history of science in the last half of the twentieth century. For more detailed source material, see the references at the end of each entry.

Abbate, Janet. *Inventing the Internet*. Cambridge: MIT Press, 2000.

Alcock, John. *The Triumph of Sociobiology*. New York: Oxford University Press, 2001.

Aczel, Amir D. *Fermat's Last Theorem: Unlocking the Secret of an Ancient Mathematical Problem*. New York: Four Walls Eight Windows, 1996.

Adams, Steve. *Frontiers: Twentieth-Century Physics*. London: Taylor and Francis, 2000.

Aldersey-Williams, Hugh. *The Most Beautiful Molecule: The Discovery of the Buckyball*. New York: Wiley, 1997.

Aldridge, Susan. *The Thread of Life: The Story of Genes and Genetic Engineering*. New York: Cambridge University Press, 1996.

Alley, Richard B. *The Two-Mile Time Machine: Ice Cores, Abrupt Climate Change, and Our Future*. Princeton: Princeton University Press, 2000.

Alpher, Ralph A., and Robert Herman. *Genesis of the Big Bang*. New York: Oxford University Press, 2001.

Alvarez, Walter. *T. Rex and the Crater of Doom*. Princeton: Princeton University Press, 1997.

Appel, Toby A. *Shaping Biology: The National Science Foundation and American Biological Research, 1945–1975*. Baltimore: Johns Hopkins University Press, 2000.

Aspray, William. *John von Neumann and the Origins of Modern Computing*. Cambridge: MIT Press, 1990.

Bailey, George. *Galileo's Children: Science, Sakharov, and the Power of the State*. New York: Arcade, 1990.

Ballard, Robert D., with Will Hively. *The Eternal Darkness: A Personal History of Deep-Sea Exploration*. Princeton: Princeton University Press, 2000.

Barbour, Ian G. *When Science Meets Religion: Enemies, Strangers, or Partners?* San Francisco: HarperSanFrancisco, 2000.

Beatty, J. Kelly, Carolyn Collins Peterson, and Andrew Chaikin, editors. *The New Solar System*. Fourth edition. New York: Cambridge University Press, 1999.

Beeman, Randal S., and James A. Pritchard. *A Green and Permanent Land: Ecology and Agriculture in the Twentieth Century*. Lawrence: University Press of Kansas, 2001.

Ben-David, Joseph. *The Scientist's Role in Society: A Comparative Study.* Upper Saddle River, NJ: Prentice-Hall, 1971.

Berstein, Vadim J. *The Perversion of Knowledge: The True Story of Soviet Science.* Boulder, CO: Westview, 2001.

Bloom, Samuel William. *The Word as Scalpel: A History of Medical Sociology.* New York: Oxford University Press, 2002.

Bloor, David. *Knowledge and Social Imagery.* Second edition. Chicago: University of Chicago Press, 1991.

Blum, Deborah. *Love at Goon Park: Harry Harlow and the Science of Affection.* Cambridge, MA: Perseus, 2002.

Bocking, Stephen. *Ecologists and Environmental Politics: A History of Contemporary Ecology.* New Haven: Yale University Press, 1997.

Boss, Alan. *Looking for Earths: The Race to Find New Solar Systems.* New York: Wiley, 1998.

Bowler, Peter J. *The Norton History of the Environmental Sciences.* New York: Norton, 1993.

Brockman, John, editor. *The Next Fifty Years: Science in the First Half of the Twenty-First Century.* New York: Vintage, 2002.

Bronowksi, Jacob. *Science and Human Values.* Revised edition. New York: Julian Messner, 1965.

Brooke, John Hedley. *Science and Religion: Some Historical Perspectives.* New York: Cambridge University Press, 1991.

Brown, Laurie, Abraham Pais, and Brian Pippard, editors. *Twentieth Century Physics.* 3 volumes. New York and Philadelphia: American Institute of Physics, 1995.

Brown, Lester R. *Seeds of Change: The Green Revolution and Development in the 1970's.* New York: Praeger, 1970.

Bullough, Vern L. *Science in the Bedroom: A History of Sex Research.* New York: Basic, 1994.

Burrows, William E. *Exploring Space: Voyages in the Solar System and Beyond.* New York: Random House, 1990.

———. *This New Ocean: The Story of the First Space Age.* New York: Random House, 1998.

Cagin, Seth, and Phillip Dray. *Between Earth and Sky: How CFCs Changed Our World and Endangered the Ozone Layer.* New York: Pantheon, 1993.

Chadarevian, Soraya de. *Designs for Life: Molecular Biology after World War II.* Cambridge: Cambridge University Press, 2002.

Chaikin, Andrew. *A Man on the Moon: The Voyages of the Apollo Astronauts.* New York: Viking, 1994.

Chaisson, Eric J. *The Hubble Wars: Astrophysics Meets Astropolitics in the Two-Billion-Dollar Struggle over the Hubble Space Telescope.* New York: HarperCollins, 1994.

Clarke, Arthur C. *Greetings, Carbon-Based Bipeds! Collected Essays, 1934–1998.* New York: St. Martin's, 1999.

Clute, John, and Peter Nicholls. *The Encyclopedia of Science Fiction.* New York: St. Martin's Griffin, 1993.

Cohen, Joel. *How Many People Can the Earth Support?* New York: Norton, 1995.

Cotterell, Arthur. *The First Emperor of China: The Greatest Archeological Find of Our Time.* New York: Holt, Rinehart and Winston, 1981.

Crease, Robert P. *Making Physics: A Biography of Brookhaven National Laboratory, 1946–1972.* Chicago: University of Chicago Press, 1999.

Crease, Robert P., and Charles C. Mann. *The Second Creation: Makers of the Revolution in Twentieth-Century Physics.* New York: Macmillan, 1986.

Crevier, Daniel. *AI: The Tumultuous History of the Search for Artificial Intelligence.* New York: Basic, 1993.

Cronon, William, editor. *Uncommon Ground: Toward Reinventing Nature.* New York: Norton, 1995.

Croswell, Ken. *Planet Quest: The Epic Discovery of Alien Solar Systems.* New York: Free Press, 1997.

Danziger, Kurt. *Constructing the Subject: Historical Origins of Psychological Research.* New York: Cambridge University Press, 1990.

Darling, David. *Life Everywhere: The Maverick Science of Astrobiology.* New York: Basic, 2001.

Davidson, Keay. *Carl Sagan: A Life.* New York: Wiley, 1999.

Dawkins, Richard. *The Extended Phenotype: The Long Reach of the Gene.* Revised edition. Oxford: Oxford University Press, 1999.

Dick, Steven J. *The Biological Universe: The Twentieth-Century Extraterrestrial Life Debate and the Limits of Science.* New York: Cambridge University Press, 1996.

Drexler, K. Eric. *Engines of Creation: The Coming Era of Nanotechnology.* New York: Anchor, 1986.

Durbin, Paul T., editor. *The Culture of Science, Technology, and Medicine.* New York: Free Press, 1980.

Ehrlich, Paul R., and Anne H. Ehrlich. *The Population Explosion.* New York: Simon and Schuster, 1990.

Etheridge, Elizabeth W. *Sentinel for Health: A History of the Centers for Disease Control.* Berkeley and Los Angeles: University of California Press, 1992.

Etzkowitz, Henry, Carol Kemelgor, and Brian Uzzi. *Athena Unbound: The Advancement of Women in Science and Technology.* Cambridge: Cambridge University Press, 2000.

Eve, Raymond A., and Francis B. Harrold. *The Creationist Movement in Modern America.* Boston: Twayne, 1990.

Fagan, Brian M. *Floods, Famines, and Emperors: El Niño and the Fate of Civilizations.* New York: HarperCollins, 2000.

————. *Time Detectives: How Archaeologists Use Technology to Recapture the Past.* New York: Simon and Schuster, 1995.

Fancher, Raymond. *The Intelligence Men: Makers of the I.Q. Controversy.* New York: Norton, 1985.

Ferry, Georgina. *Dorothy Hodgkin: A Life.* London: Granta, 1998.

Feyerabend, Paul K. *Against Method: Outline of an Anarchistic Theory of Knowledge.* Third edition. New York: Verso, 1993.

Fitzgerald, Frances. *Way out There in the Blue: Reagan, Star Wars and the End of the Cold War.* New York: Simon and Schuster, 2000.

Fowler, Brenda. *Iceman: Uncovering the Life and Times of a Prehistoric Man Found in an Alpine Glacier.* Chicago: University of Chicago Press, 2001.

Fowler, T. Kenneth. *The Fusion Quest.* Baltimore: Johns Hopkins University Press, 1997.

Franklin, Allan. *Are There Really Neutrinos? An Evidential History.* Cambridge, MA: Perseus, 2001.

Friedman, Robert Mare. *Appropriating the Weather: Vilhelm Bjerknes and the Construction of Modern Meteorology.* Ithaca, NY: Cornell University Press, 1989.

Fruton, Joseph. *Proteins, Enzymes, Genes: The Interplay of Chemistry and Biology.* New Haven: Yale University Press, 1999.

Galison, Peter, and Bruce Hevly, editors. *Big Science: The Growth of Large-Scale Research.* Stanford, CA: Stanford University Press, 1992.

Gamow, George. *My World Line: An Informal Autobiography.* New York: Viking, 1970.

Gardner, Howard. *Intelligence Reframed: Multiple Intelligences for the 21st Century.* New York: Basic, 1999.

Garrett, Laurie. *The Coming Plague: Newly Emerging Diseases in a World out of Balance.* New York: Farrar, Straus and Giroux, 1994.

Geiger, Roger L. *Research and Relevant Knowledge: American Research Universities since World War II.* New York: Oxford University Press, 1993.

Gillies, James. *How the Web Was Born: The Story of the World Wide Web.* New York: Oxford University Press, 2000.

Glashow, Sheldon L., with Ben Bova. *Interactions: A Journey through the Mind of a Particle Physicist and the Matter of This World.* New York: Warner, 1988.

Gleick, James. *Chaos: Making a New Science.* New York: Penguin, 1987.

————. *Genius: The Life and Science of Richard Feynman.* New York: Pantheon, 1992.

Glen, William. *The Road to Jaramillo: Critical Years of the Revolution in Earth Science.* Stanford, CA: Stanford University Press, 1982.

Goleman, Daniel. *Emotional Intelligence: Why It Can Matter More Than IQ.* New York: Bantam, 1995.

Goodall, Jane. *Through a Window: My Thirty Years with the Chimpanzees of Gombe.* Boston: Houghton Mifflin, 1990.

Goodchild, Peter J. *Robert Oppenheimer: Shatterer of Worlds.* Boston: Houghton Mifflin, 1981.

Gould, Stephen Jay. *Full House: The Spread of Excellence from Plato to Darwin.* New York: Harmony, 1996.

————. *Wonderful Life: The Burgess Shale and the Nature of History.* New York: Norton, 1989.

Graham, Loren R. *What Have We Learned about Science and Technology from the Russian Experience?* Stanford, CA: Stanford University Press, 1998.

Greene, Brian. *The Elegant Universe: Superstrings, Hidden Dimensions, and the Quest for the Ultimate Theory.* New York: Norton, 1999.

Gribbin, John. *Q Is for Quantum: An Encyclopedia of Particle Physics.* New York: Free Press, 1998.

Gross, Paul R., Norman Levitt, Martin W. Lewis, editors. *The Flight from Science and Reason.* Baltimore: Johns Hopkins University Press, 1996.

Hacking, Ian. *Rewriting the Soul: Multiple Personality and the Sciences of Memory.* Princeton: Princeton University Press, 1995.

Hafner, Katie, and Matthew Lyon. *Where Wizards Stay Up Late: The Origins of the Internet.* New York: Simon and Schuster, 1996.

Hager, Thomas. *Force of Nature: The Life of Linus Pauling.* New York: Simon and Schuster, 1995.

Hall, Stephen S. *Invisible Frontiers: The Race to Synthesize a Human Gene.* New York: Atlantic Monthly, 1987.

Hargittai, István. *The Road to Stockholm: Nobel Prizes, Science, and Scientists.* New York: Oxford University Press, 2002.

Hawking, Stephen W. *A Brief History of Time.* New York: Bantam, 1988.

Healy, David. *The Creation of Psychopharmacology.* Cambridge: Harvard University Press, 2002.

Herken, Gregg. *Brotherhood of the Bomb: The Tangled Lives and Loyalties of Robert Oppenheimer, Ernest Lawrence, and Edward Teller.* New York: Henry Holt, 2002.

Hessenbruch, Arne. *Reader's Guide to the History of Science.* Chicago: Fitzroy Dearborn, 2000.

Hirsh, Richard F. *Glimpsing an Invisible Universe: The Emergence of X-Ray Astronomy.* New York: Cambridge University Press, 1983.

Hogan, James P. *Mind Matters: Exploring the World of Artificial Intelligence.* New York: Ballantine, 1997.

Holloway, David. *Stalin and the Bomb: The Soviet Union and Atomic Energy, 1939–1956.* New Haven: Yale University Press, 1994.

Horgan, John. *The End of Science: Facing the Limits of Knowledge in the Twilight of the Scientific Age.* Reading, MA: Addison-Wesley, 1996.

Hough, Susan Elizabeth. *Earthshaking Science: What We Know (and Don't Know) about Earthquakes.* Princeton: Princeton University Press, 2002.

Houghton, John. *Global Warming: The Complete Briefing.* Second edition. Cambridge: Cambridge University Press, 1997.

Hsu, Feng-hsiung. *Behind Deep Blue: Building the Computer That Defeated the World Chess Champion.* Princeton: Princeton University Press, 2002.

Hull, David L. *Science as a Process: An Evolutionary Account of the Social and Conceptual Development of Science.* Chicago: University of Chicago Press, 1988.

Hutton, Drew, and Libby Connors. *A History of the Australian Environment Movement.* Melbourne, Australia: Cambridge University Press, 1999.

Jasanoff, Sheila, and others, editors. *Handbook of Science and Technology Studies.* Thousand Oaks, CA: Sage, 1995.

Johnson, George. *Strange Beauty: Murray Gell-Mann and the Revolution in Twentieth-Century Physics.* New York: Knopf, 1999.

Jonas, Gerald. *The Circuit Riders: Rockefeller Money and the Rise of Modern Science.* New York: Norton, 1989.

Jones, James H. *Alfred C. Kinsey: A Public/Private Life.* New York: Norton, 1997.

————. *Bad Blood: The Tuskegee Syphilis Experiment.* New York: Free Press, 1981.

Judson, Horace Freeland. *The Eighth Day of Creation: Makers of the Revolution in Biology.* Expanded edition. Woodbury, NY: Cold Spring Harbor Laboratory, 1996.

Kaku, Michio. *Visions: How Science Will Revolutionize the 21st Century.* New York: Anchor, 1997.

Kay, Lily E. *Who Wrote the Book of Life? A History of the Genetic Code.* Stanford, CA: Stanford University Press, 2000.

Keller, Evelyn Fox. *A Feeling for the Organism: The Life and Work of Barbara McClintock.* San Francisco: W. H. Freeman, 1983.

Kevles, Daniel J. *In the Name of Eugenics: Genetics and the Uses of Human Heredity.* New York: Knopf, 1986.

————. *The Physicists: The History of a Scientific Community in Modern America.* Revised edition. Cambridge: Harvard University Press, 1995.

Kevles, Daniel J., and Leroy Hood, editors. *The Code of Codes: Scientific and Social Issues in the Human Genome Project.* Cambridge: Harvard University Press, 1992.

Koertge, Noretta, editor. *A House Built on Sand: Exposing Postmodernist Myths about Science.* New York: Oxford University Press, 1998.

Koppes, Clayton R. *JPL and the American Space Program: A History of the Jet Propulsion Laboratory.* New Haven: Yale University Press, 1982.

Kragh, Helge. *Cosmology and Controversy: The Historical Development of Two Theories of the Universe.* Princeton: Princeton University Press, 1996.

————. *An Introduction to the Historiography of Science.* New York: Cambridge University Press, 1987.

————. *Quantum Generations: A History of Physics in the Twentieth Century.* Princeton: Princeton University Press, 1999.

Lawrence, David M. *Upheaval from the Abyss: Ocean Floor Mapping and the Earth Science Revolution.* New Brunswick, NJ: Rutgers University Press, 2002.

Leakey, Richard E., and Roger Lewin. *Origins Reconsidered: In Search of What Makes Us Human.* New York: Doubleday, 1992.

Lear, Linda. *Rachel Carson: Witness for Nature.* New York: Henry Holt, 1997.

LeDoux, Joseph. *Synaptic Self: How Our Brains Become Who We Are.* New York: Viking, 2002.

Lee, Kelley. *Historical Dictionary of the World Health Organization.* Lanham, MD: Scarecrow, 1998.

Leslie, Stuart W. *The Cold War and American Science: The Military-Industrial-Academic Complex at MIT and Stanford.* New York: Columbia University Press, 1994.

Levy, David H. *Impact Jupiter: The Crash of Shoemaker-Levy 9.* New York: Plenum, 1995.

————. *Shoemaker by Levy: The Man Who Made an Impact.* Princeton: Princeton University Press, 2000.

Levy, Stuart. *The Antibiotic Paradox: How Miracle Drugs Are Destroying the Miracle.* New York: Plenum, 1992.

Lindberg, David C., and Ronald L. Numbers, editors. *God and Nature: Historical Essays on the Encounter between Christianity and Science.* Berkeley: University of California Press, 1986.

Livi-Bacci, Massimo. *A Concise History of World Population.* Second edition. Oxford: Blackwell, 1997.

Lourie, Richard. *Sakharov: A Biography.* Hanover, NH: Brandeis University Press, 2002.

Lovelock, James. *Gaia: A New Look at Life on Earth.* Oxford: Oxford University Press, 1979.

————. *Homage to Gaia: The Life of an Independent Scientist.* Oxford: Oxford University Press, 2000.

MacKinnon, Barbara, editor. *Human Cloning: Science, Ethics, and Public Policy.* Urbana: University of Illinois Press, 2000.

Macrina, Francis L. *Scientific Integrity: An Introductory Text with Cases.* Washington, DC: American Society for Microbiology, 1995.

Maddox, Brenda. *Rosiland Franklin: The Dark Lady of DNA.* New York: HarperCollins, 2002.

Maddox, John. *What Remains to Be Discovered: Mapping the Secrets of the Universe, the Origins of Life, and the Future of the Human Race.* New York: Free Press, 1998.

Magee, Bryan. *Karl Popper.* New York: Viking, 1973.

Magner, Lois N. *A History of the Life Sciences.* Second edition. New York: Marcel Dekker, 1994.

Margulis, Lynn. *Symbiotic Planet: A New Look at Evolution.* New York: Basic, 1998.

Margulis, Lynn, and Dorion Sagan. *Acquiring Genomes: A Theory of the Origins of Species.* New York: Basic, 2002.

————. *Microcosmos: Four Billion Years of Evolution from Our Microbial Ancestors.* New York: Summit, 1986.

Marks, Lara V. *Sexual Chemistry: A History of the Contraceptive Pill.* New Haven: Yale University Press, 2001.

Martino, Joseph P. *Science Funding: Politics and Porkbarrel.* New Brunswick, NJ: Transaction, 1992.

Mather, John C., and John Boslough. *The Very First Light: The True Inside Story of the Scientific Journey back to the Dawn of the Universe.* New York: Basic, 1996.

Matricon, Jean, and Georges Waysand. *The Cold Wars: A History of Superconductivity.* New Brunswick, NJ: Rutgers University Press, 2003.

Mayewski, Paul Andrew, and Frank White. *The Ice Chronicles: The Quest to Understand Global Climate Change.* Hanover, NH: University Press of New England, 2002.

Mayr, Ernst. *The Growth of Biological Thought: Diversity, Evolution, and Inheritance.* Cambridge: Harvard University Press, 1982.

McElheny, Victor K. *Watson and DNA: Making a Scientific Revolution.* Cambridge, MA: Perseus, 2003.

McGee, Glenn, editor. *Pragmatic Bioethics.* Second edition. Cambridge: MIT Press, 2003.

McGrayne, Sharon Bertsch. *Nobel Prize Women in Science.* Second edition. Secaucus, NJ: Carol, 1998.

McHughen, Alan. *Pandora's Picnic Basket: The Potential and Hazards of Genetically Modified Foods.* New York: Oxford University Press, 2000.

McKeown, Thomas. *The Role of Medicine: Dream, Mirage, or Nemesis?* Oxford: Blackwell, 1979.

McNeill, John Robert. *Something New under the Sun: An Environmental History of the Twentieth Century.* New York: Norton, 2000.

McRae, Murdo William, editor. *The Literature of Science: Perspectives on Popular Science Writing.* Athens: University of Georgia Press, 1993.

Mehra, Jagdish, and Kimball A. Milton. *Climbing the Mountain: The Scientific Biography of Julian Schwinger.* Oxford: Oxford University Press, 2000.

Minsky, Marvin. *The Society of Mind.* New York: Simon and Schuster, 1988.

Moir, Anne, and David Jessel. *Brain Sex: The Real Difference between Men and Women.* New York: Penguin, 1989.

Montgomery, Sy. *Walking with the Great Apes: Jane Goodall, Dian Fossey, Biruté Galdikas.* Boston: Houghton Mifflin, 1991.

Morrell, Virginia. *Ancestral Passions: The Leakey Family and the Quest for Humankind's Beginnings.* New York: Simon and Schuster, 1995.

Mould, Richard F. *Chernobyl Record: The Definitive History of the Chernobyl Catastrophe.* Bristol, UK: Institute of Physics, 2000.

Mowat, Farley. *Woman in the Mists: The Story of Dian Fossey and the Mountain Gorillas of Africa.* New York: Warner, 1987.

Munson, Richard. *Cousteau: The Captain and His World.* New York: William Morrow, 1989.

Nasar, Sylvia. *A Beautiful Mind: A Biography of John Forbes Nash, Jr., Winner of the Nobel Prize in Economics, 1994.* New York: Simon and Schuster, 1998.

Nelkin, Dorothy. *Selling Science: How the Press Covers Science and Technology.* Revised edition. San Francisco: W. H. Freeman, 1995.

Newton, David E. *The Ozone Dilemma: A Reference Handbook.* Santa Barbara, CA: ABC-CLIO, 1995.

Numbers, Ronald L. *The Creationists: The Evolution of Scientific Creationism.* New York: Knopf, 1992.

Olby, R. C., and others, editors. *Companion to the History of Modern Science.* New York: Routledge, 1990.

Oreskes, Naomi, editor. *Plate Tectonics: An Insider's History of the Modern Theory of the Earth.* Boulder, CO: Westview, 2002.

Overbye, Dennis. *Lonely Hearts of the Cosmos: The Story of the Scientific Quest for the Secret of the Universe.* New York: HarperCollins, 1991.

Paul, John R. *A History of Poliomyelitis.* New Haven: Yale University Press, 1971.

Popper, Karl. *Unended Quest: An Intellectual Autobiography.* La Salle, IL: Open Court, 1982.

Porter, Roy. *The Greatest Benefit to Mankind: A Medical History of Humanity.* New York: Norton, 1997.

Poundstone, William. *Carl Sagan: A Life in the Cosmos.* New York: Henry Holt, 1999.

Pyenson, Lewis, and Susan Sheets-Pyenson. *Servants of Nature: A History of Scientific Institutions, Enterprises, and Sensibilities.* New York: Norton, 1999.

Raup, David M. *The Nemesis Affair: A Story of the Death of Dinosaurs and the Ways of Science.* Revised and expanded edition. New York: Norton, 1999.

Reid, T. R. *The Chip: How Two Americans Invented the Microchip and Launched a Revolution.* New York: Random House, 2001.

Rhodes, Richard. *Dark Sun: The Making of the Hydrogen Bomb.* New York: Simon and Schuster, 1995.

———. *Deadly Feasts: Tracking the Secrets of a Terrifying New Plague.* New York: Simon and Schuster, 1997.

Ridley, Rosalind M., and Harry F. Baker. *Fatal Protein: The Story of CJD, BSE, and Other Prion Diseases.* Oxford: Oxford University Press, 1998.

Rossiter, Margaret W. *Women Scientists in America: Before Affirmative Action, 1940–1972.* Baltimore: Johns Hopkins University Press, 1995.

Sagan, Carl. *The Demon-Haunted World: Science as a Candle in the Dark.* New York: Random House, 1996.

Sapp, Jan. *Evolution by Association: A History of Symbiosis.* New York: Oxford University Press, 1994.

Schiebinger, Londa. *Has Feminism Changed Science?* Cambridge: Harvard University Press, 1999.

Schopf, J. William. *Cradle of Life: The Discovery of Earth's Earliest Fossils.* Princeton: Princeton University Press, 1999.

Selye, Hans. *The Stress of My Life: A Scientist's Memoirs.* Toronto: McClelland and Stewart, 1977.

Shannon, Thomas A. *Genetic Engineering: A Documentary History.* Westport, CT: Greenwood, 1999.

Shepard, Alan, and Deke Slayton. *Moon Shot: The Inside Story of America's Race to the Moon.* Atlanta: Turner, 1994.

Shermer, Michael. *The Borderlands of Science: Where Sense Meets Nonsense.* New York: Oxford University Press, 2001.

Simon, Bart. *Undead Science: Science Studies and the Afterlife of Cold Fusion.* New Brunswick, NJ: Rutgers University Press, 2002.

Smith, Jane S. *Patenting the Sun: Polio and the Salk Vaccine.* New York: William Morrow, 1990.

Söderqvist, Thomas, editor. *Historiography of Contemporary Science and Technology.* New York: Harwood Academic, 1997.

Soyfer, Valery N. *Lysenko and the Tragedy of Soviet Science.* New Brunswick, NJ: Rutgers University Press, 1994.

Starzl, Thomas E. *The Puzzle People: Memoirs of a Transplant Surgeon.* Pittsburgh: University of Pittsburgh Press, 1992.

Straus, Eugene. *Rosalyn Yalow, Nobel Laureate: Her Life and Work in Medicine. A Biographical Memoir.* New York: Plenum, 1998.

Tattersall, Ian, and Jeffrey H. Schwartz. *Extinct Humans.* Boulder, CO: Westview, 2000.

Taylor, Nick. *Laser: The Inventor, the Nobel Laureate, and the Thirty-Year Patent War.* New York: Simon and Schuster, 2000.

Teitelman, Robert. *Profits of Science: The American Marriage of Business and Technology.* New York: Basic, 1994.

Teller, Edward, with Judith Shoolery. *Memoirs: A Twentieth-Century Journey in Science and Politics.* Cambridge, MA: Perseus, 2001.

Thomas, David Hurst. *Skull Wars: Kennewick Man, Archaeology, and the Battle for Native American Identity.* New York: Basic, 2000.

Thomas, Patricia. *Big Shot: Passion, Politics, and the Struggle for an AIDS Vaccine.* New York: Public Affairs, 2001.

Thompson, Dick. *Volcano Cowboys: The Rocky Evolution of a Dangerous Science.* New York: St. Martin's, 2000.

Thorne, Brian. *Carl Rogers.* Thousand Oaks, CA: Sage, 1992.

Toumey, Christopher P. *God's Own Scientists: Creationists in a Secular World.* New Brunswick, NJ: Rutgers University Press, 1994.

Townes, Charles H. *How the Laser Happened: Adventures of a Scientist.* New York: Oxford University Press, 1999.

Tucker, Jonathan B. *Scourge: The Once and Future Threat of Smallpox.* New York: Atlantic Monthly, 2001.

Tucker, Wallace, and Karen Tucker. *The Cosmic Inquirers: Modern Telescopes and Their Makers.* Cambridge, MA: Harvard University Press, 1986.

———. *Revealing the Universe: The Making of the Chandra X-ray Observatory.* Cambridge: Harvard University Press, 2001.

Van Dover, Cindy Lee. *The Octopus's Garden: Hydrothermal Vents and Other Mysteries of the Deep Sea.* Reading, MA: Addison-Wesley, 1996.

Verschuur, Gerrit L. *Impact! The Threat of Comets and Asteroids.* New York: Oxford University Press, 1996.

Wakeford, Tom. *Liaisons of Life: From Hornworts to Hippos, How the Unassuming Microbe Has Driven Evolution.* New York: Wiley, 2001.

Walter, Malcolm. *The Search for Life on Mars.* Cambridge, MA: Perseus, 1999.

Wang, Jessica. *American Science in an Age of Anxiety: Scientists, Anticommunism, and the Cold War.* Durham: University of North Carolina Press, 1999.

Watkins, Elizabeth Siegel. *On the Pill: A Social History of Oral Contraceptives, 1950–1970.* Baltimore: Johns Hopkins University Press, 1998.

Watson, James D. *Genes, Girls, and Gamow: After the Double Helix.* London: Oxford University Press, 2001.

Weatherall, Miles. *In Search of a Cure: A History of Pharmaceutical Discovery.* New York: Oxford University Press, 1991.

Weiner, Jonathan. *The Beak of the Finch: A Story of Evolution in Our Time.* New York: Knopf, 1994.

Wheeler, John Archibald, with Kenneth Ford. *Geons, Black Holes and Quantum Foam: A Life of Physics.* New York: Norton, 1998.

Wiener, Daniel N. *B. F. Skinner: Benign Anarchist.* Boston: Allyn and Bacon, 1996.

Wilmut, Ian, Keith Campbell, and Colin Tudge. *The Second Creation: Dolly and the Age of Biological Control.* New York: Farrar, Straus and Giroux, 2000.

Wilson, David Sloan. *Darwin's Cathedral: Evolution, Religion, and the Nature of Society.* Chicago: University of Chicago Press, 2002.

Wilson, Edward O. *Naturalist.* Washington, DC: Island, 1994.

————. *Sociobiology: The New Synthesis.* Twentieth-Fifth Anniversary Edition. Cambridge: Belknap Press of Harvard University Press, 2000.

Worster, Donald. *Nature's Economy: A History of Ecological Ideas.* Second edition. New York: Cambridge University Press, 1994.

Wulff, David M. *Psychology of Religion: Classic and Contemporary Views.* New York: Wiley, 1991.

Index

Page numbers in **boldface** denote main entries.

About the Author

Eric G. Swedin is on the faculty at Weber State University in Ogden, Utah. Dr. Swedin has also written *Healing Souls: Psychotherapy in the Latter-day Saint Community* (2003).